International
REVIEW OF
Neurobiology
Volume 92

International

REVIEW OF

Neurobiology

Volume 92

SERIES EDITORS

RONALD J. BRADLEY

Department of Psychiatry, College of Medicine
The University of Tennessee Health Science Center
Memphis, Tennessee, USA

R. ADRON HARRIS

Waggoner Center for Alcohol and Drug Addiction Research
The University of Texas at Austin
Austin, Texas, USA

PETER JENNER

Division of Pharmacology and Therapeutics
GKT School of Biomedical Sciences
King's College, London, UK

EDITORIAL BOARD

Dreams and Dreaming

EDITED BY

ANGELA CLOW

Department of Psychology
University of Westminster
London
UK

and

PATRICK McNAMARA

Department of Neurology and Behavioral Neurosciences
Boston University School of Medicine
Boston
Massachusetts, USA

ELSEVIER

AMSTERDAM • BOSTON • HEIDELBERG • LONDON
NEW YORK • OXFORD • PARIS • SAN DIEGO
SAN FRANCISCO • SINGAPORE • SYDNEY • TOKYO
Academic Press is an imprint of Elsevier

Academic Press is an imprint of Elsevier
360 Park Avenue South, New York, NY 10010-1700
525 B Street, Suite 1900, San Diego, California 92101-4495, USA
32 Jamestown Road, London NW1 7BY, UK

Permissions may be sought directly from Elsevier's Science & Technology Rights
Department in Oxford, UK: phone: (þ44) 1865 843830, fax: (þ44) 1865 853333,
E-mail: permissions@elsevier.com. You may also complete your request on-line
via the Elsevier homepage (http://elsevier.com), by selecting "Support & Contact"
then "Copyright and Permission" and then "Obtaining Permissions."

For information on all Elsevier publications
visit our website at books.elsevier.com

ISBN: 978-0-12-381322-0

PRINTED AND BOUND IN THE UNITED STATES OF AMERICA
10 11 12 13   9 8 7 6 5 4 3 2 1

CONTENTS

REM and NREM Sleep Mentation

PATRICK MCNAMARA, PATRICIA JOHNSON, DEIRDRE MCLAREN, ERICA HARRIS,
CATHERINE BEAUHARNAIS AND SANFORD AUERBACH

Neuroimaging of Dreaming: State of the Art and Limitations

CAROLINE KUSSÉ, VINCENZO MUTO, LAURA MASCETTI, LUCA MATARAZZO,
ARIANE FORET, ANAHITA SHAFFII-LE BOURDIEC AND PIERRE MAQUET

Memory Consolidation, The Diurnal Rhythm of Cortisol, and The Nature of Dreams: A New Hypothesis

JESSICA D. PAYNE

Characteristics and Contents of Dreams

MICHAEL SCHREDL

Trait and Neurobiological Correlates of Individual Differences in Dream Recall and Dream Content

MARK BLAGROVE AND EDWARD F. PACE-SCHOTT

Consciousness in Dreams

DAVID KAHN AND TZIVIA GOVER

The Underlying Emotion and the Dream: Relating Dream Imagery to the Dreamer's Underlying Emotion can Help Elucidate the Nature of Dreaming

ERNEST HARTMANN

Dreaming, Handedness, and Sleep Architecture: Interhemispheric Mechanisms

STEPHEN D. CHRISTMAN AND RUTH E. PROPPER

To What Extent Do Neurobiological Sleep-Waking Processes Support Psychoanalysis?

CLAUDE GOTTESMANN

The Use of Dreams in Modern Psychotherapy

CLARA E. HILL AND SARAH KNOX

CONTRIBUTORS

Numbers in parentheses indicate the pages on which the authors contributions begin.

Sanford Auerbach (69), Sleep Disorders Center, Boston Medical Center, Boston University School of Medicine, Boston, MA, USA

Catherine Beauharnais (69), Department of Neurology, Boston University School of Medicine and VA New England Healthcare System, Boston, MA, USA

Mark Blagrove (155), Department of Psychology, School of Human and Health Sciences, Swansea University, Wales, UK

Kelly Bulkeley (31), Visiting Scholar, The graduate Theological Union, Berkeley, California, USA

Melissa M. Burnham (47), Department of Educational Psychology, Counseling, & Human Development, University of Nevada, Reno, NV, USA

Stephen D. Christman (215), Department of Psychology, University of Toledo, Toledo, OH, USA

Christian Conte (47), Department of Educational Psychology, Counseling, & Human Development, University of Nevada, Reno, NV, USA

Ariane Foret (87), Cyclotron Research Centre, University of Liège, Liège, Belgium

Claude Gottesmann (1, 233), Département de Biologie, Faculté des Sciences, Université de Nice-Sophia Antipolis, Nice, France

Tzivia Gover (181), Holyoke Community College, Holyoke, MA, USA

Erica Harris (69), Department of Neurology, Boston University School of Medicine and VA New England Healthcare System, Boston, MA, USA

Ernest Hartmann (197), Department of Psychiatry, Tufts University School of Medicine, Boston, MA, USA

Clara E. Hill (291), Department of Psychology, University of Maryland, College Park, MD, USA

Patricia Johnson (69), Department of Neurology, Boston University School of Medicine and VA New England Healthcare System, Boston, MA, USA

David Kahn (181), Department of Psychiatry, Harvard Medical School, Boston, MA, USA

Sarah Knox (291), College of Education, Marquette University, Milwaukee, WI, USA

Caroline Kussé (87), Cyclotron Research Centre, University of Liège, Liège, Belgium

Pierre Maquet (87), Cyclotron Research Centre, University of Liège, Liège, Belgium

Laura Mascetti (87), Cyclotron Research Centre, University of Liège, Liège, Belgium

Luca Matarazzo (87), Cyclotron Research Centre, University of Liège, Liège, Belgium

Deirdre McLaren (69), Department of Neurology, Boston University School of Medicine and VA New England Healthcare System, Boston, MA, USA

Patrick McNamara (69), Department of Neurology, Boston University School of Medicine and VA New England Healthcare System, Boston, MA, USA

Vincenzo Muto (87), Cyclotron Research Centre, University of Liège, Liège, Belgium

Edward F. Pace-Schott (155), Department of Psychology, University of Massachusetts Amherst, Amherst, MA, USA

Jessica D. Payne (101), Department of Psychology, University of Notre Dame, Notre Dame, IN, USA

Ruth E. Propper (215), Psychology Department, Merrimack College, North Andover, MA, USA

Michael Schredl (135), Sleep laboratory, Central Institute of Mental Health, Mannheim, Germany

Anahita Shaffii-Le Bourdiec (87), Cyclotron Research Centre, University of Liège, Liège, Belgium

PREFACE

This special issue presents 13 papers on breakthrough research in dreams and dreaming. Gottesman recounts the history neurobiologic approaches to dreams. Bulkeley surveys the anthropologic literature on traditional conceptions of dreams. Burnham and Conte review developmental data while McNamara *et al.* identify REM vs NREM content differences. Kussé *et al.* review neuroimaging studies of brain correlates of dreams, while Payne summarizes evidence for memory processing in dreams. Schredl reviews content of typical dreams, while Blagrove and Pace-Schott identify individual differences constraints on dream recall. Christman and Propper identify links between handedness and dream content, and Kahn and Gover review the nature of dream consciousness. The final chapters by Hartman, Gottesman, and Hill and Knox demonstrate emotional and therapeutic effects of dream content. Dreams and dreaming can no longer be considered mere epiphenomena of the sleeping brain. Dreaming, dream recall, and dream content all appear to have functional effects, some identified and some yet to be identified.

ANGELA CLOW
PATRICK McNAMARA

THE DEVELOPMENT OF THE SCIENCE OF DREAMING

Claude Gottesmann

Département de Biologie, Faculté des Sciences, Université de Nice-Sophia Antipolis, Nice, France

Although the main peripheral features of dreaming were identified two millennia ago, the neurobiological study of the basic and higher integrated processes underlying rapid eye movement (REM) sleep only began about 70 years ago. Today, the combined contributions of the successive and complementary methods of electrophysiology, imaging, pharmacology, and neurochemistry have provided a good level of knowledge of the opposite but complementary activating and inhibitory processes which regulate waking mentation and which are disturbed during REM sleep, inducing a schizophrenic-like mental activity.

Dreaming has undoubtedly fascinated man since the first appearance of reflective consciousness in our species. The search for dream interpretations was already underlined in the Talmud (V century): "an uninterpreted dream is like an unread letter" (Fromm, 1953), and today we still focus on the same search for the "Interpretation of dreams" (Freud, 1900). While the meaning of dreams is still being actively pursued, for more than two centuries now the similarity between dreams and madness has been recognized: "the madman is a waking dreamer" (Kant), "Dreams are short madness and madness a long dream" (Schopenhauer). In the last century, the well-known neurophysiopathologist Hughlings Jackson beautifully anticipated—in spite of his era's only rudimentary awareness of neurophysiology—that if you "find out all about dreams . . . you will find out all about insanity." This assertion has been more recently developed by neuropsychiatrists such as Henri Ey (1967), who stated, with better knowledge of the neurobiological basis of dreaming, "It is obvious, it cannot be but obvious that the dream and madness spurt from the same source." This latter assertion is of particular interest since, as we will see, it is supported by recent findings in both psychiatry and neurobiology.

Consequently, despite the persistence of serious gaps in our knowledge of the precise neurobiological mechanisms that underlie dreaming as well as of its psychological signification, we will examine the extraordinary adventure of the discovery of the main dreaming sleep stage, i.e., rapid eye movement (REM) sleep, also referred to as paradoxical sleep.

I. Results

A. PREHISTORY OF REM SLEEP DISCOVERIES

The behavioral features of dreaming are already found in Lucretius (circa 98–55 BC). In *De Rerum Natura*, he described in detail the behavioral correlates of dreaming in humans and animals:

> In truth you will see strong horses when their limbs lie at rest, yet sweat in their sleep, and go on panting, and strain every nerve as though for victory, or else as though the barriers were opened (struggle to start). And hunters' dogs often in their soft sleep yet suddenly toss their legs, and all at once give tongue, and again and again sniff the air with their nostrils, as if they had found and were following the tracks of wild beasts, yea, roused from slumber they often pursue empty images of stags, as though they saw them in eager flight, until they shake off the delusion and return to themselves. But the fawning brood of pups brought up in the house, in a moment shake their body and lift it from the ground, just as if they beheld unknown forms and faces. And the wilder any breed may be, the more must it need rage in its sleep. But the diverse tribes of birds fly off, and on a sudden in the night time trouble the peace of the groves of the gods with the whirr of wings, as if in their gentle sleep they have seen hawks, flying in pursuit, offer fight and battle. Moreover, the minds of men, which with mighty movements bring forth mighty deeds, often in sleep do and dare just the same; they storm kings, are captured, join battle, raise a loud cry, as though being murdered—all without moving. Many men fight hard, and utter groans through their pain, and, as though the teeth of a panther or savage lion bit them, fill all around them with loud cries. Many in their sleep discourse of high affairs, and very often have been witness to their own guilt. Many meet death; many as though they were falling headlong with all their body from high mountains to the earth, are beside themselves

with fear, and, as though bereft of reason, scarcely recover themselves from sleep, quivering with the turmoil of their body. Likewise, a thirsty man sits down beside a stream or a pleasant spring, and gulps almost the whole river down his throat. Cleanly persons often, if bound fast in slumber they think they are lifting their dress at a latrine or a shallow pot, pour forth the filtered liquid from their whole body, and the Babylonian coverlets of rich beauty are soaked. Later on those, into the channel of whose life the vital seed is passing for the first time, when the ripeness of time has created it in their limbs, there come from without idols from every body, heralding a glorious face or beautiful colouring, which stir and rouse their members swelling with much seed, and often, as though all were over, they pour forth huge floods of moisture and soil their clothes. (lines 987–1036)

Much more recently, Fontana (1765) described convulsion-like movements and constantly restricted pupils in cats during sleep. A century later, a major book appeared on the characteristics of dreaming, although it left out its physiological correlates (Maury, 1861). Soon afterward, Hervey de Saint Denys (1867) described erection during sleep and observed that in the first period of night sleep there is no dreaming and that "there is no sleep of mind. Consciousness ignores sleep" (p. 165), an observation which has been confirmed today (Cavallero, 2003; Cavallero et al., 1992; Foulkes, 1962; Tracy and Tracy, 1974). The eye movements that occur during dreaming were described by Ladd (1892), after having been more generally discovered during sleep by Raehlmann and Witkowsky (1877, 1878); the latter also noted fissurated pupils during sleep, whose insensitivity to light was underlined by Gernet (1889). Complementary to premonitory finding, Berger and Loewy (1898) observed pupillary dilations during sleep and mentioned their relation to dreaming: the linkage of pupillary dilation with REM sleep eye movements during which dreams occur was more recently confirmed (Berlucchi, 1965; Berlucchi et al., 1964). Finally, also highly interesting observation from the 19th century, Weed and Hallam (1896) reported that "The dreams occurring from 5:00 to 6:30 are the most frequent, most interesting, most vivid and most varied" (p. 405). Moreover, more than half of the dreams contain "disagreeable" emotions, an observation that was recently recalled by Revonsuo (2003) after having been also quantified by Manacéine (29 pleasant vs 57 unpleasant dreams in a study of 381 dreams) (Manacéine, 1897) (quoted by Kleitman, 1939).

The last century began with Freud's tremendous "The interpretation of dreams" (Freud, 1900). An examination of the modern neurobiological support for his theory will be undertaken in the separate chapter related to psychoanalysis. It will only be mentioned here from a neurobiological standpoint that, although sexuality and sexual symbols are almost overrepresented in his dream interpretations, there is no report in his book of the erections that are observed almost systematically upon awakening from REM sleep. This is rather unexpected, since

Freud had an extraordinary knowledge of classical literature (see his powerful first chapter in which he mentions Hervey de Saint Denys); also, since his theory encountered strong resistance in the scientific community at the time (as it partly still does now!), recalling this peripheral characteristic of sexuality would have been a strong argument in favor of his theory. However, current knowledge has established that these erections are not related to libido, as they occur in neonates, in the elderly, and in animals (Schmidt *et al.*, 1994, 1999, 2000).

With Hans Berger (1929), the extraordinary contemporary history of the electrophysiological study of sleep-related phenomena began. Caton (1875) had previously detected galvanometric variations during brain activity in rabbits and monkeys (an experimental feat at that time): "the grey matter appears to have a relation to its function. When any part of the grey matter is in a state of functional activity, its electrical current usually exhibits negative variations" (p. 278). Berger, however, was the first to record electroencephalograms (EEGs), placing electrodes on the scalp of his son and on the cortex of patients. He clearly distinguished waking from sleep-related electrocortical activities.

In relation to REM sleep, Jacobson (1930) recorded eye movements by "connecting one electrode near the orbital bridge and another behind the ear ... this arrangement tended to yield ... marked galvanometric variations" (p. 694). He also wrote, as quoted by Kleitman (1965): "when a person dreams ... most of his eyes are active. When the sleeper whose eyes move under his closed lids ... awakes ... you are likely to find ... that he had seen something in a dream" (Jacobson, 1938).

During this same rich decade prior to the true discovery of REM sleep, Derbyshire *et al.* (1936) observed that, during sleep in cats, in addition to the slow, high-amplitude electrocortical waves, "when sleep was apparently less tranquil judging by twitching of the vibrissae, there were only small rapid waves as in the alert waking state" (p. 582). This result was premonitory of the later discovered REM sleep characteristics. Similarly, Blake and Gerard (1937) observed periods in human sleep with "feeble irregular potentials ... yet the test sound (auditory stimuli threshold) evokes no response" (p. 696). Thus, the most important criterion of REM sleep (low-voltage EEG) was coupled with increased arousal threshold, which was to be confirmed later by Dement (1958).

Two other papers from 1937 were crucial in leading the way toward the discovery of REM sleep. The first presented research was performed by Loomis *et al.* (1937). Following other earlier papers on sleep (Loomis et al., 1935a, 1935b) that clearly distinguished waking from sleep EEGs, in the 1937 paper the authors presented, for the first time, a night hypnogram. They also showed in a figure, and described in detail, that dreaming occurred during the low-voltage EEG stage "B." Another paper focusing on cats from the same year was crucial. Indeed, Klaue (1937) described a deep sleep stage (tiefer Schlaf) with "quiet electrical current," "complete muscular relaxation ... and numerous jerks of single extremities" (p. 514). He also showed a figure with a clear-cut rapid and low-amplitude electrocorticogram. Finally, in the last relevant report of the decade, Blake *et al.* (1939), with the participation of Nathaniel Kleitman, confirmed the existence of a period during sleep with low-amplitude EEG and with beta waves, "called 'null' (or low voltage)" (p. 50).

In 1942, McGlade (1942) observed a relationship between twitches and dreaming. "Two subjects . . . were awakened on 6 occasions by a gentle shaking at 1 a.m. before the group of foot twitches had begun. Neither subject recalled a dream . . . When they were again awakened, after the first large body movement at the end of the group of twitches, they invariably stated they had dreamed" (p. 139).

Although Ohlmeyer had earlier mentioned (in 1936) that he observed penile erection during sleep with a mean duration of 25.5 minutes and with a 79.7 minute periodicity of occurrence, he and his colleagues (Ohlmeyer *et al.*, 1944) later presented more complete results obtained on six subjects. He found that penile erection lasted for a mean duration of 25.3 minutes and with a periodicity of 85 minutes. Three years later, Ohlmeyer and Brilmayer (1947), in a paper originally submitted in September 1944, described that erection was delayed in the case of delayed sleep, with and without hypnotic administration. Moreover, they considered to be well known the fact that morning erection occurs independently of a full bladder.

Gibbs and Gibbs (1950), in their EEG atlas, showed in one figure "early morning sleep" with "a pattern . . . difficult to distinguish from normal waking activity."

Finally, Humphrey and Zangwill (1951) wrote that parieto-occipital lesions suppress dreaming. This conclusion, which was based on observations in several WWII soldiers and in one case following an operated abscess, has now been well-confirmed through the study of lesions at the parieto-temporo-occipital junction (Solms, 2000) and of transcranial direct current stimulation of the posterior parietal cortex (Jakobson *et al.*, 2009).

B. Definitive Establishment of REM Sleep

In 1953, the year that is now considered to be the date of the first discovery of REM sleep (Aserinsky and Kleitman, 1953), research into sleep had already established the existence of a stage with low-voltage cortical activity (Blake and Gerard, 1937; Blake *et al.*, 1939; Derbyshire *et al.*, 1936; Gibbs and Gibbs, 1950; Klaue, 1937; Loomis *et al.*, 1937) and that dreaming occurs during this stage (Loomis *et al.*, 1937). Moreover, this stage was known to be associated with atonia (Klaue, 1937), jerks–twitches (Derbyshire *et al.*, 1936; Klaue, 1937; McGlade, 1942), and increased arousal threshold (Blake and Gerard, 1937). Further, one author had clearly written that eyeball movement recordings strongly suggested that when a subject is dreaming, his eyeballs are active under the eyelids (Jacobson, 1938). Finally, one team had reported erection during sleep (Ohlmeyer and Brilmayer, 1947; Ohlmeyer *et al.*, 1944), although classical authors had already mentioned such peripheral activities as being related to dreaming (Hervey de Saint Denys, 1867; Lucrece, 1900).

Consequently, when in the 1950s Nathaniel Kleitman sent Eugene Aserinsky to Jacobson's laboratory to learn how to record eye movements during sleep (Lufkin, 1968) and he decided to record them in association with EEG and with respiration, he certainly already anticipated at least in part what his student would find. Indeed, in his masterly first edition of "Sleep and Wakefulness," which already had 1434 references (Kleitman, 1939), he analyzed the papers of all the previously mentioned authors, although only with respect to the 1937 Loomis *et al.* paper did he mention the low-voltage EEG observed during dreaming (p. 156).

The research of Aserinsky and Kleitman (1953) showed that "slow rolling or pendular eye movements such as have been observed in sleeping children and adults" could be recorded. However, they also recorded "rapid, jerky and binocularly symmetrical movements." In 20 of 27 cases, the subjects reported dreaming with visual content, whereas outside of the eye movement periods 19 of 23 did not report dreams. The first observation of eye movements occurred 1 hour 40 minutes after sleep onset. They were associated with body movements, which also occurred independently of eye movements. Moreover, the EEG was described "as invariably of low amplitude in the frontal as well as in the occipital level." Finally, the respiratory rate was increased during these periods. As already mentioned (Gottesmann, 2001), when I once met Aserinsky, he told me that at his dissertation defense at the University of Chicago in 1953 the members of the committee told him that doctorates are awarded not for obtained results but in the hope of future results!

The next year, a largely forgotten paper—which very probably guided the future research of Jouvet (1962; Jouvet and Michel, 1959, 1960)—showed that REM sleep-generating processes particularly involve the brainstem. Indeed, Rioch (1954), whom I visited once, observed that brainstem-transected cats and dogs (high mesencephalic transection) "occasionally show running movements and respiratory changes similar to those interpreted as 'dreaming' in sleeping normal animals" (p. 133–134). This was the first description of dreaming-like behavior in brainstem-limited preparations.

Another paper of Aserinsky and Kleitman (1955) provided a description of slow eye movements that appear at sleep onset and disappear at the onset of stages III and IV, and which are very different from the REMs that occur every 90 minutes, with associated increased respiratory and heart rate. The following years witnessed the emergence of William C. Dement's powerful results. He first studied schizophrenic patients (Dement, 1955) who, horrifyingly for us today, underwent lobotomies and the more accepted convulsive therapy. Dement clearly confirmed rapid low-amplitude EEG in these patients as well as in controls: "There was no apparent change in the schizophrenic's ability to recall dreams" (p. 265). Two years later, he published two major papers with Kleitman (Dement and Kleitman, 1957a, 1957b) confirming the low-voltage EEG activity of REM sleep and, importantly for our topic, a relationship between eye movements and dream content (see also Berger and Oswald, 1962; Dement and Wolpert, 1958). This was later confirmed by two papers from Roffwarg's group (Herman et al., 1984; Roffwarg et al., 1962), although it has since been challenged because eye movements still occur in the blind despite the absence of corneo-retinal potential (Amadeo and Gomez, 1966; Gross et al., 1965); further, several other groups were unable to confirm the relationship (Jacobs et al., 1972; Moskowitz and Berger, 1969).

To better concentrate our attention on the neurobiological processes supporting dreaming, this chapter will not address all the major results on the basic brainstem mechanisms underlying REM sleep.

C. Forebrain Processes Involved in Dreaming

This chapter will only address REM sleep. Indeed, although dreams have been described as occurring during slow-wave sleep (Bosinelli, 1995; Bosinelli and Cigogna, 2003; Cavallero, 2003; Foulkes, 1962; Tracy and Tracy, 1974), recent research suggests that dreaming can only occur within the neurobiological processes of REM sleep (Nielsen, 2000) and that dreams occurring during slow-wave sleep contain "covert" REM sleep activities. This has since been confirmed, particularly for dreaming that occurs shortly after sleep onset (called SOREM: sleep-onset REM), where the EEG is like during REM sleep and during which there are slow eye movements as during REM sleep (Takeuchi et al., 1999, 2001). Dement wrote once that REM sleep is like a symphony: Some instruments can be absent, but the music persists.

1. REM Sleep-Related Electrophysiological Activities in the Cortex

During REM sleep, the cortex is activated. This has been shown by the activation of neurons, particularly of pyramidal neurons either directly (Evarts, 1962, 1964) or indirectly (Arduini et al., 1963; Morrison and Pompeiano, 1965) recorded or antidromically stimulated (Pisano et al., 1962). This activation is sustained by brainstem influences (Datta and Siwek, 1997, 2002; Sakai, 1985,

1988; Steriade and McCarley, 1990). Above the classic low-voltage rapid EEG of up to 25 Hz, the higher frequency gamma rhythm centered around 40 Hz (Bouyer *et al.*, 1981; Ribary *et al.*, 1991) was later discovered to also occur during REM sleep (Llinas and Ribary, 1993). However, during this sleep stage the gamma rhythm becomes uncoupled between different brain areas (Cantero *et al.*, 2004; Corsi-Cabrera *et al.*, 2003; Massimini *et al.*, 2005; Perez-Garci *et al.*, 2001), reflecting a disturbance in the intracerebral connections; this uncoupling is also observed during waking in schizophrenia (Kubicki *et al.*, 2008; Meyer-Lindenberg *et al.*, 2001, 2005; Peled *et al.*, 2000; Tononi and Edelman, 2000; Young *et al.*, 1998). The function of the higher EEG high-frequency (nearly 140 Hz) forebrain rhythms that have been observed in animals is open to discussion (Hunt *et al.*, 2009).

Forebrain-activating processes were also shown during REM sleep by virtue of a high thalamocortical excitability level (Cordeau *et al.*, 1964, 1965; Favale *et al.*, 1965; Okuma and Fujimori, 1963; Palestini *et al.*, 1964), even though the cortex is nearly disconnected from the periphery during REM sleep (Weitzman and Kremen, 1965; Williams *et al.*, 1962). The recorded negative surface potential (Wurtz, 1965, 1966) is also an indication of major depolarizing currents taking place in the cortex during REM sleep. Finally, the detected ponto-geniculo-occipital (PGO) and eye movement potentials in the cortex (McCarley *et al.*, 1983; Miyauchi *et al.*, 1987; Salzarulo *et al.*, 1975), which are related to phasic activations (Kiyono and Iwama, 1965; Satoh, 1971) lasting up to about 100 milliseconds (Miyauchi *et al.*, 1987), also participate in the increased cortical excitability during this sleep stage when compared with slow-wave sleep.

However, the activating processes that occur at the cortical level during REM sleep are not the only ones that regulate its function. True inhibitory processes also take place at the cortical level during REM sleep. These inhibitory processes had already been shown in dogs in the 19th century by Bubnoff and Heidenhain (Bubnoff and Heidenhain, 1881), who suppressed cortical-induced movements by peripheral and cortical stimulation. Much more recently, Creutzfeldt *et al.* (1956) showed that cortical neurons can be inhibited by local electrical stimulation, a finding that was subsequently confirmed by geniculate radiation stimulation (Evarts *et al.*, 1960). Indeed, Evarts (Evarts *et al.*, 1960) showed that the cortical recovery cycle of evoked potentials after radiation stimulation was shorter during sleep than it was during waking. Evarts (1964, 1965) was also the first to clearly show, through the study of pyramidal neurons, that cortical inhibitory (control) influences disappear during REM sleep; this was soon confirmed by Demetrescu *et al.* (1966) by examining thalamocortical responsiveness. It must be added that Rossi *et al.* (1965) and Allison (1965, 1968) also observed a significantly shorter recovery cycle (disinhibition) of cortically evoked potentials during REM sleep. In humans as well, Kisley *et al.* (2003) found a shortening of the recovery cycle of auditory-evoked potentials. It is of importance that this disinhibition at the cortical level is identical to that observed in the same paradigm in waking schizophrenics. Together with the previously noted forebrain disconnection processes, this is a second similarity in terms of cortical dysfunction between REM sleep and schizophrenia.

Therefore, cortical-activating and inhibitory influences, as identified by electrophysiology, are antagonistic but obligatory complementary processes that act together to ensure the proper functioning of the waking brain. This equilibrated functioning is, however, exhausted during REM sleep.

2. *Results Involving Positron Emission Tomography, Glucose Uptake, Magnetic Resonance Imaging, and Blood Oxygen Level*

The cortical activation that takes place during REM sleep, as evidenced by increased blood flow, was observed in the visual associative areas by Madsen *et al.* (1991); Madsen also established a relationship between the activation and the dreaming. Several years later, Maquet *et al.* (1996) observed increased blood flow in several structures including the anterior cingulate cortex, the posterior part of the right operculum, the right amygdala, and the surrounding entorhinal cortex. "The amygdalo-cingulate coactivation could account for the emotional and affective aspects of dreams" (p. 165). Braun *et al.* (1997) found a similar activation of the neo- and paleocortex and the particular activation of the associative visual cortex (Braun *et al.*, 1998). Hong *et al.* (1995) observed that the eye movements seen during REM sleep are associated with an increased activation of the frontal eye field, and speculated that these eye movements could be "saccadic scans of targets in the dream scene" (p. 570). This result is very interesting because it was believed for decades that the eye movements of REM sleep are induced by the PGO waves and eye movement potentials originating from the pontine (Gottesmann, 1967; Michel *et al.*, 1964; Pompeiano and Morrison, 1965, 1966a, 1966b; Vanni-Mercier and Debilly, 1998; Vanni-Mercier *et al.*, 1996) and third nucleus (Gottesmann, 1969) levels. However, one must remember that the function of the cortex in the simultaneous occurrence of eye movements and dreaming was already hypothesized by Dement and Kleitman (1957b), Roffwarg *et al.* (1962), and Herman *et al.* (1984), and that occipital lesions decrease REM sleep eye movements whereas frontal ablation increases them (Jeannerod *et al.*, 1965; Mouret *et al.*, 1965). These results suggest that there are two kinds of eye movements occurring during REM sleep: reflex movements, which persist in pontine animals (Jouvet, 1962), and a second type that is related to dreams and that is of cortical origin. Finally, different results have shown that the primary visual cortex is activated during REM sleep eye movements (Miyauchi *et al.*, 2009), and even independently of them (Hong *et al.*, 2009).

Finally, also with respect to prefrontal activation processes, it must be underlined that the dorsolateral prefrontal cortex recovers its waking activation level later than other cortical areas upon arousal from REM sleep (Balkin *et al.*, 2002), thus possibly participating in the forgetting of dreams.

Nevertheless, some forebrain areas involved in higher integrated functions have been shown to be deactivated during REM sleep. For example, this is the case with the dorsolateral prefrontal cortex, a more recently evolved brain structure that is involved in higher intellectual functions (anticipation processes, long-term memory, etc.) (Braun *et al.*, 1997; Lövblad *et al.*, 1999; Maquet *et al.*, 1996, 2004). It is of interest that this same deactivation is another specific characteristic of schizophrenia, particularly when cognitive functions are disturbed (Buschbaum *et al.*, 1982; Fletcher *et al.*, 1998; Weinberger *et al.*, 1986). Moreover, there is one occasion when the dorsolateral prefrontal cortex and the posterior cingulate cortex (which is not part of the limbic system) are deactivated together, as they are in REM sleep: it is when pianists are so involved in their playing that they lose contact with the environment (Parsons *et al.*, 2005), like in schizophrenia.

The deactivation of the primary visual cortex (Braun *et al.*, 1998) is now a matter of discussion (Hong *et al.*, 2009). However, it has long been established that the subject is sensorily disconnected from the periphery (Blake and Gerard, 1937; Dement, 1958; Williams *et al.*, 1964). Another line of evidence supporting this notion is the lack of resetting of the gamma rhythm by peripheral stimulation during REM sleep (Llinas and Ribary, 1993). Even if the primary visual cortex deactivation is uncertain, the at least partial disconnection from the periphery is obvious at the thalamic level. Indeed, while the thalamic postsynaptic neurons are activated during REM sleep, as shown by the thalamocortical responsiveness during REM sleep (Albe-Fessard *et al.*, 1964; Favale *et al.*, 1965; Gandolfo *et al.*, 1980; Rossi *et al.*, 1965; Steriade, 1970), there is a presynaptic inhibition during the eye movement bursts (that occur during dreaming), as shown by the depolarization of thalamic afferents (Dagnino *et al.*, 1969; Gandolfo *et al.*, 1980; Ghelarducci *et al.*, 1970; Iwama *et al.*, 1966; Sakakura and Iwama, 1965). This presynaptic inhibition could be related to GABAergic influences originating in the thalamic reticular nucleus (Steriade *et al.*, 1985, 1987) or to vestibular ascending influences (Ghelarducci *et al.*, 1970; Morrison and Pompeiano, 1966). Indeed, contrary to the waking presynaptic inhibition, the presynaptic inhibition that takes place during the eye movement bursts is not of cortical origin, since they persist after homolateral decortication (Gandolfo and Gottesmann, 1982). Here also, it must be underlined that the lowering of sensory constraints, a consequence of the partial central sensory deafferentation, is also a criterion of schizophrenia (Behrendt and Young, 2005), and primary visual cortex deactivation has been observed in at least one patient with visual and auditory hallucinations (Silbersweig *et al.*, 1995) (personal communication, 2007). This functional sensory deafferentation could also explain the increased pain threshold that is observed during acute episodes of this disease (Griffin and Tyrrell, 2003).

Finally, a recently published biophysical hypothesis suggests that both dreaming imagery and psychotic hallucinations could be the consequence of hologram-like representations induced in the primary and secondary visual areas by biophoton activation related to redox molecular processes (Bokkon *et al.*, 2010).

3. *Neurochemistry and Pharmacology*

The one transmitter for which the activating properties are without question is glutamate, whose action increases sodium channel permeability. Its influence on the basic and higher integrated processes of dreaming is beginning to be well established. Concerning the basic support of REM sleep, various studies have shown that glutamate promotes REM sleep-on processes (Datta *et al.*, 2001, 2002; Onoe and Sakai, 1995; Sakai and Crochet, 2003). In the prefrontal cortex, glutamate is produced from collaterals of pyramidal neurons (Levy *et al.*, 2006), astrocytes (Ye *et al.*, 2003), and subcortical neurons (Wang *et al.*, 2005). Its prefrontal levels do not differ between REM sleep and waking (Dash *et al.*, 2009; Léna *et al.*, 2005); similarly, there is no difference in glutamate transporter mRNA expression in the prefrontal and primary visual cortex in schizophrenic patients compared with healthy controls (Lauriat *et al.*, 2005). In contrast, glutamate levels are significantly lower in the nucleus accumbens during REM sleep than they are during waking (Léna *et al.*, 2005), although the aspartate level remains unchanged. This is the consequence of lowered hippocampal glutamate release, which prevents its release in the nucleus accumbens from neurons originating in the prefrontal cortex (Grace, 2000). This decrease in glutamate is in accordance with the theory of schizophrenia (Grace, 2000; Heresco-Levy, 2000). Indeed, glutamate antagonists induce psychotic symptoms and, at the same time, vivid dreaming (Reeves *et al.*, 2001). It is now nearly entirely accepted that the decrease in hippocampal and prefrontal glutamatergic afferents to the nucleus accumbens favors the occurrence of hallucinations, delusions, and bizarre thought processes, all of which are common to both dreaming and schizophrenia. Moreover, this double loss of hippocampal and prefrontal afferents to the nucleus accumbens disinhibits the glutamatergic afferents coming from the amygdala (which are strongly activated during REM sleep (Maquet and Franck, 1997)) and contributes to the affective disturbances observed in schizophrenia (Grace, 2000).

Acetylcholine was shown to be capable of inducing REM sleep by brainstem microinjection as early as 1963 (Cordeau *et al.*, 1963), and its involvement in this sleep stage was definitively established by injection of the agonist carbachol the following year (George *et al.*, 1964). That acetylcholine activates cortical processes had already been indirectly shown in 1949 through the use of inhibitors of acetylcholine destruction (Bremer and Chatonnet, 1949), and then later with atropine (a muscarinic receptor blocker), which suppressed cortical waking patterns (Wikler, 1952). Acetylcholine activates cortical neurons by decreasing K^+ currents and blocking the Ca^{2+}-dependent component of the slow hyperpolarization that follows the action potential. Acetylcholine also blocks M currents, a nonactivating current which is active at resting potentials and which turns on slowly with depolarization (Nicoll *et al.*, 1990). The cortical-activating influences of acetylcholine have been definitively established by numerous studies (Mednikova *et al.*, 1998; Szymusiak *et al.*, 1990; Vanderwolf, 1988). The cortical-activating acetylcholine originates from intracortical neurons and from the nucleus basalis (Meynert nucleus in humans) (Buzsaki and Gage, 1989; Buzsaki *et al.*, 1988), which is activated during REM sleep as evidenced by the maximal release of acetylcholine during this sleep phase (Vasquez and Baghdoyan, 2001). The cortical release occurs mainly at varicosities (diffuse release); indeed, only 14% of the involved terminals give rise to synaptic junctions (Descarries *et al.*, 1997). Thus, acetylcholine acts more as a neuromodulator than as a neurotransmitter. Its level was first reported to be slightly higher during REM sleep than during waking (Jasper and Tessier, 1971). However, a more recent study on cats (Marrosu *et al.*, 1995) found that its release was highest during active waking, lowest during slow-wave sleep, and intermediate during REM sleep, when its level only reached that of quiet waking.

Acetylcholine also has inhibitory influences at the cortical level (Muzur *et al.*, 2002). "Inhibitory responses ... had shorter latency and faster recovery than the excitatory ones. The threshold of both inhibitory and excitatory responses were almost the same. (In some cases) responses to acetylcholine reversed from excitatory to inhibitory with increasing dose" (p. 123) (Nelson *et al.*, 1973). Recent findings have confirmed these inhibitory influences on pyramidal neurons (Giulledge and Stuart, 2005; Levy *et al.*, 2006). It is worth noting that acetylcholine is crucial for cognitive function (Perry and Piggott, 2003; Perry *et al.*, 1999; Sarter and Bruno, 2000) and that lowering its level in the forebrain promotes the occurrence of hallucinations during waking; this is what occurs in schizophrenia (Collerton *et al.*, 2005), a psychosis in which there is a lowering of muscarinic receptors in the cortex (Raedler *et al.*, 2003).

Serotonin was shown early on to have no significant influence on REM sleep, as this sleep stage did not recover after the administration of 5-hydroxytryptophan to reserpinized cats (Matsumoto and Jouvet, 1964). The neurons which innervate the cortex originate from the dorsal and medial mesencephalic raphe nuclei (Dahlstrom and Fuxe, 1964; Fuxe, 1965). They become silent as soon as REM sleep begins (McGinty and Harper, 1976; McGinty *et al.*, 1974; Rasmussen *et al.*, 1984), and serotonin release is decreased at the cortical level during REM sleep (Cespuglio *et al.*, 1992). Although serotonin that is released at the varicosity level (Descarries *et al.*, 1975; Nelson *et al.*, 1973) can depolarize some cortical neurons through 5-HT$_2$ receptors, this activation seems to partly act on GABAergic interneurons, resulting in the hyperpolarization of pyramidal neurons

(Araneda and Andrade, 1991). Serotonin directly inhibits the majority of cortical neurons (Krnjevic and Phillis, 1963; Reader *et al.*, 1979) through 5-HT$_{1A}$ receptors, which increase K$^+$ conductance. However, serotonin also increases the signal/noise ratio (McCormick *et al.*, 1993), thereby increasing neuron performance. Although the raphe nuclei act as a PS-off system and have no significant function in REM sleep-generating processes (Sakai and Crochet, 2001), the decrease in forebrain serotonin content during REM sleep certainly participates in the abnormal mental activity seen in dreaming; consistent with this, reuptake inhibitors are used as adjuvants of antischizophrenia molecules (Silver *et al.*, 2000; Van Hes *et al.*, 2003).

A positive influence of catecholamines on REM sleep was shown as early as 1964, in experiments showing that dihydroxyphenylalanine increased this sleep stage in reserpinized cats (Matsumoto and Jouvet, 1964). However, soon afterward it was shown that noradrenergic neurons of the locus coeruleus, which innervate the cortex (Dahlstrom and Fuxe, 1964), become silent during REM sleep (Aston-Jones and Bloom, 1981a; Hobson *et al.*, 1975; Rasmussen *et al.*, 1986) and that the level of noradrenaline in the cortex is thus decreased during REM sleep relative to during waking and slow-wave sleep (Léna *et al.*, 2005). This is also the case in the nucleus accumbens (Léna *et al.*, 2005), which is mainly innervated by medulla oblongata A$_2$ neurons (Delfs *et al.*, 1998).

Noradrenaline is released at the varicosity level (Descarries *et al.*, 1977; Fuxe *et al.*, 1968) and binds to several brain receptors with differing effects: postsynaptic excitatory α_1 and β_1 receptors, the mostly presynaptic inhibitory α_2 receptors, and inhibitory β_2 receptors (Langer, 2008). Although noradrenaline has mainly inhibitory influences at the cortical level (Foote *et al.*, 1975; Frederickson *et al.*, 1971; Krnjevic and Phillis, 1963; Manunta and Edeline, 1999; Nelson *et al.*, 1973; Reader *et al.*, 1979)—and α_2 agonists reinforce the effects of anesthetics (Crassous *et al.*, 2007)—it increases the signal/noise ratio of neuronal functioning (Aston-Jones and Bloom, 1981b; Berlucchi, 1997; Foote *et al.*, 1975; Warren and Dykes, 1996; Waterhouse *et al.*, 1990).

Noradrenaline is crucial for cognitive processes. In aging primates, there is a parallel decrease in cortical catecholamine levels and performance (Arnsten and Goldman-Rakic, 1987; Goldman-Rakic and Brown, 1981). More generally, NA depletion increases the frequency of wrong responses to irrelevant stimuli while decreasing the frequency of correct responses to relevant stimuli (Milstein *et al.*, 2007; Selden *et al.*, 1990). Increases in cognitive performance are also associated with increases in prefrontal noradrenaline release (Berridge *et al.*, 2006). Moreover, stimulation of the locus coeruleus for therapeutic purposes in humans induces "well-being (and) improves clarity of ... thinking" (p. 179) (Libet and Gleason, 1994). The silencing of noradrenergic neurons during REM sleep certainly contributes to the cognitive attenuation (Johnson, 2003) and explains at least in part the discontinuity of dreams and particularly the "loss of self-conscious awareness" (Hofle *et al.*, 1997) that is characteristic of this sleep stage as well as of schizophrenia (Gottesmann and Gottesman, 2007). The similarity between the mental disturbances seen in dreaming and in schizophrenia related to noradrenaline deficits is reinforced by the fact that noradrenaline reuptake inhibitors are used as adjuvants in the treatment of this psychosis (Friedman *et al.*, 1999; Linner *et al.*, 2002). Finally, the reappearance of noradrenaline function in the few seconds preceding behavioral arousal upon awakening (Aston-Jones and Bloom, 1981a), which represents the precocious recovery of brain-waking properties except in the prefrontal cortex (Balkin *et al.*, 2002), probably promotes the forgetting of dreams (Gottesmann, 2006). A disturbance in this forgetting process could favor a schizophrenic belief in the reality of hallucinations (Gottesmann, 2008; Kelly, 1998).

Dopamine was first identified in the brain by Dahlström and Fuxe (1964), and its presence in the cortex was shown by Thierry and colleagues (1973). Subsequently, Hökfelt *et al.* (1974) showed that there are two kinds of dopaminergic forebrain terminals, one in the frontal cortex and one in the limbic system, and Fuxe *et al.* (1974) distinguished mesocortical and mesolimbic terminals for them, both of which originate from the A_{10} ventral tegmental area. Dopamine at the prefrontal cortex level is mostly released from varicosities, with apposed synapses involved in only 40% of cases (Smiley and Goldman-Rakic, 1993), meaning that its release is most often diffuse, as is the case with other monoamines. Dopamine inhibits the occurrence of REM sleep (Crochet and Sakai, 1999, 2003), see Gottesmann (2010). It also inhibits cortical neurons, which was shown long ago (Krnjevic and Phillis, 1963; Reader *et al.*, 1979), by increasing the release of GABA from interneurons via D_2 receptors (Grobin and Deutch, 1998) as well as by directly inhibiting pyramidal neurons through D_1 receptors (Abi-Dargham and Moore, 2003; Pirot *et al.*, 1992; Rétaux *et al.*, 1991). At the same time, dopamine, like noradrenaline, also increases the signal-to-noise ratio of synaptic afferent influences (Luciana *et al.*, 1998). Further, its importance for cognitive processes in the prefrontal cortex (Nieoullon, 2002) is underscored by the fact that D_1 receptor gene expression is highest in adolescence and early adulthood, a critical life-stage in schizophrenia. Moreover, the expression of both D_1 and D_2 receptors decreases significantly in old age (Weickert *et al.*, 2007).

Dopaminergic neurons fire continuously during sleep-waking stages (Miller *et al.*, 1983; Trulson and Preussler, 1984), and it was recently confirmed that during REM sleep they fire in bursts (Dahan *et al.*, 2007; Miller *et al.*, 1983) as they do during active waking, thereby releasing more dopamine (Gonon, 1988; Suaud-Chagny *et al.*, 1992). However, although a maximal amount of dopamine release has indeed been demonstrated in the nucleus accumbens during REM sleep (Léna *et al.*, 2005)—which is in accordance with the abnormally high dopaminergic function observed in schizophrenia (MacKay *et al.*, 1982) as well as with the hallucinatory activity and vivid dreaming that are induced by dopamine agonists (Arnulf *et al.*, 2000; Buffenstein *et al.*, 1999; Grace, 1991)—a decrease in dopamine concentration has been observed in the prefrontal cortex during REM sleep in rats (Léna *et al.*, 2005). The imagery shown deactivation in the prefrontal cortex (Braun *et al.*, 1997; Lövblad *et al.*, 1999; Madsen *et al.*, 1991; Maquet *et al.*, 1996, 2004) could be behind the opposing levels of dopamine release that have been observed in the prefrontal cortex (Takahata and Moghaddam, 2000, 2003) and the nucleus accumbens (Brake *et al.*, 2000; Jackson *et al.*, 2001). Finally, a decrease in D_1 receptor activation in the prefrontal cortex, which appears to be outside a narrow window showing an inverted U-shape relationship with cognitive performance (Meyer-Lindenberg and Weinberger, 2006), could explain, together with the noradrenaline decrease as well as with other possible disturbed influences (see below), the reduction or loss of reflectiveness that is encountered during REM sleep as well as in schizophrenia.

The influence of GABA during REM sleep is, unfortunately, still open to speculation, as the release of this purely inhibitory transmitter in the forebrain during REM sleep has not been studied to date, and more generally has not been examined during the sleep–waking cycle. However, several intriguing results have been published in connection to the decreased dopamine release that takes place in the prefrontal cortex during REM sleep: (1) GABAergic interneurons are mainly (60%) situated in layers three and five and impinge the pyramidal neurons (Gabbott *et al.*, 1997), and the pyramidal neurons are highly activated during REM sleep (Arduini *et al.*, 1963; Evarts, 1964; Morrison and Pompeiano, 1965); (2) the stimulation of mesocortical neurons from the ventral tegmental area activates all fast-spiking GABAergic interneurons and inhibits the majority of pyramidal cells (Tseng *et al.*, 2006); in addition, dopamine release is decreased during REM sleep compared with waking in the medial prefrontal cortex (Léna *et al.*, 2005); (3) dopamine increases interneuron excitability (Gorelova *et al.*, 2002); (4) dopamine promotes GABA release in the prefrontal cortex (Grobin and Deutch, 1998); (5) amphetamines induce c-Fos-mediated activation of prefrontal cortex interneurons (Morshedi and Meredith, 2007). All of these results strongly suggest that there should be a decrease in GABA levels during REM sleep. Moreover, it has been shown that, in addition to a direct inhibition of pyramidal cells through $5-HT_{1A}$ receptor activation, serotonin activates cortical GABAergic interneurons through $5-HT_2$ receptors, which in turn inhibit pyramidal neurons (Araneda and Andrade, 1991; Davies *et al.*, 1987; Sheldon and Aghajanian, 1990) (see above). The silence of serotonergic neurons during REM sleep should induce a complementary GABAergic disinhibition. Finally, referring again to the parallels between REM sleep and schizophrenic processes, there is also a GABAergic deficit in this disease (Lewis, 2000; Lewis and Hashimoto, 2007; Lewis *et al.*, 2005; Maldonado-Aviles *et al.*, 2009).

II. Conclusion

The progressive steps from the discovery of REM sleep-generating processes to the successive findings regarding the neurobiological basis of dreaming underscore the crucial importance of the inseparable progress of technical methods.

It is of interest that two millennia separated the first detailed description of the major peripheral characteristics of dreaming from the first contemporary experimental results of brain research in this field, while only about 60 years were necessary to establish relatively solid knowledge of the basic and higher integrated neurobiological processes underlying REM sleep.

It should also be highlighted that, today, numerous neurobiological results on REM sleep tend to confirm the observations of philosophers, writers, and neuropsychiatrists from previous centuries (see *Introduction*) regarding a strong relationship between dreaming and schizophrenia-like mentation. This is the consequence of the breakdown of the complex equilibrium between forebrain-activating and inhibitory influences which regulate waking mental activities.

More generally, the neurobiological properties of REM sleep constitute a fascinating brain state whose characteristics are candidate endophenotypes of not only schizophrenia but also depression, mental retardation, and dementia (Gottesmann and Gottesman, 2007).

One question remains open to discussion: Does the particular REM sleep mentation have a meaning, even a symbolic signification in spite of its psychotic-like properties? This is the fundamental assertion of psychoanalysis, a theory which is still heavily disputed today.

Acknowledgments

I thank Dr. Peter Follette for his improvement of the English.

References

Abi-Dargham, A., and Moore, H. (2003). Prefrontal DA transmission at D1 receptors and the pathology of schizozphrenia. *Neuroscientist* **9**, 404–416.

Albe-Fessard, D., Massion, J., Hall, R., and Rosenblith, W. (1964). Modifications au cours de la veille et du sommeil des valeurs moyennes de réponses nerveuses centrales induites par des stimulations somatiques chez le Chat libre. *C. R. Acad. Sci.* **258**, 353–356.

Allison, T. (1965). Cortical and subcortical evoked responses to central stimuli during wakefulness and sleep. *Electroenceph. Clin. Neurophysiol.* **18**, 131–139.

Allison, T. (1968). Recovery cycles of primary evoked potentials in cats sensorimotor cortex. *Experentia* **24**, 240–241.

Amadeo, M., and Gomez, A. (1966). Eye movements, attention and dreaming in the congenetically blind. *Canad. Psychiat. Ass.* **11**, 501–507.

Araneda, R., and Andrade, R. (1991). 5-hydroxytryptamine2 and 5-hydroxytryptamine 1A receptors mediate opposing responses on membrane excitability in the association cortex. *Neuroscience* **40**, 399–412.

Arduini, A., Berlucchi, G., and Strata, P. (1963). Pyramidal activity during sleep and wakefulness. *Arch. ital. Biol.* **101**, 530–544.

Arnsten, A. F., and Goldman-Rakic, P. S. (1987). Noradrenergic mechanisms in age-related cognitive deficit. *J. Neur. Transm. Suppl.* **24**, 317–324.

Arnulf, I., Bonnet, A. M., Damier, P., Bejjani, B. P., Seilhean, D., Derenne, J. P., and Agid, Y. (2000). Hallucinations, REM sleep and Parkinson's disease: a medical hypothesis. *Neurology* **55**, 281–288.

Aserinsky, E., and Kleitman, N. (1953). Regularly occurring periods of eye motility, and concomitant phenomena during sleep. *Science* **118**, 273–274.

Aserinsky, E., and Kleitman, N. (1955). Two types of ocular motility occurring during sleep. *J. Appl. Physiol.* **8**, 1–10.

Aston-Jones, G., and Bloom, F. E. (1981a). Activity of norepinephrine-containing neurons in behaving rats anticipates fluctuations in the sleep-waking cycle. *J. Neurosci.* **1**, 876–886.

Aston-Jones, G., and Bloom, F. E. (1981b). Norepinephrine-containing locus coeruleus neurons in behaving rat exhibit pronounced responses to non-noxious environmental stimuli. *J. Neurosci.* **1**, 887–900.

Balkin, T. J., Braun, A. R., Wesensten, N. J., Jeffries, K., Varga, M., Baldwin, P., Belenky, G., and Herscovitch, P. (2002). The process of awakening: a PET study of regional brain activity patterns mediating the re-establishment of alertness and consciousness. *Brain* **125**, 2308–2319.

Behrendt, R. P., and Young, C. (2005). Hallucinations in schizophrenia, sensori impairment and brain disease: an unified model. *Behav. Brain Sci.* **27**, 771–787.

Berger, E., and Loewy, R. (1898). L'état des yeux pendant le sommeil et la théorie du sommeil. *C. R. Soc. Biol* 448–450.

Berger, H. (1929). Ueber das electroenkephalogram des Menschen. *Arch. Psychiat. Nervenkrank.* **87**, 527–570.

Berger, R. J., and Oswald, I. (1962). Eye movements during active and passive dreams. *Science* **137**, 601.

Berlucchi, G. (1997). One or many arousal system? Reflections on some of Guiseppe Moruzzi's foresights and insights about intrinsic regulation of brain activity. *Arch. Ital. Biol.* **135**, 5–14.

Berlucchi, G., Moruzzi, G., Salvi, G., and Strata, P. (1964). Pupil behavior and ocular movements during desynchronized and synchronized sleep. *Arch. Ital. Biol.* **102**, 230–244.

Berlucchi, G., and Strata., P.(1965). Ocular phenomena during synchronized and desynchronized sleep. Aspects Anatomo-Fonctionnels de la physiologie du sommeil. CNRS, Paris, pp.285–307.

Berridge, C. W., Devilbiss, D. M., Andrzejewski, M. E., Arnsten, A. F., Kelley, A. E., Schmeichel, B., Hamilton, C., and Spencer, R. C. (2006). Methylphenidate preferentially increases catecholamine neurotransmission within the prefrontal cortex at low doses that enhance cognitive function. *Biol. Psychiat.* **60**, 1111–1120.

Blake, H., and Gerard, R. W. (1937). Brain potentials during sleep. *Am. J. Physiol.* **119**, 692–703.

Blake, H., Gerard, R. W., and Kleitman, N. (1939). Factors influencing brain potentials during sleep. *J. Neurophysiol.* **2**, 48–60.

Bokkon, I., Dai, J., and Antal, I. (2010). Picture representation during REM dreams: a redox molecular hypothesis. *Biosystems* **100**, 79–86.

Bosinelli, M. (1995). Mind and consciousness during sleep. *Brain Res.* **69**, 195–201.

Bosinelli, M., and Cigogna, P. C. (2003). REM and NREM mentation: Nielsen's model once again supports the supremacy of REM. In: Sleep and Dreaming. Scientific advances and reconsiderations (E. Pace-Schott, M. Solms, M. Blagrove, and S. Harnard, eds.), Cambridge University Press, Cambridge, pp. 124–125.

Bouyer, J. J., Montaron, M. F., and Rougeul, A. (1981). Fast fronto-parietal rhythms during combined focused attentive behavior and immobility in cat: cortical and thalamic localizations. *Electroenceph. Clin. Neurophysiol.* **51**, 244–252.

Brake, W. G., Flores, G., Francis, D., Meaney, M. J., Srivastava, L. K., and Gratton, A. (2000). Enhanced nucleus accumbens dopamine and plasma corticosterone stress responses in adult rats with neonatal excitotoxic lesions to the medial prefrontal cortex. *Neuroscience* **96**, 687–695.

Braun, A. R., Balkin, T. J., Wesensten, N. J., Carson, R. E., Varga, M., Baldwin, P., Selbie, S., Belenky, G., and Herscovitch, P. (1997). Regional cerebral blood flow throughout the sleep-wake cycle: an ^{15}0 PET study. *Brain* **120**, 1173–1197.

Braun, A. R., Balkin, T. J., Wesensten, N. J., Gwardry, F., Carson, R. E., Varga, M., Baldwin, P., Belenky, G., and Herscovitch, P. (1998). Dissociated pattern of activity in visual cortices and their projections during human rapid eye movement sleep. *Science* **279**, 91–95.

Bremer, F., and Chatonnet, J. (1949). Acetylcholine et cortex cérébral. *Arch. Int. Physiol.* **57**, 106–109.

Bubnoff, N., and Heidenhain, R. (1881). Ueber Erregungs-Hemmungsvorgänge innerhalb der motorischen Hirncentren. In: Arch. Gesam. Physiol. (E. F. W. Pflüger, ed.), pp. 137–202.

Buffenstein, A., Heaster, J., and Ko, P. (1999). Chronic psychotic illness from amphetamine. *Am. J. Psychiat* **156**, 662.

Buschbaum, M. S., Ingvar, D. H., Kessler, R., Waters, R. N., Cappelletti, J., Van Kammen, D. P., King, A. C., Johnson, J. L., Manning, R. G., Flynn, R. W., Bunney, W. E. J., and Sokoloff, L. (1982). Cerebral glucography with positron tomography, use in normal subjects and in patients with schizophrenia. *Arch. Gen. Psychiat.* **39**, 251–259.

Buzsaki, G., Bickford, R. G., Ponomareff, G., Thal, L. J., Mandel, R., and Cage, F. H. (1988). Nucleus basalis and thalamic control of neocortical activity in the freely moving rat. *J. Neurosci.* **8**, 4007–4026.

Buzsaki, G., and Gage, F. H. (1989). The nucleus basalis: a key structure in neocortical arousal. In: Central Cholinergic Synaptic Transmission (M. Froescher and U. Misgeld, eds.), Kirhäuser Verlag, Basel, pp. 159–171.

Cantero, J. L., Atienza, M., Madsen, J. R., and Stickgold, R. (2004). Gamma EEG dynamics in neocortex and hippocampus during human wakefulness and sleep. *NeuroImage* **22**, 1271–1280.

Caton, R. (1875). The electrical currents of the brain. *Br. Med. J.* **278**.

Cavallero, C. (2003). REM sleep dreaming: the never-ending story. In: Sleep and Dreaming. Scientific Advances and Reconsideration (E. Pace-Schott, M. Solms, M. Blagrove, and S. Harnard, eds.), Cambridge University Press, Cambridge, pp. 127–128.

Cavallero, C., Cicogna, P., Natale, V., Occhionero, M., and Zito., A. (1992). Slow wave sleep dreaming. *Sleep* **15**, 562–566.

Cespuglio, R., Houdouin, F., Oulerich, M., El Mansari, M., and Jouvet, M. (1992). Axonal and somato-dendritic modalities of serotonin release: their involvement in sleep regulation, triggering and maintenance. *J. Sleep Res.* **1**, 150–156.

Collerton, D., Perry, E., and McKeith, I. (2005). A novel perception and attention deficit model for recurrent visual hallucinations. *Brain Behav. Sci.* **28**, 737–757.

Cordeau, J. P., Moreau, A., Beaulnes, A., and Laurin, C. (1963). EEG and behavioral changes following microinjections of acetylcholine and adrenaline in the brain stem of cats. *Arch. Ital. Biol.* **101**, 30–47.

Cordeau, J. P., Walsh, J., and Mahut, H. (1964). Sensory transmission during various stages of sleep and waking. *Electroenceph. Clin. Neurophysiol.* **17**, 442.

Cordeau, J. P., Walsh, J., and Mahut, H. (1965). Variations dans la transmission des messages sensoriels en fonction des différents états d'éveil et de sommeil. In: Aspects Anatomo-Fonctionnels de la physiologie du sommeil. (CNRS, ed.), CNRS, Paris, pp. 477–507.

Corsi-Cabrera, M., Miro, E., del Rio Portilla, Y., Perez-Garci, E., Villanueva, Y., and Guevara, M. (2003). Rapid eye movement sleep dreaming is characterized by uncoupled EEG activity netween frontal abd perceptual cortical regions. *Brain Cogn.* **51**, 337–345.

Crassous, P. A., Denis, C., Paris, H., and Sénard, J. M. (2007). Interest of alpha2-adrenergic agonists and antagonists in clinical practice: background facts and perspectives. *Curr. Top. Med. Chem* **7**, 187–194.

Creutzfeldt, O., Baumgartner, G., and Schoen, L. (1956). Reaktionen einzelner Neurons des senso-motorischen cortex nach elektrischen Reizen. I Hemmung und Erregung nach direkten und contralateralen Einzelreizen. *Arch. Psychiat. Nervenkrank.* **194**, 597–619.

Crochet, S., and Sakai, K. (1999). Effects of microdialysis application in monoamines on the EEG and behavioral states in the cat mesopontine tegmentum. *Eur. J. Neurosci.* **11**, 3738–3752.

Crochet, S., and Sakai, K. (2003). Dopaminergic modulation of behavioral states in mesopontine tegmentum: a reverse microdialysis study in freely moving cats. *Sleep* **26**, 801–806.

Dagnino, N., Favale, E., Loeb, C., Manfredi, M., and Seitun, A. (1969). Presynaptic and postsynaptic changes in specific thalamic nuclei during deep sleep. *Arch. Ital. Biol.* **107**, 668–684.

Dahan, L., Astier, B., Vautrelle, N., Urbain, N., Kocsis, B., and Chouvet, G. (2007). Prominent burst firing of dopaminergic neurons in the vental tegmental area during paradoxical sleep. *Neuropschopharmacology* **32**, 1232–1241.

Dahlstrom, A., and Fuxe, K. (1964). Evidence for existence of monoamine-containing neurons in the central nervous system. *Acta Physiol. Scand.* **62** (Suppl. 232), 3–55.

Dash, M. B., Douglas, C. L., Vyazovskiy, V., Cirelli, C., and Tononi, G. (2009). Long-term homeostasis of extracellular glutamate in the rat cerebral cortex across sleep and waking states. *J. Neurosci.* **29**, 620–629.

Datta, S., and Siwek, D. F. (1997). Excitation of the brainstem pedunculopontine tegmentum cells induces wakefulness and REM sleep. *J. Neurophysiol.* **77**, 2975–2988.

Datta, S., and Siwek, D. F. (2002). Single cell activity patterns of pedunculopontine tegmentum neurons across the sleep-wake cycle in freely moving rats. *J. Neurosci. Res.* **70**, 611–621.

Datta, K., Spoley, E. E., Mavanji, V., and Patterson, E. H. (2002). A novel role of pedunculopontine tegmental kainate receptors: a mechanism of rapid eye movement sleep generation in the rat. *Neuroscience* **114**, 157–164.

Datta, S., Spoley, E. E., and Patterson, E. H. (2001). Microinjection of glutamate into the pedunculopontine tegmentum induces REM sleep and wakefulness in the rat. *Am. J. Physiol.* **280**, R752–R759.

Davies, M. F., Deisz, R. A., Prince, D. A., and Peroutka., S. J. (1987). Two distinct effects of 5-hydroxytryptamine on single cortical neurons. *Brain Res.* **423**, 347–352.

Delfs, J. M., Zhu, Y., Druhan, J. P., and Aston-Jones, G. S. (1998). Origin of noradrenergic afferents to the shell subregion of the nucleus accumbens: anterograde and retrograde tract-tracing studies in the rat. *Brain Res.* **806**, 127–140.

Dement, W. (1955). Dream recall and eye movements during sleep in schizophrenic and normals. *J. Neural Eng.* **122**, 263–269.

Dement, W. C. (1958). The occurrence of low voltage fast electroencephalogram patterns during behavioral sleep in the cat. *Electroenceph. Clin. Neurophysiol.* **10**, 291–296.

Dement, W., and Kleitman, N. (1957a). Cyclic variations in EEG during sleep and their relation to eye movements, body motility, and dreaming. *Electroenceph. Clin. Neurophysiol.* **9**, 673–690.

Dement, W., and Kleitman, N. (1957b). The relation of eye movements during sleep to dream activity: an objective method for the study of dreaming. *J. Exp. Psychol.* **53**, 339–346.

Dement, W., and Wolpert, E. A. (1958). The relation of eye movements, body motility, and external stimuli to dream content. *J. Exp. Psychol.* **55**, 543–553.

Demetrescu, M., Demetrescu, M., and Iosif, G. (1966). Diffuse regulation of visual thalamo-cortical responsiveness during sleep and wakefulness. *Electroenceph. Clin. Neurophysiol.* **20**, 450–469.

Derbyshire, A. J., Rempel, B., Forbes, A., and Lambert, E. F. (1936). The effects of anesthetics on action potentials in the cerebral cortex of the cat. *Am. J. Physiol.* **116**, 577–596.

Descarries, L., Beaudet, A., and Watkins, K. C. (1975). Serotonin nerve terminals in the adult rat neocortex. *Brain Res.* **100**, 563–588.

Descarries, L., Gisinger, V., and Steriade, M. (1997). Diffuse transmission by acetylcholine in the CNS. *Prog. Neurobiol.* **53**, 603–625.

Descarries, L., Watkins, K. C., and Lapierre, Y. (1977). Noradrenergic axon terminals in the cerebral cortex of rats. III. Topometric ultrastructural analysis. *Brain Res.* **133**, 197–222.

Evarts, E. V. (1962). Activity of neurons in visual cortex of the cat during sleep with low voltage EEG activity. *J. Neurophysiol.* **25**, 812–816.

Evarts, E. V. (1964). Temporal patterns of discharge of pyramidal tract neurons during sleep and waking in the monkey. *J. Neurophysiol.* **27**, 152–171.

Evarts, E. V. (1965). Neuronal activity in visual and motor cortex during sleep and waking. In: Aspects Anatomo-fonctionnels de la physiologie du sommeil. (CNRS, ed.), Paris, pp. 189–212.

Evarts, E. V., Fleming, T. C., and Huttenlocher, P. R. (1960). Recovery cycle of visual cortex of the awake and sleeping cat. *Am. J. Physiol.* **199**, 373–376.

Ey, H. (1967). La dissolution du champ de la conscience dans le phénomène sommeil-veille et ses rapports avec la psychopathologie. *Pres. Med.* **75**, 575–578.

Favale, E., Loeb, C., Manfredi, M., and Sacco, G. (1965). Somatic afferent transmission and cortical responsiveness during natural sleep and arousal in the cat. *Electroenceph. Clin. Neurophysiol.* **18**, 354–368.

Fletcher, P. C., McKenna, P. J., Frith, C. D., Grasby, P. M., Friston, K. J., and Dolan, R. J. (1998). Brain activation in schizophrenia during a graded memory task studied with functional neuroimaging. *Arch. Gen. Psychiat.* **55**, 1001–1008.

Fontana, F. (1765). Dei moti dell'ride. *Stamperia Jacopo* Giusti, Lucca.

Foote, S. L., Freedman, R., and Oliver, A. P. (1975). Effects of putative neurotransmitters on neuronal activity on monkey auditory cortex. *Brain Res.* **86**, 229–242.

Foulkes, W. D. (1962). Dream reports from different stages of sleep. *J. Abnorm. Soc. Psychol.* **65**, 14–25.

Frederickson, R. C. A., Jordan, L. M., and Phillis, J. W. (1971). The action of noradrenaline on cortical neurons: effects of pH. *Brain Res.* **35**, 556–560.

Freud, S. (1900). The Interpretation of Dreams, The Hogart Press, London.

Friedman, J. I., Adler, D. N., and Davis, K. L. (1999). The role of norepinephrine in the physiopathology of cognitive disorders: potential applications to the treatment of cognitive dysfunction in schizophrenia and Alzheimer's disease. *Biol. Psychiat.* **46**, 1243–1252.

Fromm, E. (1953). Le langage oublié. Payot, Paris.

Fuxe, K. (1965). evidence for the existence of monoamine neurons in the central nervous system. *Acta Physiol. Scand. Suppl.* **247**, 37–85.

Fuxe, K., Hamberger, B., and Hökfelt, T. (1968). Distribution of noradrenaline nerve terminals in cortical areas in the rat. *Brain Res.* **8**, 125–131.

Fuxe, K., Hökfelt, T., Johanson, O., and Lidbrink, P. (1974). The origin of the dopamine nerve terminals in limbic and frontal cortex. Evidence for meso-cortico dopamine neurons. *Brain Res.* **82**, 349–355.

Gabbott, P. L. A., Dickie, B. G. M., Vaid, R. R., Headlam, A. J. N., and Bacon, S. J. (1997). Local-circuit neurones in the medial prefrontal cortex (areas 25, 32 and 24b) in the rat: morphology and quantitative distribution. *J. Comp. Neurol.* **377**, 465–499.

Gandolfo, G., Arnaud, C., and Gottesmann, C. (1980). Transmission in the ventrobasal complex of rat during the sleep-waking cycle. *Brain Res. Bull.* **5**, 921–927.

Gandolfo, G., and Gottesmann, C. (1982). Transmission in the ventrobasal complex of thalamus during rapid sleep and wakefulness in the homolaterally neodecorticated eat. *Acta Neurobiol. Exp.* **42**, 443–455.

George, R., Haslett, W. L., and Jenden, D. J. (1964). A cholinergic mechanism in the brainstem reticular formation: induction of paradoxical sleep. *Int. J. Neuropharmacol.* **3**, 541–552.

Gernet, R. (1889). Das Verhalten der Augen im Schlaf, Gustav Shade, Berlin.

Ghelarducci, B., Pisa, M., and Pompeiano, M. (1970). Transformation of somatic afferent volleys across the prethalamic and thalamic components of the lemniscal system during the rapid eye movements of sleep. *Electroenceph. Clin. Neurophysiol.* **29**, 348–357.

Gibbs, F. A., and Gibbs, E. L. (1950). Atlas of Electroencephalography. Addison-Wesley Press, Cambridge.

Giulledge, A. T., and Stuart, G. J. (2005). Cholinergic inhibition of neocortical pyramidal neurons. *J. Neurosci.* **25**, 10308–10320.

Goldman-Rakic, P. S., and Brown, R. M. (1981). Regional changes of monoamines in cerebral cortex and subcortical structures of aged rhesus monkeys. *Neuroscience* **6**, 177–187.

Gonon, F. G. (1988). Nonlinear relationship between impulse flow and dopamine released by rat midbrain dopaminergic neurons as studied by *in vivo* electrochemistry. *Neuroscience* **24**, 19–28.

Gorelova, N., Seamans, J. K., and Yang, C. R. (2002). Mechanisms of dopamine activation of fast-spiking interneurons that exert inhibition in rat prefrontal cortex. *J. Neurophysiol.* **88**, 3150–3166.

Gottesmann, C. (1967). Recherche sur la psychophysiologie du sommeil chez le Rat. Presses du Palais Royal, discussion and summary in English. Paris (still available).

Gottesmann, C. (1969). Etude Ser Les Activites Electrophysiologiques Phasiques Chez lerat. *Physiol. Behav.* **4**, 495–504.

Gottesmann, C. (2001). The golden age of rapid eye movement sleep discoveries. I. Lucretius-1964. *Prog. Neurobiol.* **65**, 211–287.

Gottesmann, C. (2006). The dreaming sleep stage: a new neurobiological model of schizophrenia? *Neuroscience* **140**, 1105–1115.

Gottesmann, C. (2008). Noradrenaline involvement in basic and higher integrated REM sleep processes. *Prog. Neurobiol.* **82**, 237–272.

Gottesmann, C. (2010). Aminergic influences in the regulation of basic REM sleep processes. In: Rapid Eye Movement Sleep: Regulation and Function. (B. N. Mallick, S. R. Pandi-Perumal, A. R. Morrison, and R. McCarley, eds.), Cambridge University Press, Cambridge.

Gottesmann, C., and Gottesman, I. I. (2007). The neurobiological characteristics of the rapid eye movement (REM) dreaming sleep stage are candidate endophenotypes of depression, schizophrenia, mental retardation and dementia. *Prog. Neurobiol.* **81**, 237–250.

Grace, A. A. (1991). Phasic versus tonic dopamine release and the modulation of dopamine system responsivity: a hypothesis for the etiology of schizophrenia. *Neuroscience* **41**, 1–24.

Grace, A. A. (2000). Gating information flow within the limbic system and the pathophysiology of schizophrenia. *Brain Res. Rev.* **31**, 330–341.

Griffin, J., and Tyrrell, I. (2003). Human Givens: A New Approach to Emotional Health and Clear Thinking, HG Publishing, New York.

Grobin, A. C., and Deutch, A. Y. (1998). Dopaminergic regulation of extracellular g-aminobutyric acid levels in the prefrontal cortex of the rat. *J. Pharmacol. Exp. Ther.* **285**, 350–357.

Gross, J., Byrne, J., and Fisher, C. (1965). Eye movements during emergent stage 1 EEG in subjects with lifelong blindness. *J. Neural Eng.* **141**, 365–370.

Heresco-Levy, U. (2000). N-methyl-D-aspartate (NMDA) receptor-based treatment approaches in schizophrenia: the first decade. *Int. J. Neuropharmacol.* **3**, 243–258.

Herman, J. H., Erman, M., Boys, R., Peiser, L., Taylor, M. E., and Roffwarg, H. (1984). Evidence for a directional correspondence between eye movements and dream imagery in REM sleep. *Sleep* **7**, 52–63.

Hervey de Saint Denys, M. J. L. (1867). Les rêves et les moyens de les diriger. Amyot, Paris (republished (1964) Tchou, Paris).

Hobson, J. A., McCarley, R. W., and Wyzinski, P. W. (1975). Sleep cycle oscillation: reciprocal discharge by two brainstem neuronal groups. *Science* **189**, 55–58.

Hofle, N., Paus, T., Reutens, D., Fiset, P., Gotman, J., Evans, A. C., and Jones, B. E. (1997). Regional cerebral blood flow changes as a function of delta and spindle activity during slow wave sleep. *J. Neurosci.* **17**, 4800–4808.

Hökfelt, T., Ljungdahl, A., Fuxe, K., and Johanson, O. (1974). Dopamine nerve terminals in the rat limbic cortex. Aspects of the dopamine hypothesis of schizophrenia. *Science* **184**, 177–179.

Hong, C. C. H., Gillin, J. C., Dow, B. C., Wu, J., and Buschbaum, M. S. (1995). Localized and lateralized cerebral glucose metabolism associated with eye movements during REM sleep and wakefulness: a positron emission tomography (PET) study. *Sleep* **18**, 570–580.

Hong, C. C. H., Harris, J. C., Pearlson, G. D., Kim, J. -S., Calhoun, V. C., Fallon, J. H., Golay, X., Gillen, J. S., Simmonds, D. J., van Zijl, P. C. M., Zee, D. S., and Pekar, J. J. (2009). fMRI evidence for multisensory recruitment associated with rapid eye movements during sleep. *Hum. Brain Map.* **30**, 1705–1722.

Humphrey, M. E., and Zangwill, O. L. (1951). Cessation of dreaming after brain injury. *J. Neurol. Neurosurg. Psychiat.* **14**, 322–325.

Hunt, M. J., Matulewicz, P., Gottesmann, C., and Kasacki, S. (2009). State-dependent changes in high frequency oscillations recorded in the rat nucleus accumbens. *Neuroscience* **164**, 380–386.

Iwama, K., Kawamoto, T., Sakkakura, H., and Kasamatsu, T. (1966). Responsiveness of cat lateral geniculate at pre- and postsynaptic levels during natural sleep. *Physiol. Behav.* **1**, 45–53.

Jackson, M. E., Frost, A. S., and Moghaddam, B. (2001). Stimulation of prefrontal cortex at physiologically relevant frequencies inhibits dopamine release in nucleus accumbens. *J. Neurochem.* **78**, 920–923.

Jacobs, B. L., Feldman, M., and Bender, M. B. (1972). Are the eye movements of dreaming sleep related to the visual images of the dreams. *Psychophysiology* **9**, 393–401.

Jacobson, A. (1930). Electrical measurements of neuromuscular states during mental activities. III. Visual imagination and recollection. *Am. J. Physiol.* **95**, 694–702.

Jacobson, A. (1938). You can Sleep Well. The ABC's of Restful Sleep for the Average Person. Whittley House, New York.

Jakobson, A. J., Fitzgerald, P., and Conduit, R. (2009). The effect of transcanial direct current stimulation to the right posterior parietal cortex on dream recall from REM sleep. *Sleep Biol. Rhyt.* **7**, A5.

Jasper, H. H., and Tessier, J. (1971). Acetylcholine liberation from cerebral cortex during paradoxical (REM sleep). *Science* **172**, 601–602.

Jeannerod, M., Mouret, J., and Jouvet, M. (1965). Etude de la motricité oculaire au cours de la phase paradoxale du sommeil chez le Chat. *Electroenceph. Clin. Neurophysiol.* **18**, 554–566.

Johnson, J. D. (2003). Noradrenergic control of cognition: global attenuation and an interrupt function. *Med. Hypoth.* **60**, 689–692.

Jouvet, M. (1962). Recherches sur les structures nerveuses et les mécanismes responsables des différentes phases du sommeil physiologique. *Arch. Ital. Biol.* **100**, 125–206.

Jouvet, M., and Michel, F. (1959). Corrélations électromyographiques du sommeil chez le Chat décortiqué et mésencéphalique chronique. *C. R. Soc. Biol.* **153**, 422–425.

Jouvet, M., and Michel, F. (1960). Mise en évidence d'un "centre hypnique" au niveau du rhombencéphale chez le Chat. *C. R. Acad. Sci.* **251**, 1188–1190.

Kelly, P. H. (1998). Defective inhibition of dream event memory formation: a hypothesized mechanism in the onset and progression of symptoms of schizophrenia. *Brain Res. Bull.* **46**, 189–197.

Kisley, M. A., Olincy, A., Robbins, E., Polk, S. D., Adler, L. E., Waldo, M. C., and Freedman, R. (2003). Sensory gating impairment associated with schizophrenia persists into REM sleep. *Psychophysiology* **40**, 29–38.

Kiyono, S., and Iwama, K. (1965). Phasic activity of cat's cerebral cortex during paradoxical sleep. *Med. J. Osa. Univers.* **16**, 149–159.

Klaue, R. (1937). Die bioelektrische Tätigkeit der Grosshirnrinde im normalen Schlaf und in der Narkose durch Schlafmittel. *J. Psychol. Neurol.* **47**, 510–531.

Kleitman, N. (1939). Sleep and Wakefulness. As Alternating Phases in the Cycle of Existence, The University Of Chicago Press, Chicago.

Kleitman, N. (1965). Sleep and Wakefulness, University Chicago Press, Chicago, p. 552.

Krnjevic, K., and Phillis, J. W. (1963). Actions of certain amines on cerebral cortex neurons. *Brit. J. Pharmacol.* **20**, 471–490.

Kubicki, M., Styner, M., Gerig, G., Markant, D., Smith, K., MacCarley, R. W., and Shenton, M. E. (2008). Reduced interhemispheric connectivity in schizophrenic-tractography based segmentation of the corpus callosum. *Schizophr. Res.* **106**, 125–131.

Ladd, G. T. (1892). Contribution to the psychology of visual dreams. *Mind* **1**, 299–304.

Langer, S. Z. (2008). Presynaptic autoreceptors regulating transmitter release. *Neurochem. Int.* **52**, 26–30.

Lauriat, T. L., Dracheva, S., Chin, B., Schmeidler, J., McInnes, L. A., and Haroutunian, V. (2005). Quantitative analysis of glutamate transporter mRNA expression in prefrontal and primary visual cortex in normal and schizophrenic brain. *Neuroscience* **137**, 843–851.

Léna, I., Parrot, S., Deschaux, O., Muffat, S., Sauvinet, V., Renaud, B., Suaud-Chagny, M. F., and Gottesmann, C. (2005). Variations in the extracellular levels of dopamine, noradrenaline, glutamate and aspartate across the sleep-wake cycle in the medial prefrontal cortex and nucleus accumbens of freely moving rats. *J. Neurosci. Res.* **81**, 891–899.

Levy, R. B., Reves, A. D., and Aoki, C. (2006). Nicotinic and muscarinic reduction of unitary excitatory post synaptic potentials in sensory cortex: dual intracellular recording *in vitro*. *J. Neurophysiol.* **95**, 2155–2166.

Lewis, D. A. (2000). GABAergic local circuit neurons and prefrontal cortical dysfunction in schizophrenia. *Brain Res. Rev.* **31**, 270–276.

Lewis, D. A., and Hashimoto, H. (2007). Deciphering the disease process of schizophrenia: the contribution of cortical GABA neurons. *Intern. Rev. Neurobiol.* **78**, 109–131.

Lewis, D. A., Hashimoto, T., and Volk, D. W. (2005). Cortical inhibitory neurons in schizophrenia. *Nat. Rev. Neurosci.* **6**, 312–324.

Libet, B., and Gleason, G. A. (1994). The human locus coeruleus and anxiogenesis. *Brain Res.* **634**, 178–180.

Linner, L., Wiker, C., Wadenberg, M. L., Schalling, M., and Svensson, T. H. (2002). Noradrenaline reuptake inhibition enhances the antipsychotic-like effect of raclopride and potentiates D2-blockade-induced dopamine release in the medial prefrontal cortex of the rat. *Neuropsychpharmacology* **27**, 691–698.

Llinas, R., and Ribary, U. (1993). Coherent 40 Hz oscillation characterizes dream state in humans. *Proc. Nat. Acad. Sci. USA* **90**, 2078–2081.

Loomis, A. L., Harvey, E. N., and Hobart, G. A. I. (1935a). Further observations on the potential rhythms of the cerebral cortex during sleep. *Science* **82**, 198–200.

Loomis, A. L., Harvey, E. N., and Hobart, G. A. I. (1935b). Potential rhythms of the cerebral cortex during sleep. *Science* **81**, 597–598.

Loomis, A. L., Harvey, E. N., and Hobart, G. A. I. (1937). Cerebral states during sleep, as studied by human brain potentials. *J. Exp. Psychol.* **21**, 127–144.

Lövblad, K. O., Thomas, R., Jakob, P. M., Scammel, T., Bassetti, C., Griswold, M., Ives, J., Matheson, J., Edelman, R. R., and Warach, S. (1999). Silent functional magnetic resonance imaging demonstrates focal activation in rapid eye movement sleep. *Neurology* **53**, 2193–2195.

Luciana, M., Collins, P. F., and Depue, R. A. (1998). Opposing role for dopamine and serotonin in the modulation of human spatial working memory functions. *Cer. Cort.* **8**, 218–226.

Lucrece, T.C. (1900). De rerum natura OXONII. E Translation Baily, C. Topographeo Clarendoniano. 178 pp.

Lufkin, B. (1968). Letter to the editor. *Psychophysiology* **5**, 449–450.

MacKay, A. V., Iversen, L. L., Rossor, M., Spokes, E., Bird, E., Arregui, A., and Snyder, S. (1982). Increased brain dopamine and dopamine receptors in schizophrenia. *Arch. Gen. Psychiat.* **39**, 991–997.

Madsen, P. L., Holm, S., Vorstrup, S., Friberg, L., Lassen, N. A., and Wildschiodtz, G. (1991). Human regional cerebral blood flow during rapid-eye-movement sleep. *J. Cerbr. Blood Flow Met.* **11**, 502–507.

Maldonado-Aviles, J. G., Curley, A. A., Hashimoto, T., Morrow, A. L., Ramsey, A. J., O'Donnell, P., Volk, D. W., and Lewis, D. A. (2009). Altered markers of tonic inhibition in the dorsolateral prefrontal cortex of subjects with schizophrenia. *Am. J. Psychiat.* **166**, 450–459.

Manacéine, M. D. (1897). Sleep: Its Physiology, Pathology, Hygiene, and Psychology, Walter Scott, London.

Manunta, Y., and Edeline, J. M. (1999). Effects of noradrenaline on frequency tuning of auditory cortex neurons during wakefulness and slow wave sleep. *Eur. J. Neurosci.* **11**, 2134–2150.

Maquet, P., and Franck, G. (1997). REM sleep and amygdala. *Mol. Psychiat.* **2**, 195–196.

Maquet, P., Peters, J. M., Aerts, J., Delfiore, G., Degueldre, C., Luxen, A., and Franck, G. , (1996). Functional neuroanatomy of human rapid-eye-movement sleep and dreaming. *Nature* **383**, 163–166.

Maquet, P., Ruby, P., Schwartz, S., Laurey, S., Albouy, T., Dang-Vu, T., Desseilles, M., Boly, M., Melchior, G., and Peigneux, P. (2004). Regional organisation of brain activity during paradoxical sleep. *Arch. Ital. Biol.* **142**, 413–419.

Marrosu, F., Portas, C., Mascia, M. F., Casu, M. A., Fa, M., Giagheddu, M., Imperato, A., and Gessa, G. L. (1995). Microdialysis measurement of cortical and hippocampal acetylcholine release during sleep-wake cycle in freely moving cats. *Brain Res.* **671**, 329–332.

Massimini, M., Ferrarelli, F., Huber, R., Esser, S. K., Singh, H., and Tononi, G. (2005). Breakdown of cortical effective connectivity during sleep. *Science* **309**, 2228–2232.

Matsumoto, J., and Jouvet, M. (1964). Effets de la réserpine, DOPA et 5 HTP sur les deux états de sommeil. *C. R. Soc. Biol.* **158**, 2137–2140.

Maury, F. (1861). Le sommeil et les rêves. Didier (ed.), 156.

McCarley, R. W., Winkelman, J. W., and Duffy, F. H. (1983). Human cerebral potentials associated with REM sleep rapid eye movements: links to PGO waves and waking potentials. *Brain Res.* **274**, 359–364.

McCormick, D. A., Wang, Z., and Huguenard, J. (1993). Neurotransmitter control of neocortical neuronal activity and excitability. *Cereb. Cort.* **3**, 387–398.

McGinty, D. J., and Harper, R. M. (1976). Dorsal raphe neurons: depression of firing during sleep in cats. *Brain Res.* **101**, 569–575.

McGinty, D. J., Harper, R. M., and Fairbanks, M. K. (1974). Neuronal unit activity and the control of sleep states. In: Advances in Sleep Research, Vol. 1 (E. D. Weitzman, ed.), Spectrum, New York, pp. 173–216.

McGlade, H. B. (1942). The relationship between gastric motility, muscular twitching during sleep and dreaming. *Am. J. Digest. Dis.* **9**, 137–140.

Mednikova, Y. S., Karnup, S. V., and Loseva, E. V. (1998). Cholinergic excitation of dendrites in neocortical neurons. *Neuroscience* **87**, 783–796.

Meyer-Lindenberg, A., Olsen, R. K., Kohn, P. D., Brown, T., Egan, M. F., Weinberber, D. R., and Berman, K. F. (2005). Regionally specific disturbance of dorsolateral prefrontal-hippocampal function connectivity in schizophrenia. *Arch. Gen. Psychiat.* **62**, 379–386.

Meyer-Lindenberg, A., Poline, J. B., Kohn, P. D., Holt, J. L., Egan, M. F., Weinberger, D. R., and Berman, K. F. (2001). Evidence for abnormal cortical functional connectivity during working memory in schizophrenia. *Am. J. Psychiat.* **158**, 1809–1817.

Meyer-Lindenberg, A., and Weinberger, D. R. (2006). Intermediate phenotypes and genetic mechanisms of psychiatric disorders. *Nat. Neurosci. Rev.* **7**, 818–827.

Michel, F., Jeannerod, M., Mouret, J., Rechtschaffen, A., and Jouvet, M. (1964). Sur les mécanismes de l'activité de pointes au niveau du système visuel au cours de la phase paradoxale du sommeil. *C. R. Soc. Biol.* **158**, 103–106.

Miller, J. D., Farber, J., Gatz, P., Roffwarg, H., and German, D. (1983). Activity of mesencephalic dopamine and non-dopamine neurons across stages of sleep and waking in the rat. *Brain Res.* **273**, 133–141.

Milstein, J. A., Lehmann, O., Theobald, D. E., Dalley, J. W., and Robbins, E. (2007). Selective depletion of cortical noradrenaline by anti-domain beta-hydroxylase-saporin impairs attentional function and enhances the effect of guanfacine in the rat. *Psychopharmacology* **190**, 51–63.

Miyauchi, S., Misaki, M., Kan, S., Fukunaga, T., and Koike, T. (2009). Human brain activity time-locked to rapid eye movements during REM sleep. *Exp. Brain Res.* **192**, 657–667.

Miyauchi, S., Takino, R., Fukuda, H., and Torii, S. (1987). Electrophysiological evidence for dreaming: human cerebral potentials associated with rapid eye movements during REM sleep. *Electroenceph. Clin. Neurophysiol.* **66**, 383–390.

Morrison, A. R., and Pompeiano, O. (1965). Pyramidal discharge from somatosensory cortex and cortical control of primary afferents during sleep. *Arch. Ital. Biol.* **103**, 538–568.

Morrison, A. R., and Pompeiano, O. (1966). Vestibular influences during sleep. IV. Functional relations between vestibular nuclei and lateral geniculate nucleus during desynchronized sleep. *Arch. Ital. Biol.* **104**, 425–458.

Morshedi, M. M., and Meredith, G. E. (2007). Differential laminar effects of amphetamine on prefrontal parvalbumin interneurons. *Neuroscience* **149**, 617–624.

Moskowitz, E., and Berger, R. J. (1969). Rapid eye movements and dream imagery: are they related. *Nature* **244**, 613–614.

Mouret, J., Jeannerod, M., and Jouvet, M. (1965). Mise en jeu du système oculomoteur pendant le sommeil. *Confin. Neurol.* **25**, 291–299.

Muzur, A., Pace-Schott, E., and Hobson, J. A. (2002). The prefrontal cortex in sleep. *Trends Cogn. Sci.* **6**, 475–481.

Nelson, C. N., Hoffer, B. J., Chu, N. S., and Bloom, F. E. (1973). Cytochemical and pharmacological studies on polysensory neurons in the primate frontal cortex. *Brain Res.* **62**, 115–133.

Nicoll, R. A., Malenka, R. C., and Kauer., J. A. (1990). Functional comparison of neurotransmitter receptor subtypes in mammalian central nervous system. *Physiol. Rev.* **70**, 513–566.

Nielsen, T. (2000). cognition in REM and NREM sleep: a review and possible reconciliation of two models of sleep mentation. *Behav. Brain Sci.* **23**, 851–866.

Nieoullon, A. (2002). Dopamine and the regulation of cognition and attention. *Prog. Neurobiol.* **67**, 53–83.

Ohlmeyer, P., and Brilmayer, B. (1947). Periodische Vörgange im Schlaf. II Mitteilung. *Pflüg. Arch.* **249**, 50–55.

Ohlmeyer, P., Brilmayer, H., and Hüllstrung, H. (1944). Periodische Vorgänge im Schlaf. *Pflüg. Arch.* **248**, 559–560.

Okuma, T., and Fujimori, M. (1963). Electrographic and evoked potential studies during sleep in the cat (The study of sleep I. *Fol. Psychiat. Neurol. Jap.* **17**, 25–50.

Onoe, H., and Sakai, K. (1995). Kainate receptors: a novel mechanism in paradoxical (REM) sleep generation. *NeuroReport* **6**, 353–356.

Palestini, M., Pisano, M., Rosadini, G., and Rossi, G. F. (1964). Visual cortical responses evoked by stimulating lateral geniculate body and optic radiations in awake and sleeping cats. *Exp. Neurol.* **9**, 17–30.

Parsons, L. M., Sergent, J., Hodges, D. A., and Fox, P. T. (2005). The brain basis of piano performance. *Neuropsychologia* **43**, 199–215.

Peled, A., Geva, A. B., Kremen, W. S., Blankfeld, H. M., Esfandiarfard, R., and Nordahl, T. E. (2000). Functional connectivity and working memory in schizophrenia: an EEG study. *Int. J. Neurosci.* **106**, 47–61.

Perez-Garci, E., del Rio-Portilla, Y., Guevara, M. A., Arce, C., and Corsi-cabrera, M. (2001). Paradoxical sleep is characterized by uncoupled gamma activity between frontal and perceptual cortical regions. *Sleep* **24**, 118–126.

Perry, E. K., and Piggott, M. A. (2003). Neurotransmitter mechanisms of dreaming: implication of modulatory systems based on dream intensity. In: Sleep and Dreaming. Scientific Advances and Reconsiderations (E. Pace-Schott, M. Solms, M. Blagrove, and S. Harnad, eds.), Cambridge University Press, Cambridge, pp. 202–204.

Perry, E., Walker, M., Grace, J., and Perry, R. (1999). Acetylcholine in mind: a neurotransmitter of consciousness? *Trends Neurosci.* **22**, 273–280.

Pirot, S., Godbout, R., Mantz, J., Tassin, J. P., Glowinski, J., and Thierry, A. M. (1992). Inhibitory effects of ventral tegmental area stimulation on the activity of the prefrontal cortex neurons:

evidence for the involvement of both dopaminergic and GABAergic components. *Neuroscience* **49**, 857–865.

Pisano, M., Rosadini, G., and Rossi, G. F. (1962). Riposte corticali evocate da stimoli dromici ed antidromici durante il sonno e la veglia. *Riv. Neurobiol.* **8**, 414–426.

Pompeiano, O., and Morrison, A. R. (1965). Vestibular influences during sleep. I. Abolition of the rapid eye movements of desynchronized sleep following vestibular lesions. *Arch. Ital. Biol.* **103**, 569–595.

Pompeiano, O., and Morrison, A. R. (1966a). Vestibular input to the lateral geniculate nucleus during desynchronized sleep. *Pflüg. Arch.* **290**, 272–274.

Pompeiano, O., and Morrison, A. R. (1966b). Vestibular origin of the rapid eye movements during desynchronized sleep. *Experentia* **22**, 60–61.

Raedler, T., Knable, M., Jones, D., Urbina, R., Gorey, J., Lee, K., Egan, M., Coppola, R., and Weinberger, D. (2003). *In vivo* determination of muscarinic acetylcholine receptor availability in schizophrenia. *Am. J. Psychiatry* **160**, 118–127.

Raehlmann, E., and Witkowski, L. (1877). Ueber atypische Augenbewegungen. *Arch. Anat. Physiol.* **S, 454—471**.

Raehlmann, E., and Witkowski, L. (1878). Ueber das Verhalten der pupilen während des Schlafes nebst Bemerkungen zur innervation der Iris. *Arch. Anat. Physiol.* **S, 109—121**.

Rasmussen, K., Heym, J., and Jacobs, B. L. (1984). Activity of serotonin-containing neurons in nucleus centralis superior of freely moving cats. *Exp. Neurol.* **83**, 302–317.

Rasmussen, K., Morilak, D. A., and Jacobs, B. L. (1986). Single unit activity of locus coeruleus neurons in the freely moving cat. I. Naturalistic behaviors and in response to simple and complex stimuli. *Brain Res.* **371**, 324–334.

Reader, T. A., Ferron, A., Descarries, L., and Jasper, H. H. (1979). Modulatory role for biogenic amines in the cerebral cortex. Microiontopheric studies. *Brain Res.* **160**, 219–229.

Reeves, M., Lindholm, D. E., Myles, P. S., Fletcher, H., and Hunt, J. O. (2001). Adding ketamine to morphine for patient-controlled analgesia after major abdominal surgery: a double-blind, randomized trial. *Anesth. Analg.* **93**, 116–120.

Rétaux, R., Besson, M. J., and Penit-Soria, J. (1991). Opposing effects of dopamine D2 receptor stimulation on the spontaneous and electrically-evoked release of 3H GABA on rat prefrontal cortex slices. *Neuroscience* **42**, 61–72.

Revonsuo, A. (2003). The reinterpretation of dreams: an evolutionary hypothesis of the function of dreaming. In: Sleep and Dreaming. Scientific Advances and Reconsiderations (E. Pace-Schott, M. Solms, M. Blagrove, and S. Harnad, eds.), Cambridge University Press, Cambridge, pp. 85–109.

Ribary, U., Ionnides, A. A., Singh, K. D., Hasson, R., Bolton, J., Lado, F., Mogilner, A., and Llinas, R. (1991). Magnetic field tomography of coherent thalamocortical 40 Hz oscillations in humans. *Proc. Nat. Acad. Sci. USA* **88**, 11037–11041.

Rioch, M. D. (1954). Discussion of W.R. Hess's paper. In: Brain Mechanisms and Consciousness (J. F. Delafresnaye, ed.), Blackwell Scientific Publications, Oxford, pp. 133–136.

Roffwarg, H. P., Dement, W. C., Muzio, J. N., and Fisher, C. (1962). Dream imagery: relationship to rapid eye movements of sleep. *Arch. Gen. Psychiat.* **7**, 235–258.

Rossi, G. F., Palestini, M., Pisano, M., and Rosadini, G. (1965). An experimental study of the cortical reactivity during sleep and wakefulness. In: Aspects anatomo-fonctionnels de la physiologie du sommeil. CNRS, Paris, pp. 509–532.

Sakai, K. (1985). Anatomical and physiological basis of paradoxical sleep. In: Brain Mechanisms of Sleep. (R. drucker-Colin, A. R. Morrison, P. L. Parmeggiani, eds.), Raven press, New York, pp. 111–137.

Sakai, K. (1988). Executive mechanisms of paradoxical sleep. *Arch. Ital. Biol.* **126**, 239–257.

Sakai, K., and Crochet, S. (2001). Role of dorsal raphe neurons in paradoxical sleep generation in the cat: no evidence for serotonergic mechanisms. *Eur. J. Neurosci.* **13**, 103–112.

Sakai, K., and Crochet, S. (2003). A neural mechanism of sleep and wakefulness. *Sleep Biol. Rhyt.* **1**, 29–42.

Sakakura, H., and Iwama, K. (1965). Presynaptic inhibition and postsynaptic facilitation in lateral geniculate body and so-called deep sleep wave activity. *Tohok. J. Exp. Med.* **87**, 40–51.

Salzarulo, P., Lairy, G. C., Bancaud, J., and Munari, C. (1975). Direct depth recording of the striate cortex during REM sleep in man: are there PGO potentials?. *Electroenceph. Clin. Neurophysiol.* **38**, 199–202.

Sarter, M., and Bruno, J. P. (2000). Cortical cholinergic inputs mediating arousal, attentional processing and dreaming: differential afferent regulation of the basal forebrain by telencephalic and brainstem afferents. *Neuroscience* **95**, 933–952.

Satoh, T. (1971). Direct cortical response and PGO spike during paradoxical sleep of the cat. *Brain Res.* **28**, 576–578.

Schmidt, M. H., Sakai, K., Valatx, J. L., and Jouvet, M. (1999). The effects of spinal or mesencephalic transections on sleep-related erections and ex-copula penile reflexes in the rat. *Sleep* **22**, 409–418.

Schmidt, M. H., Valatx, J. L., Sakai, K., Fort, P., and Jouvet, M. (2000). Role of the lateral preoptic area in sleep-related erectile mechanisms and sleep generation in the rat. *J. Neurosci.* **20**, 6640–6647.

Schmidt, M. H., Valatx, J. L., Schmidt, H. S., Wauquier, A., and Jouvet, M. (1994). Experimental evidence of penile erections during paradoxical sleep in the rat. *NeuroReport* **5**, 561–564.

Selden, N. R. W., Robbins, T. W., and Everitt, B. J. (1990). Enhanced behavioral conditioning to context and impaired behavioral and neuroendocrine responses to conditioned stimuli following ceruleocortical noradrenergic lesions: support for an attentional hypothesis of central noradrenergic function. *J. Neurosci.* **10**, 531–539.

Sheldon, P. W., and Aghajanian, G. K. (1990). Serotonin (5-HT) induces IPSPs in pyramidal layer cells of rat piriform cortex: evidence for the involvement of a $5-HT_2$-activated interneuron. *Brain Res.* **506**, 62–69.

Silbersweig, D. A., Stern, E., Frith, C. D., Cahill, C., Holmes, A., Grootoonk, S., Seaward, J., McKenna, P., Chua, S. E., Schnorr, L., Jones, T., and Frackowiak, R. S. J. (1995). A functional neuroanatomy of hallucinations in schizophrenia. *Nature* **378**, 176–179.

Silver, H., Barash, I., Aharon, N., Kaplan, A., and Poyurovsky, M. (2000). Fluvoxamine augmentation of antipsychotics improves negative symptoms in psychotic chronic schizophrenic patients: a placebo-controlled study. *Int. Clin. Psychopharmacol.* **15**, 257–261.

Smiley, J. F., and Goldman-Rakic, P. S. (1993). Heterogenous targets of dopamine synapses in monkey prefrontal cortex demonstrated by serial section electron microscopy: a laminar analysis using the silver-enhanced diaminobenzidine sulfite (SEDS) immunolabeling technique. *Cereb. Cort.* **3**, 223–238.

Solms, M. (2000). Dreaming and REM sleep are controlled by different brain mechanisms. *Behav. Brain Sci.* **23**, 843–850.

Steriade, M. (1970). Ascending control of thalamic and cortical responsiveness. *Int. Rev. Neurobiol.* **12**, 87–144.

Steriade, M., Deschenes, M., Domich, L., and Mulle., C. (1985). Abolition of spindle oscillations in thalamic neurons disconnected from nucleus reticulari thalami. *J. Neurophysiol.* **54**, 1473–1497.

Steriade, M., Domich, L., Oakson, G., and Deschenes., M. (1987). The deafferented reticular thalamic nucleus generates spindle rhythmicity. *J. Neurophysiol.* **57**, 260–273.

Steriade, M., and McCarley, R. W. (1990). Brainstem Control of Wakefulness and Sleep, Plenum Press, New York, p. 267.

Suaud-Chagny, M. F., Chergui, K., Chouvet, G., and Gonon, F. (1992). Relationship between dopamine release in the rat nucleus accumbens and the discharge activity of dopaminergic neurons during local *in vivo* application of amino acids in the ventral tegmental area. *Neuroscience* **49**, 63–72.

Szymusiak, R., McGinty, D., Shepard, D., shouse, M. N., and Sterman, M. (1990). Effects of systemic atropine sulfate administration on the frequency content of the cat sensorimotor EEG during sleep and waking. *Behav. Neurosci.* **104**, 217–225.

Takahata, R., and Moghaddam, B. (2000). Target-specific glutamatergic regulation of dopamine neurons in the ventral tegmental area. *J. Neurochem.* **75**, 1775–1778.

Takahata, R., and Moghaddam, B. (2003). Activation of glutamate neurotransmission in the prefrontal cortex sustains the motoric and dopaminergic effects of phencyclidine. *Neuropsychopharmacol.* **28**, 117–1124.

Takeuchi, T., Miyasita, A., Inugami, M., and Yamamoto, Y. (2001). Intrinsic dreams are not produced without REM sleep mechanisms: evidence through elicitation of sleep onset periods. *J. Sleep Res.* **10**, 43–52.

Takeuchi, T., Ogilvie, R. D., Ferrelli, A. V., Murphy, T., Yamamoto, Y., and Inugami, M. (1999). Dreams are not produced without REM sleep mechanisms. *Sleep Res. Online* **2** (Suppl. 1), 279.

Thierry, A. M., Blanc, G., Sobel, A., Stinus, L., and Glowinski, J. (1973). Dopaminergic terminals in the rat cortex. *Science* **182**, 499–501.

Tononi, G., and Edelman, G. M. (2000). Schizophrenia and the mechanism of conscious integration. *Brain Res. Rev.* **31**, 391–400.

Tracy, R. L., and Tracy, L. N. (1974). Reports of mental activity from sleep stages 2 and 4. *Perc. Mot. Skills* **38**, 647–648.

Trulson, M. E., and Preussler, D. W. (1984). Dopamine-containing ventral tegmental area neurons in freely moving cats: activity during the sleep-waking cycle and effects of stress. *Exp. Neurol.* **83**, 367–377.

Tseng, K. Y., Mallet, N., Toreson, K. L., Le Moine, C., Gonon, F., and O'Donnell, P. (2006). Excitatory response of prefrontal cortical fast-spiking interneurons to ventral tegmental area stimulation *in vivo. Synapse* **59**, 412–417.

Van Hes, R., Smid, P., Stroomer, C. N., Tipker, K., Tulp, M. T., Van der Heyden, J. A., McCreary, A. C., Hesselink, M. B., and Kruse, C. G. (2003). SL V310, a novel, potential antipsychotic, combining potent dopamine d2 receptor antagonism with serotonin reuptake inhibition. *Bioorg. Med. Chem. Lett.* **13**, 405–408.

Vanderwolf, C. H. (1988). Cerebral activity and behavior control by central cholinergic and serotonergic systems. *Int. Rev. Neurobiol.* **30**, 225–340.

Vanni-Mercier, G., and Debilly, G. (1998). A key role for the caudoventral pontine tegmentum in the simultaneous generation of eye saccades in bursts and associated ponto-geniculo-occipital waves during paradoxical sleep in the cat. *Neuroscience* **86**, 571–585.

Vanni-Mercier, G., Debilly, G., Lin, J. S., and Pélisson, D. (1996). The caudo ventral tegmentum is involved in the generation of high velocity eye saccades in bursts during paradoxical sleep in the cat. *Neurosci. Lett.* **213**, 127–131.

Vasquez, J., and Baghdoyan, H. A. (2001). Basal forebrain acetylcholine release during REM sleep is significantly greater than during waking. *Am. J. Physiol.* **280**, R598–R601.

Wang, X., Ai, J., Hampson, D. R., and Snead, I.I.I.O.C. (2005). Altered glutamate and GABA release within thalamocortical circuitry in metabotropic glutamate receptor 4 knockout mice. *Cellular Neurosci.* **134**, 1195–1203.

Warren, R. A., and Dykes, R. W. (1996). Transient and long-lasting effects of iontophoretically administered norepinephrine on somatosensory cortical neurons in halothane-anesthetized cats. *Can. J. Physiol. Pharmacol.* **74**, 38–57.

Waterhouse, B. D., Azizi, S. A., Burne, R. A., and Woodward, D. (1990). Modulation of rat cortical area 17 neuronal responses to moving visual stimuli during norepinephrine and serotonin microiontophoresis. *Brain Res.* **514**, 276–292.

Weed, S. C., and Hallam, F. M. (1896). A study of the dream-consciousness. *Am. J. Psychol.* **7**, 405–411.

Weickert, C. S., Webster, M. J., Gondipalli, P., Rothmond, D., Fatula, R. J., Herman, M. M., Kleinman, J. E., and Akil, M. (2007). Postnatal alterations in dopaminergic markers in the human prefrontal cortex. *Neuroscience* **144**, 1109–1119.

Weinberger, D. R., Berman, K. F., and Zec, R. F. , (1986). Physiological dysfunction of dorsolateral prefrontal cortex in schizophrenia. 1. Regional cerebral blood flow evidence. *Arch. Gen. Psychiat.* **43**, 114–124.

Weitzman, E. D., and Kremen, H. (1965). Auditory evoked responses during different stages of sleep in man. *Electroenceph. Clin. Neurophysiol.* **18**, 65–70.

Wikler, A. (1952). Pharmacological dissociation of behavior and EEG "sleep patterns" in dogs: morphine, N-allylmorphine and atropine. *Proc. Soc. exp. Biol. Med.* **79**, 261–265.

Williams, H. L., Hammack, J. T., Daly, R. L., Dement, W. C., and Lubin, A. (1964). Responses to auditory stimulation, sleep loss and the EEG stages of sleep. *Electroenceph. Clin. Neurophysiol.* **16**, 269–279.

Williams, H. L., Tepas, D. I., and Morlock, J. H. C. (1962). Evoked responses to clicks and electroencephalographic stages of sleep in man. *Science* **138**, 685–686.

Wurtz, R. H. (1965). Steàdy potential shifts in the rat during desynchronized sleep. *Electroenceph. Clin. Neurophysiol.* **19**, 521–523.

Wurtz, R. H. (1966). steady potential fields during sleep and wakefulness in the cat. *Exp. Neurol.* **15**, 274–292.

Ye, Z., Wyeth, M. S., Baltan-Tekkok, S., and Ransom, B. R. (2003). Functional hemichannels in astrocytes: a novel mechanism of glutamate release. *J. Neurosci.* **23**, 3588–3596.

Young, C. E., Beach, T. G., Falkai, P., and Honer, W. G. (1998). SNAP-25 deficit and hippocampal connectivity in schizophrenia. *Cer. Cort.* **8**, 261–268.

DREAMING AS INSPIRATION: EVIDENCE FROM RELIGION, PHILOSOPHY, LITERATURE, AND FILM

Kelly Bulkeley

Visiting Scholar, The Graduate Theological Union, Berkeley, California, USA

This paper presents evidence from the history of religion, philosophy, literature, and film to suggest that dreaming is a primal wellspring of creative inspiration. Powerful, reality-bending dreams have motivated the cultural creativity of people all over the world and throughout history. Examples include the dream revelations of Egyptian Pharaohs, the philosophical insights of Socrates, the dark literary themes of Fyodor Dostoevsky, and the cinematic artistry of Akira Kurusawa. Although the conclusions that can be drawn from these sources are limited by several methodological factors, the evidence gives contemporary researchers good reasons to explore the creative potentials of dreaming and the impact on waking life behavior of certain types of extraordinary dream experience.

I. Introduction

In recent years several researchers have identified meaningful patterns in dream content (Domhoff, 1996; Foulkes, 1999; Hartmann, 1998; Hunt, 1989; Strauch and Meier, 1996; Zadra and Domhoff, 2010). Their findings refute the notion that dreams are nothing but random neural nonsense. The latter idea, while often repeated in public discourse about dreaming, lacks any actual data to support it. The frontier of scientific dream research is moving beyond the question of whether dreams have meaning, to the much more interesting question of what

31

kinds of meanings can be accurately discerned in dream content. Among other things, these meanings relate to the aspects of waking personality, relationships, activities, and cultural preferences (Bulkeley and Domhoff, 2010). At the very least, then, we can confidently say that dreaming is meaningful in the sense of accurately mirroring many of the important emotional concerns in a person's waking life.

This chapter explores the meaningfulness of dreaming at a different level of analysis. To what extent do dreams not simply reflect waking life, but actively motivate and inspire people to undertake creative endeavors in their waking lives? Rather than looking at how waking influences dreaming, the question here is to what extent does dreaming influence waking. Here the research findings are less empirically substantial, though still intriguing in terms of hinting at deeper powers and unconscious intentions in the human brain–mind system. To facilitate future studies in this area, this chapter reviews several sources of evidence for the idea that dreaming serves as a wellspring of creative inspiration in human culture, from religion and philosophy to literature and film. This chapter does not set out to prove that idea in a complete and systematic fashion, but rather to review some of the best available data that other researchers can use in pursuing more detailed investigations in the future.

Examples from four broad realms of cultural activity will be discussed in roughly chronological order: religion, philosophy, literature, and film. Of course this leaves out of consideration the inspirational impact of dreaming on other forms of cultural expression such as music (see Grace, 2001), painting (see Coxhead and Hiller, 1976; Platow, 2005), and graphic novels (e.g., the works of Neil Gaiman and Jesse Reklaw). But given the limitations of space, the focus on these four areas provides a good, quick introduction to the general subject of inspiring dreams because they offer an abundance of cross-cultural and historical information.

II. Quality of Evidence

All forms of dream research face the question of how a *report* of a dream relates to the *experience* of a dream. Does the waking description of what happened in a dream accurately and completely represent the person's feelings, thoughts, and sensations within the dream-as-dreamt? Put in the sharpest and most methodologically troubling terms, how do we ever know a dream report has not been edited, revised, embellished, or completely fabricated by the dreamer? This question becomes even more difficult to answer when the dream reports are translated from different languages, drawn from ancient historical texts, and/or explicitly connected to a person's cultural activities, adding incentives to shape the dreams in accordance with waking norms and expectations.

For the most part, the evidence presented here remains vulnerable to this line of criticism. Great caution is required, therefore, in making theoretical generalizations about dreaming based on subjective accounts like these. They signal the *potentials* of dreaming creativity, nothing more and nothing less. It is up to researchers today to examine, verify, and if possible extend those potentials by developing new sources of empirical information.

III. Religion

The references in this section may all be found in *Dreaming in the World's Religions: A Comparative History* (Bulkeley, 2008).

In ancient Mesopotamian and Egyptian cultures, dream-based theophanies (revelations of the divine) were widely reported by kings and pharaohs. The messages conveyed in these dreams typically concerned support and guidance in battle, prophecies of future weal and woe, and instructions on ritual practices and temple building. The classic example comes from Thutmose IV (@1400 BCE), a New Kingdom Pharaoh who set out to revive the lost civic spirit of the earlier Egyptian dynasties and whose greatest achievement was the restoration of the Great Sphinx at Giza. He said he was inspired to undertake this literally monumental religious task by a dream:

> One of these days the King's Son Thutmose was strolling at the time of the midday and he rested in the shadow of this Great God [the Sphinx]. Slumber and sleep overcame him at the moment when the sun was at the zenith, and he found the majesty of this august god speaking with his own mouth as a father speaks to his son, as follows: 'Behold me, look upon me, my son Thutmose. I am your father Harmakhis-Khepri-Re-Atum. I will give to you my kingly office on earth as foremost of the living, and you shall wear the crown of Upper Egypt and the crown of Lower Egypt. . . .To you shall belong the earth in its length and its breadth and all that which the eye of the All-Lord illuminates. You shall possess provisions from within the Two Lands as well as the great products of every foreign country. For the extent of a long period of years my face has been turned to you and my heart devoted to you. You belong to me. Behold, my state is like that of one who is in suffering, and all my members are out of joint, for the sand of the desert, this place on which I am, presses upon me. I have waited to have you do what is in my heart, for I know that you are my son and my champion. Approach! Behold, I am with you. I am your guide.' He completed this speech. And this King's Son awoke when he heard this. . . . He recognized the words of this god and he placed silence in his heart.

Of course we can never know if Thutmose IV really had such a dream, instead of making it up to serve his political purposes. The fact that he inscribed his dream in stone next to the Great Sphinx suggests that he and the Egyptians considered such a dream both plausible and meaningful. Given Thutmose IV's enthusiasm for reviving traditional religious practices, it seems likely that a belief in the inspiring potential of dreams reaches back into the earliest history of Egyptian civilization.

Just as early, we find the awareness that some dreams might appear inspiring but actually serve to mislead and deceive people. In Homer's epic poems *The Iliad* and *The Odyssey* people receive dream visitations from gods who inspire the hapless humans in order to trick them into performing certain actions, sometimes benignly for the dreamer (the maiden Nausikaa, who receives a disguised dream from the goddess Athena meant to help Odysseus) and sometimes malevolently harmful to the dreamer (the Greek leader Agamemnon, who receives a false dream from Zeus that lures him into a deadly battlefield trap). Contrary to the widespread assumption that ancient people were gullible and naïve about their dreams, these passages from Homer suggest the ancient Greeks, at least, had a clear awareness of the difficulty of accurately discerning the true meanings of dreams and remaining vigilant against the possibility of being deceived.

A similarly skeptical idea is expressed in the Hebrew Bible in the book of the prophet Jeremiah, who warned people not to listen to rival prophets who confuse their own personal feelings with a truly divine revelation. God told Jeremiah, "I have heard what the prophets have said who prophesy lies in my name, saying, 'I have dreamed, I have dreamed!'" Just because a dream seemed inspiring, it should not be accepted unless it met the test of God's truth as revealed in other, more trustworthy sources. For Jeremiah to convey such a warning, we may surmise there were indeed people at that time who were acting on the perceived inspiration of their dreams.

Jeremiah's cautionary attitude stands in tension with the more favorable view of inspiring dreams in much of the Hebrew Bible and Christian New Testament. The lives of the patriarchs Abraham, Jacob, and Joseph were filled with God-sent dreams that motivated and guided their religious behavior. Several dreams occurred in relation to the birth of Jesus as described in the book of Matthew, with quite specific instructions about what the dreamers should do and where they should go. The apostle Paul in the book of Acts describes dreams that came to reassure him and give hope to others in times of fear and despair.

Muhammad, the founder of Islam, received his first divine revelation in the cool Arabian caves where he slept and prayed alone for long stretches of time. Whether or not his original call came in a dream or a waking vision, it seems clear that Muhammad valued his own dreams and those of others as legitimate sources of insight into God's will. He mentioned two of his own dreams in the Qur'an, and he reportedly asked his followers to describe their dreams to him each morning. The idea of dreaming as a source of religious inspiration has very strong roots in Islam and continues to this day. However, Muhammad warned his followers not to lie about their dreams, but always tell the truth about what happened in them. According to one of the sayings (*hadiths*) of the Prophet, "Whoever claims to have had a dream in which he says he saw something he did not shall be ordered [in Hell] to tie a knot between two barley grains and will not be able to do so." Again, as in the case of Jeremiah, such a stern warning suggests that some people felt a temptation to embellish their dreams in order to gain religious favor and social prestige.

Looking at Asian cultures, the earliest dynasties of China show evidence of royal dignitaries and common people alike viewing dreams as bearing religious significance and practical relevance for waking life. Rulers consulted court officials with experience in dream divination, while shamans and healers offered dream interpretation services to the general public. Given the emphasis on proper worship of one's ancestors in Chinese culture, dreams in which one's ancestors appeared took on great importance as directly inspiring factors in people's religious beliefs and practices.

For the Chinese mystical tradition of Daoism, dreaming could be a spiritual pathway to a deeper understanding of the impermanence of waking reality. Here is the famous "butterfly" dream in Zhuangzi's *Inner Chapters*, from around the 3rd century BCE:

> Long ago, a certain Zhuangzi dreamt he was a butterfly—a butterfly fluttering here and there on a whim, happy and carefree, knowing nothing of Zhuangzi. Then all of a sudden he woke up to find that he was, beyond all doubt, Zhuangzi. Who knows if it was Zhuangzi dreaming a butterfly, or a butterfly dreaming Zhuangzi? Zhuangzi and butterfly: clearly there's a difference. This is called *the transformation of things*.

In Daoism it is not any particular dream so much as dreaming itself that elicits an inspiring religious effect. Once we appreciate the vivid realism of dreaming and the authenticity of our experiences within that realm, we are prepared to understand deeper Daoist truths about "the transformation of things" as the ultimate reality.

If we turn to anthropological and ethnographic data from the indigenous cultures of Africa, Oceania, and the Americas, the results are overwhelmingly in favor of the "dreams as inspiration" thesis. Virtually every known indigenous culture has its own teachings about dreams, and in most cases there is clear evidence of dream-generated cultural creativity.

In Africa, tribal healing traditions have always included dreaming as a source of diagnostic insight and healing power via curative encounters with ancestors and other spirit beings. In present day Africa, when Christian and Muslim proselytizers compete for converts, the indigenous people often rely on dreams for creative responses to the religious tensions and conflicts that trouble their lives.

The Aboriginal Australian tradition of the "Dreamtime" revolves around a fundamental belief in dreaming as a primal force of cosmic creativity. To enter the Dreamtime via ritual, vision, or dream is to engage with the mythic powers and share in their generative vitality.

Among Native Americans, dreams were woven throughout their individual and cultural lives, inspiring everything from the designs of hunting gear to the timing and focus of group ceremonies. The goal of "vision quest" rituals practiced by the youths of many native cultures around the Great Lakes region was to evoke a special dream or vision that became a powerful source of inspiration and guidance for the rest of the person's life.

Much more could be said, of course, about the inspiring dreams of the Tibetan Buddhist sage Milarepa, the Christian mystic Emmanuel Swedenborg, the American prophet Joseph Smith and other early leaders of the Church of Latter Day Saints, and many other figures in religious history. But the basic point should already be clear: Most if not all of the world's religious traditions regard dreaming as a potential source of creative inspiration.

IV. Philosophy

The inspiring role of dreams takes a different form in the realm of philosophy. Here, it is not so much the impact of particular dreams (though that happens, too) as the challenge to waking rationality posed by the very existence and occurrence of dreaming. The skeptical theme noted in several religious traditions becomes a central preoccupation of the Western philosophical tradition, particularly after Rene Descartes and the European Enlightenment from the 17th century onward. The culmination, perhaps, of this rationalist strain of Western philosophy came with Norman Malcolm's 1964 assertion that consciousness within dreaming is impossible since dreaming by definition lacks consciousness, and Daniel Dennett's 1977 claim that we have no real experience of dreaming and thus dream recall is an exercise in self-delusion. For these philosophers, the universal phenomenon of dreaming serves as an inspiration to advance a rationalist world-view that denies virtually all significance to dreams.

It was not always so. A brief look at the teachings of the ancient Greek philosopher Socrates suggests that dreaming could play a more constructive role in the process of philosophical reasoning.

Socrates (470–399 BCE) left no writings of his own, but through his student Plato a fairly clear portrait comes through of an engaging teacher who considered it his mission to question people's assumptions and decenter their ordinary beliefs and attitudes. In Plato's dialogue *Theaetetus* Socrates asks a new student several questions to test his aptitude for training in philosophy. One of his questions regarded dreaming:

Socrates: Have you not taken note of another doubt that is raised in these cases, especially about sleeping and waking?

Theaetetus: What is that?

Socrates: The question I imagine you have often heard asked—what evidence could be appealed to, supposing we were asked at this very moment whether we are asleep or awake, dreaming all that passes through our minds or talking to one another in the waking state?" (Plato, 1961b)

Note that Socrates assumed this to be a commonly-asked question; this gives us another historical glimpse of the dream beliefs of people 2400 years ago. The young man confessed that his inability to answer the question left him feeling surprised and amazed at his own lack of sure knowledge. Socrates, however, interpreted his startled reaction as a promising sign. Such disoriented feelings indicate that old barriers were being removed and new insights were emerging: "This sense of wonder is the mark of the philosopher. Philosophy indeed has no other origin." For Socrates, the ontological uncertainty generated by the powerful realism of dreaming experience can be an inspiration to pursue the path of philosophical discovery and self-knowledge.

Plato's dialogue *Crito* described the last days of Socrates, while he was in prison awaiting execution by the city of Athens. When his friends arrived to visit Socrates they were surprised to find the great thinker passing his time by translating some of Aesop's fables into verse. They asked why he was doing this, and he answered:

I did it in the attempt to discover the meaning of certain dreams, and to clear my conscience, in case this was the art which I had been told to practice. It is like this, you see. In the course of my life I have often had the same dream, appearing in different forms at different times, but always saying the same thing, 'Socrates, practice and cultivate the arts.' In the past I used to think that it was impelling me and exhorting me to do what I was actually doing; I mean that the dream, like a spectator encouraging a runner in a race, was urging me on to do what I was doing already, that is, practicing the arts, because philosophy is the greatest of the arts, and I was practicing it. But ever since my trial, while the festival of the god has been delaying my execution, I have felt that perhaps it might be this popular form of art that the dream intended me to practice, in which case I ought to practice it and not disobey. I thought it would be safer not to take my departure before I had cleared my conscience by writing poetry and so obeying the dream. (Plato, 1961a)

His friends apparently had nothing to say in response to this, and the rest of the dialogue concerned other philosophical issues. But in this story Socrates shared several fascinating pieces of information. First was his recurrent dream, which he reported as having encouraged and inspired his life as a philosopher. Second was his willingness to question his lifelong interpretation of that dream and consider the possibility that it had a more literal message, i.e., to practice a conventional art form such as poetry. Third was his evident desire to respond properly in waking life to his understanding of the dream's meaning. Instead of escaping Athens to the safety of another city (which his friends were desperately trying to persuade him to do), Socrates chose to stay in his prison cell and write children's verses. He preferred to sacrifice his life and question his whole lifelong vocation as a philosopher rather than fail to obey the guiding intention of his dreams.

It might seem this philosophical openness to dreaming was finally and completely eradicated during the Enlightenment, when Descartes sharply distinguished waking rationality from dreaming illusion in *Discourse on the Method* (1637). But the great irony of Cartesian philosophy is that it was inspired, in the fullest sense of that term, by a series of three dreams the 23-year-old Descartes experienced on a particularly significant night in his life. As recounted in John Cole's *The Olympian Dreams of Rene Descartes* (1992), the young man wrote these dreams down in a private journal and interpreted them as announcing his vocation as a philosopher, which he then pursued with well-known results. He never mentioned these dreams, however, and the true extent of their influence on his philosophy remains a mystery. We are left with the paradox that Cartesian rationalism is grounded in the "irrational" experience of dreaming (see Bulkeley, 2005).

V. Literature

With this information on dreams in religion and philosophy providing a degree of historical and intellectual context, we may now consider the various roles of dreaming in literary creativity. In some cases writers are inspired to write certain stories or poems because of a particular dream. In other cases a writer may create a work with a dream episode for one of the characters. Or a writer may try to recreate aspects of dream experience for the reader, with "dreamy" elements of form and content. Finally, a writer may be inspired to think more imaginatively in general by his or her dreams, contributing artistic vitality but not necessarily specific details to the creative process.

One could turn in many directions for examples of these kinds of literary dream inspirations. The following merely represents my perspective on the easily accessed and theoretically interesting sources for further study.

The Russian novelist Fyodor Dostoevsky (1821–1881) wrote several novels and short stories in which characters experience dreams that reveal important aspects of their personality and motivation. For example, in *Crime and Punishment* (1866) the murderer Raskolnikov has several vivid dreams and nightmares that chronicle his tortured journey from criminal madness to spiritual salvation. Given that Dostoevsky drew on his own experiences with epilepsy to write about the seizure-prone prince in *The Idiot* (1869), it seems likely he drew on his personal dreams as inspiration for writing about his characters' dreams.

Samuel Taylor Coleridge (1772–1834), English writer and literary critic, also wove elements of dreaming through his two most famous poems, "The Rime of the Ancient Mariner" and "Kubla Khan, or, A Vision in a Dream, a Fragment." The latter, first published in 1816, is presented as the virtual transcript of a dream (probably stimulated to some degree by opium) that was interrupted by an unbidden visitor:

> On awakening he appeared to himself to have a distinct recollection of the whole, and taking his pen, ink, and paper, instantly and eagerly wrote down the lines that are here preserved. At this moment he was unfortunately called out by a person on business from Porlock, and detained by him above an hour, and on his return to his room, found, to his no small surprise and mortification, that though he still retained some vague and dim recollection of the general purport of the vision, yet, with the exception of some eight or ten scattered lines and images, all the rest had passed away like the images on the surface of a stream into which a stone has been cast, but, alas! without the after restoration of the latter! (Coleridge, preface to the 1816 edition)

This account has long since been regarded as emblematic of the ephemeral nature of dream recall and the danger of forgetting even the most vividly meaningful dreams if one is distracted by something else upon awakening. In 1997 two special issues of the journal *Dreaming* were devoted Coleridge's treatment of dreams in his literary and philosophical works. These articles highlight Coleridge in relation to the Romantic movement, which took dreams as seriously as the Enlightenment had disparaged them.

The genres of fantasy, horror, and science fiction have naturally provided opportunities for exploring the literary potentials of dreaming. The writings of Edgar Allen Poe (1809–1849) deliberately aimed at eliciting nightmarish feelings of horror and dread in his readers, as seen in *A Descent into the Maelstrom*, *The Murders in the Rue Morgue*, and the poem "A Dream Within a Dream." Robert Louis Stevenson (1850–1894) used his own dreams and nightmares to develop the plot of *The Strange Case of Dr. Jekyll and Mr. Hyde*, a terrifying tale of split personality, and he attributed much of his literary inspiration in general to the "brownies," the Scottish fairies who visited him at night in his sleep. H.P. Lovecraft (1890–1937) wrote numerous short stories in which dreaming features as a frightening portal between the normal world of sanity and the unnamable horrors that lurk in every shadow. According to Robert Bloch, an early protégé of Lovecraft's and later an accomplished science fiction writer in his own right, said this about him:

> The one theme incontrovertibly constant in both his life and his work is a preoccupation with dreams. From earliest childhood on, Lovecraft's sleep ushered him into a world filled with vivid visions of alien and exotic landscapes that at times formed a background for terrifying nightmares.... Gradually he built up a rationale for both reality and dreams, nothing less than a history of the entire universe. (Bloch, 1963, p 6–7)

Another prominent writer in the fantasy and science fiction realm, Ursula K. LeGuin (1929–present), has made dreams and the creative potentials of dreaming a recurrent theme in her works. Perhaps because her father was A.L. Kroeber, a prominent anthropologist with expertise in the Native American cultures of the Pacific Northwest, LeGuin shows in her writing a deep respect for the power of dreaming and an urgent concern about the threats to that power by Western materialism. *The Word for World Is Forest* (1976) envisions an alien planet of forest-dwelling creatures whose waking lives are thoroughly integrated with lucid dreaming experiences; the brutal human colonists who enslave and abuse these creatures seem to represent all the forces of modernity that conspire to destroy dreaming. In *The Lathe of Heaven* (1971) a man living in a dystopic near-future America discovers he can change reality with his dreams, an ability which a sleep researcher tries to use to improve the world, with horrible results. In these and other writings, LeGuin displays an appreciation for dreaming as a profound source of inspiration and creative power. She also evokes in her readers an acute sense of the vulnerability of dreaming to manipulation and violence.

Some authors have published their personal dream journals, offering tremendous possibilities for correlating a writer's dreaming patterns with the literary qualities of his or her published works. American poet and novelist Jack Kerouac (1922–1969), an early innovator in the iconoclastic Beat movement, wrote out his dreams with a passionate determination to give them the freest possible expression: "I wrote non-stop so that the subconscious could speak for itself in its own form, that is, uninterruptedly flowing & rippling" (Kerouac, 1961, p 4). Several of the characters from Kerouac's novels *On the Road* and *The Subterraneans* appear in his dreams, suggesting many possible connections between his life and works. "Everybody interested in their dreams," he said, "should use the method of fishing their dreams out *in time* before they disappear forever" (4, italics in original).

One of the final books published by William S. Burroughs (1914–1997), another highly influential figure in the Beat movement, was *My Education: A Book of Dreams*, in which he offered a provocative, nonlinear meditation on dreams, dreaming, and the literary process. Burroughs made a basic distinction between two types of dreams:

> The conventional dream, approved by the psychoanalyst, clearly, or by obvious association, refers to the dreamer's waking life, the people and places he knows, his desires, wishes, and obsessions. Such dreams radiate a special disinterest. They are as boring and as commonplace as the average dreamer. There is a special class of dreams, in my experience, that are not dreams at all but quite as real as so-called waking life and, in the two examples I will relate, completely unfamiliar as regards my waking experience but, if one can specify degrees of reality, more real by the impact of unfamiliar scenes, places, personnel, even odors. (Burroughs, 1995, p 2)

This distinction between ordinary and extraordinary dreams, common among many cultures through history, gives us a more precise insight into which kinds of dream experiences are likely to play a role in artistic creativity.

British novelist and critic Graham Greene (1904–1991) allowed his executors to publish soon after his death *A World of My Own: A Dream Diary*, in which Greene reflects on the interweaving of his dreams through the course of his writing career (he kept a meticulously detailed dream journal for 25 years). The book is a remarkable document that testifies to the stimulating role of dreaming in a writer's life, enabling him to move freely and creatively between what he called the "World of My Own" and the "Common World."

These examples could be multiplied many times over. Two of the most spectacularly popular book series of recent times, J.K. Rowling's *Harry Potter* stories about a magical school for young wizards and witches and Stephanie Meyer's *Twilight* saga about vampires, werewolves, and the human girl who loves them both, are filled with significant dreams that influence, warn, and inspire the characters. The millions of people, mostly children and teenagers, who have read these books have been well primed to accept the possibility that unusually powerful, revelatory, "magical" dreams can break through the boundaries of ordinary life to motivate acts of great courage and creativity.

Researchers interested in pursuing more detailed studies on the literary dimensions of dreaming can find additional information in Carol Schreier Rupprecht's *The Dream and the Text: Essays on Literature and Language* (1994), Bert States' *The Rhetoric of Dreams* (1988) and *Dreaming and Storytelling* (1993), Richard Russo's *Dreams Are Wiser Than Men* (1987), Marjorie and Jon Ford's *Dreams and Inward Journeys: A Reader for Writers* (1990), Nicholas Royle's *The Tiger Garden: A Book of Writers' Dreams* (1996), Roderick Townley's *Night Errands: How Poets Use Dreams* (1998), Stephen Brooks' *The Oxford Book of Dreams* (2003), Marjorie Garber's *Dream in Shakespeare: From Metaphor to Metamorphosis* (1974), and Naomi Epel's *Writer's Dreaming: 25 Writers Talk About Their Dreams and the Creative Process* (1994).

VI. Film

The invention of motion pictures in the early 20th century enabled the creation of a remarkably dream-like form of cultural experience. The audience sits quiet and motionless in a darkened space while a vivid, emotionally arousing story full of imagery, sound, and action unfolds before them—it should be no surprise that psychoanalysts have called cinema a "dream screen," or that Hollywood has been nicknamed "the dream factory." A great deal of research has been done over the years using movies as emotional presleep stimuli (see Koulack, 1993), indicating that some aspects of film can have a strong impact on dream content. Yet very little research has been done focusing on the cultural dimensions of movie incorporations into dream content. Anyone who has gathered survey data from college and university students in recent years has probably noticed the frequency of references in their dreams to contemporary movies (along with television shows and video games). This is surely a promising area for future investigation.

As with literature, films and filmmakers are inspired by dreaming in several different ways. A good contemporary example is the American director David Lynch (1946–present). He has used the uncertainty between dreaming and waking as the overarching narrative structure of *Lost Highway* (1997) and *Mulholland Drive* (2001), both of which radically and abruptly shift the audience's perception of what is and is not "real." Lynch has created numerous scenes in which characters experience elaborate dreams, most prominently in his television series *Twin Peaks* (1990–1991) when FBI special agent Dale Cooper has a bizarre dream of a midget who becomes a vital clue in his murder investigation. Lynch has also drawn on his own dreams to guide his filmmaking, including *Blue Velvet* (1986):

> I was sitting on a bench and suddenly I remembered this dream that I'd had the night before. And the dream was the ending to *Blue Velvet*. The dream gave me the police radio; the dream gave me Frank's disguise; the dream gave me the gun in the yellow man's jacket; the dream gave me the scene where Jeffrey was in the back of Dorothy's apartment, sending the wrong message, knowing Frank would hear it. I don't know how it happened, but I just had to plug and change a few things to bring it together. Everything else had been done except that. (Rodley, 1997, p 136)

The visual and emotional themes of Lynch's movies bear obvious similarities with the surrealist movement, particularly in Luis Bunuel and Salvador Dali's short film *Un chien andalou* (1929) with its jarringly violent imagery and Freudian preoccupations. Alfred Hitchcock's psychological thriller *Spellbound* (1949), with its Dali-designed dream sequence, also paved the way for Lynch's creative blending of dreaming and waking realities.

Perhaps the most famous and influential "dream movie" is *The Wizard of Oz* (1939), with its dramatic shift from the black-and-white waking world of Kansas to the vividly colorful dreaming world of Oz. This widely beloved film drew upon several features of common dream experience, including vivid sensations of flying and falling, strange transformations of people and animals, and scenes of indescribable beauty, in order to intensify the audience's immersion in the story.

At the other end of the emotional scale, the *Nightmare on Elm Street* series of horror movies uses many of the recurrent features of nightmare experiences to create a sense of frightening vulnerability and supernatural dread. Wes Craven, the director of the first film in the series (1984), acknowledged the influence of dream experience on his creative work: "I first became interested in dreams in college. When I delved into the subject, I discovered that the Russians were performing experiments in dream manipulation and were actually working on assassination through psychotic intervention. Lately, research has shown that people can actually die during a nightmare. So the idea of being accosted in your dreams became the backbone of the story" (Cooper, 1987, p 10). His 1994 film, *Wes Craven's New Nightmare*, pushed the dream premise into new territory by intermixing the "real" world of the actors and actresses who performed in the *Nightmare* series with the "dream" world they created in the films that now threatens to break into waking reality.

One of the last films from the master Japanese director Akira Kurosawa (1910–1998) was *Dreams* (1990), which presents dramatic recreations of eight of Kurosawa's own dreams, experienced at different times of his life. The lyrical imagery, conflicting emotions, and magical ambiguity of the film expressed not only the director's personal life but also the collective life of the Japanese people through a terrible period of cultural history during and after World War II. The theme of loss and mourning pervades *Dreams*: A young boy is shut out of his family home, and then struggles to understand why a beloved cherry blossom orchard has been cut down; a military officer agonizes over the spectral return of a platoon of soldiers who died under his command; and an aspiring Japanese artist vainly chases the receding figure of Vincent Van Gogh as the famous painter (played in the film by Martin Scorsese) disappears into his colorful landscapes. Kurosawa's *Dreams* shows that the experience of mourning, while acutely painful, can inspire both haunting dreams and great works of art.

Another prominent filmmaker who has devoted careful thought to the inspirational role of dreaming is the American director John Sayles (1950–present). In films such as *Passion Fish* (1992) and *Limbo* (1999) Sayles presents characters' dreams as integral elements to his storytelling and the emotional involvement of the audience. In an interview with sleep and dream researcher James Pagel, Sayles said one of the ways dreaming inspires filmmaking is to allow a wider range of opportunities for communicating with his viewers: "A dream scene can give, sometimes the character, but certainly the audience, information that the character doesn't provide willingly and consciously. That creates a subtext. . . . A dream provides information that you can put into the story with a different weighted value than the conscious actions of the characters"

(Pagel *et al.*, 2003, p 45). These observations by Sayles add to the idea of a special bond between movies and dreaming as expressions of imaginative freedom.

Several directors have said they were directly inspired by their dreams in the creation of some of their best films, including Ingmar Bergman and *Wild Strawberries* (1957), Francis Ford Coppola and *Apocalypse Now* (1979), and Robert Altman and *3 Women* (1977). About the latter film, Altman told critic Roger Ebert what originally motivated his creative efforts:

> I dreamed of the desert, and I dreamed of how these three women, and I remember that every once in a while I'd dream that I was waking up and sending out people to scout locations and cast the thing. And when I woke up in the morning, it was like I'd *done* the picture. What's more, I *liked* it. So I decided to do it. (quoted in Gabbard and Gabbard, 1987, p 241)

A pioneering empirical study in this area was performed by James Pagel and a team of researchers who gathered surveys from participants at Sundance Film Institute Screenwriters and Directors Lab from 1995 to 1997 (Pagel *et al.*, 1999). Pagel found that the filmmakers had higher dream recall and more dream impact on waking behaviors compared with a previous general population study. Furthermore, he found the more "creative" people in the film industry (directors, screenwriters, actors) had higher dream recall and dream effects on waking behavior than did the people who serve as "workers" in the filmmaking process (production crew).

Additional information on the dream–film connection can be gathered from Bernard Welt's chapter on dreams and film in *Dreaming in the Classroom* (co-authored with Phil King and Kelly Bulkeley, in press), Robert T. Eberwein's *Film and the Dream Screen: A Sleep and a Forgetting* (1985), and Leslie Halpern's *Dreams on Film: The Cinematic Struggle Between Art and Science* (2003). An amazing resource for future research is the illustrated dream journal of Italian director Federico Fellini (1920–1993), *The Book of Dreams* (2007), which contains Fellini's dreams, fantasies, and drawings, giving new insights into his genius for representing bizarre, sensuous, and paradoxical cinematic imagery. Much like the literary dream journals of Kerouac, Burroughs, and Greene, Fellini's *Book of Dreams* enables the investigation of specific correlations and influences moving between filmmaking artistry and dreaming imagination.

VII. Conclusion

Writer and critic Jorge Luis Borges once said, "Dreams are an aesthetic work, perhaps the earliest aesthetic expression" of our species (Borges, 1984). The evidence presented here, while neither definitive nor exhaustive, suggests that

Borges was justified in identifying a core connection between the primal human experience of dreaming and the cultural wonders of creative inspiration. From ancient religious leaders and philosophers to modern writers and filmmakers, all agree in testifying to the motivational power of dreaming in their waking creativity.

Several of the examples presented here display a theme that may account for the strong experiential impact of these dreams. This theme regards a disorienting uncertainty about the boundaries between waking and dreaming, what might in cognitive scientific terms be called a temporary suspension of one's reality-testing ability. Many people are familiar with this strange feeling from their own dreams—the sense that some dreams feel "realer than real," calling into question what we think about our waking reality. What makes some dreams "inspiring" seems to be their effect of expanding a person's awareness beyond what *is* to what *could be*. The Daoists, with their enchanting butterfly dream, and Socrates, with his puzzling dream questions to Theaetetus, use this existential uncertainty to generate new levels of enlightened understanding. In the writings of Lovecraft and the films of Lynch, the characters and the audience find themselves ontologically decentered as they are drawn into dark, fantastic worlds where dreaming and waking realities merge, evoking uncanny feelings of wonder and awe. Drawing on the research of Hartmann (1998), Hunt (1989), Knudson (2001), Tonay (1995), and Nielsen (1991), one could build the case that dreams with this reality-bending quality inspire people because they activate to a very high degree a wider network of brain–mind systems than is ordinarily operative in either waking or sleeping.

References

Bloch, R. (1963). Heritage of Horror. The Best of H. P. Lovecraft. Ballantine Books, New York.

Borges, J. L. (1984). Seven Nights. (Translated by E. Weinberger). New Directions, New York.

Bulkeley, K. (2005). The Wondering Brain: Thinking about Religion with and Beyond Cognitive Neuroscience. Routledge, New York.

Bulkeley, K. (2008). Dreaming in the World's Religions: A Comparative History. New York University Press, New York.

Bulkeley, K., and Domhoff., G. W. (2010). Detecting meaning in dream reports: an extension of a word search approach. *Dreaming* **20** (2), 77–95.

Burroughs, W. S. (1995). My Education: A Book of Dreams. Viking, New York.

Cole, J. R. (1992). The Olympian Dreams and Youthful Rebellion of Rene Descartes. University of Illinois Press, Urbana.

Cooper, J. (1987). The Nightmare on Elm Street Companion. St. Martin's Press, New York.

Coxhead, D., and Hiller, S. (1976). Dreams: Visions of the Night. Crossroads, New York.

Domhoff, G. W. (1996). Finding Meaning in Dreams: A Quantitative Approach. Plenum, New York.

Fellini, F. (2007). The Book of Dreams. Rizzoli, New York.

Ford, M., and J. Ford, eds. (1990). Dreams and Inward Journeys: A Reader for Writers. Harper Collins, New York.

Foulkes, D. (1999). Children's Dreaming and the Development of Consciousness. Harvard University Press, Cambridge.

Gabbard, K., and Gabbard, G. O. (1987). Psychiatry and the Cinema. University of Chicago Press, Chicago.

Grace, N. (2001). Making dreams into music: contemporary songwriters carry on an age-old dreaming tradition. In: Dreams: A Reader on the Religious, Cultural, and Psychological Dimensions of Dreaming (K.Bulkeley., ed.), Palgrave, New York.

Halpern, L. (2003). Dreams on Film: The Cinematic Struggle between Art and Science. McFarland & Company, Jefferson, North Carolina.

Hartmann, E. (1998). Dreams and Nightmares: The New Theory on the Origin and Meaning of Dreams. Plenum, New York.

Hunt, H. (1989). The Multiplicity of Dreams: Memory, Imagination, and Consciousness. Yale University Press, New Haven.

Kerouac, J. (1961). Book of Dreams. City Lights Books, San Francisco.

King, P., Bulkeley, K., and Welt., B. (In press). Dreaming in the Classroom. State University of New York Press, Albany, New York.

Knudson, R. (2001). Significant dreams: Bizarre or beautiful? *Dreaming* **11** (4), 167–178.

Koulack, D. (1993). To Catch a Dream: Explorations in Dreaming. State University of New York Press, Albany.

Nielsen, T. (1991). Reality dreams and their effects on spiritual belief: a revision of animism theory. In: Dream Images: A Mental Call to Arms (J. Gackenbach and A. A., Sheikh., eds.), Baywood, Amityville.

Pagel, J. F., Crow, D., and Sayles., J. (2003). Filmed dreams: cinematographic and story line characteristics of the cinematic dreamscapes of John Sayles. *Dreaming* **13**(1), 43–48.

Pagel, J. F., Kwiatkowski, C., and Broyles., K. E. (1999). Dream use in film making. *Dreaming* **9**(4), 247–256.

Plato (1961a). Crito. In: Plato: Collected Dialogues (E., Hamilton and H. Cairns, eds.), Princeton University Press, Princeton.

Plato (1961b). Theaetetus. In: Plato: Collected Dialogues. (E., Hamilton and H. Cairns, eds.), Princeton University Press, Princeton.

Platow, R. ed. (2005). Catalog for the Dreaming Now Exhibit at the Rose Art Museum. Brandeis University, Waltham, Massachusetts.

Rodley, C. (1997). Lynch on Lynch. Faber and Faber, London.

Royle, Ni. ed. (1996). The Tiger Garden: A Book of Writers' Dreams. Serpent's Tail, New York.

Russo, R. A. ed. (1987). Dreams are Wiser Than Men. North Atlantic Books, Berkeley.

Strauch, I., and Meier., B. (1996). In Search of Dreams: Results of Experimental Dream Research. State University of New York Press, Albany.

Tonay, V. (1995). The Art of Dreaming. Celestial Arts, Berkeley.

Townley, R. ed. (1998). Night Errands: How Poets Use Dreams. University of Pittsburgh Press, Pittsburgh.

Zadra, A. L., and William Domhoff., G. (2010). The content of dreams: methods and findings. In: Principles and Practices of Sleep Medicine (M. Kryger, T. Roth and W. Dement eds.), Elsevier Saunders, Philadelphia.

DEVELOPMENTAL PERSPECTIVE: DREAMING ACROSS THE LIFESPAN AND WHAT THIS TELLS US

Melissa M. Burnham and Christian Conte

Department of Educational Psychology, Counseling, & Human Development, University of Nevada, Reno, NV, USA

This chapter takes on the ambitious goal of describing dreaming across the lifespan, integrating both empirical dream research and clinical case examples. Each major stage of the lifespan is discussed, from infancy (where our knowledge of dreaming is speculative at best) to later adulthood. Written from the perspectives of a developmental sleep researcher and a clinician, the chapter weaves together what is known empirically with the usefulness of dreams at different stages to inform clinical understanding. We attempt to provide an integrative view of dreaming which embraces the fundamental ambiguity of dreams across the lifespan.

I. Introduction

Dreams are situated in a unique context due to their ubiquity. Dreams are the topic of everyday conversations, used to understand clients in some forms of therapy, and, to a lesser extent, have been studied scientifically. Dreaming has proven a difficult topic of scientific research at some points during the lifespan and in some individuals because dreams cannot be accessed directly. Because the study of dreams relies on personal reports, our knowledge of people's dreams depends in part on their communication ability and accuracy with regard to recalling and reporting their dreams.

When viewed with a developmental lens, the study of dreaming becomes even more complex. Although the human neonate, and, indeed, the fetus, spends the vast majority of its sleep time in rapid eye movement (REM)/active sleep, the state in which much adult dreaming occurs, it has thus far been impossible to discover whether the infant's REM/active sleep is accompanied by dreams. It is not until the early childhood years that children are linguistically competent and start reporting dreams. From early childhood on, however, dreams are known to occur, and may change as individuals develop (e.g., Foulkes, 1982).

From a developmental perspective, dreaming may develop alongside other domains, such as cognitive, social, or emotional development. Most developmental theories suggest that changes occur in these domains as individuals develop, and some purport that particular challenges emerge at different points within the lifespan (e.g., Erikson, 1959). Viewing dreaming within a developmental framework, then, one would expect that both the nature of dreaming and the content of dreams might change during the process of development. It is also important to note that the cultural context within which the person is developing will likely impact the interpretation of dreams, if not the dreaming process itself. For example, the traditional Western perspective views dreams as fictional, private, internal, and subjective (Meyer and Shore, 2001). Yet, not all societies and cultures hold the same view on dreaming. Therefore, while it is likely that dreams change as individuals develop, it is equally important to recognize that the culture within which the person is developing will likely influence both the nature of these changes and the nature of their interpretation.

In this chapter, we will present a lifespan developmental analysis of dreaming, using both science and clinical experience to frame the discussion of dreams from infancy to older adulthood. The perspectives of a researcher and a clinician will inform what we know both from research and from the experience of a clinician using a neo-Jungian point of view.

II. Background and Clinical Perspective

Before delving into the development of sleep and dreaming, some background information on theory and different perspectives on the study of dreams will be provided to give context for this review. Freud (1900) began the clinical investigation of dreams and their interpretation with *The Interpretation of Dreams*.

Freud took a dogmatic stance to his approach and seemed to disregard Karl Popper's idea of falsification (that science is differentiated from belief by its ability to be tested). Carl Jung (1961) challenged Freud on his reliance on fixed or rigid symbols in dream interpretation, and went on to develop his own theory of dreams that was much more expansive in nature. Jung was not without his own rules for dream interpretation: he believed that not only every dream, but every part of every dream, is completely unknown at the onset (Campbell, 1971). He also believed that dreams are teleological (Jung and Meyer-Grass, 2008).

A paradigm shift that occurred early in the 20th century seemed to change the approach that could be taken when working with dreams. Jung began to see dream symbols as dynamic rather than static (Jung, 1950). He began to notice that symbols which meant some things for certain patients did not mean the same thing for other patients. This deviated from the straightforward Freudian approach, and, from what Jung believed, was part of what led to the separation of the two of them (Jung, 1961). Today, over a hundred years later, we know a great deal about neurology and the science of dreaming, but we still are no closer to understanding the clinical application of dreams than they were. The most effective use of discussing dreams in clinical settings appears to occur when clinicians are non-attached. Different approaches to dream interpretation exist. Different types of dreams exist. From the anticipation of what is to come (Jung and Meyer-Grass, 2008) to the completion of the past, dreams may provide keys to the developing mind.

For psychoanalytically oriented clinicians, dreams are vital to understanding the psyche. If we do not pay attention to dreams, we miss an opportunity to expand consciousness (Mattoon, 2006). In an empirically driven world where the pendulum has swung so far to measurable events, non-measurable phenomena are often overlooked or discounted, and so it is with dreams. Mattoon (2006) noted that many times dreams are ignored because neither they are readily understood nor do they always deliver pleasant messages. Perhaps dreams are often ignored in the empirical world because they require a sophisticated level of comfort with ambiguity, a quality that is difficult to achieve. Indeed, camps of extremists range from the dedicated "black box" empiricists to the passionate believers who have taken Kierkegaard's "leap of faith." It could be, however, that the balanced approach Lao Tzu articulated for life 2500 years ago could also be the key to handling dreams. It is such a balanced approach that we seek to achieve in this chapter.

At a minimum, dreams can be used as personalized projective instruments by which clinicians can gain valuable information about clients, and by which clients can in turn gain invaluable information about themselves. At a maximum, dreams can be a gateway through which humans can peer into some unknowns that science cannot predict. When it comes to dream interpretation, sometimes it is difficult to distinguish between what is science and what is not, but dreams have been interpreted for millennia, and systematically interpreted for over a hundred years. Of course, anything we have to say about the meaning of dreams is ultimately speculation, and we may not be successful in providing a true picture of them (Jung and Meyer-Grass, 2008); but this is the nature of science: the only truth is that which is subjected to falsification.

From our earliest development we seem to rely on introjected information, because examining all data presented to us is neither feasible nor pragmatic for our survival. From accepting without question what our caretakers tell us about what is edible or not, to unquestioningly accepting what they believe about metaphysical concepts, it is in the best interest of young children to be agreeable with those who provide for them. The certainties we rely on as children continue throughout our lives until or unless we begin to question them. For many, questioning is not consciously permitted because the introjections are so power-ful. One model of the psyche proposes that introjections and other ego defense mechanisms work to push unconscious material down and keep us locked safely in our egos (Conte, 2009). The more certainties that exist for us, the more tightly we are locked in our egos. As the ego is the center of our consciousness, not the center of our psyche (Campbell, 2002), we cannot expand consciousness without exploring what may be hidden outside of it.

The psychodynamic circle presented by Conte (2009; see Fig. 1) is useful for clinicians, because it provides a physical model from which to view the meta-physical psyche. Plato used the circle to represent the psyche because it has no beginning and no end. The true self, the core of our psyche, has information for the ego; hence, it will constantly ebulliate throughout a person's life, until, theoretically, all the ideas would rise through to consciousness (Conte, 2009). This process is known as individuation (von Franz, 1964).

Dreams are innately personal because they reflect the dreamer's experiences, but dreams also may be collective in the sense that they encompass emotions, cognitions, and behaviors that derive from a pool of limited human experiences (Strauch and Meier, 1996). Themes in literature mimic themes in life, themes in life are mimicked in dreams, and dreams in turn influence literature: we act, we have feelings, we think. Experiences are simultaneously limited and infinitely varied, and so too are dreams. Dreams are thought to begin early and are experienced throughout the lifespan. The case studies presented in this chapter provide examples of dreams and their analysis from early childhood, adolescence, early adulthood, midlife, and later life.

III. Sleep and Speculations about Dreaming in Infancy

During the period of infancy, both the structure and process of sleep undergo major transformations, making sleep in this period of the lifespan quite distinct. Infants do not experience the paralysis that is typical of rapid eye movement

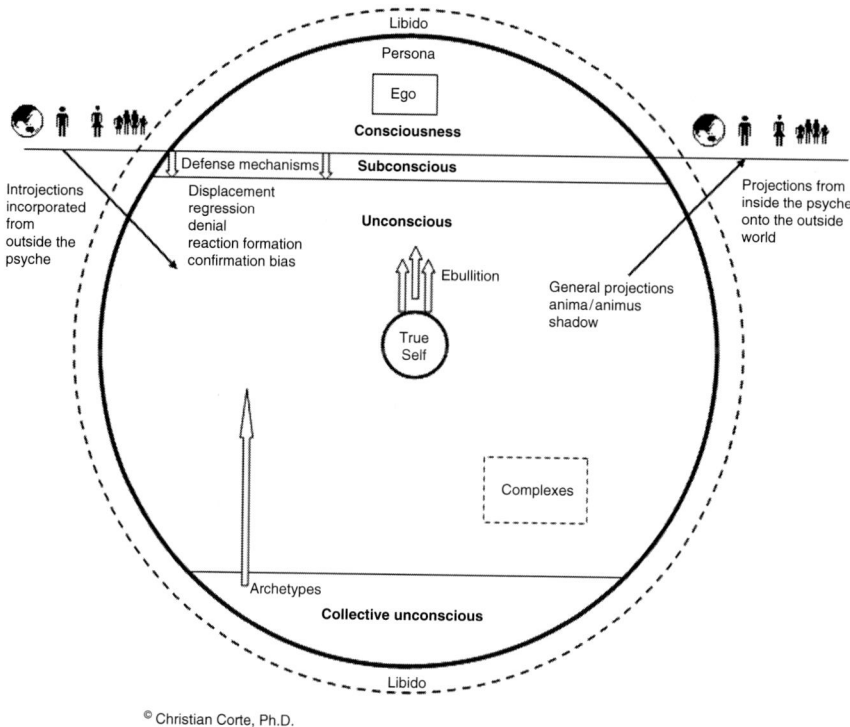

© Christian Corte, Ph.D.

FIG. 1. From Conte, 2009: *Advanced Techniques for Counseling and Psychotherapy.*

(REM) sleep in older children and adults; thus, their sleep states can be inferred by direct observation. "Active sleep" (AS) is characterized by irregular respiratory patterns, large and small skeletal muscle movement, and phasic eye movements (Montgomery-Downs, 2008). It is thought that AS is the precursor to REM sleep (although this is somewhat controversial; see Frank and Heller, 2003). The infant in "quiet sleep," on the other hand, has slow, regular respiratory patterns, no eye movements, and little gross motor movement with the exception of occasional startles (Anders and Sostek, 1976). Quiet sleep is thought to be the precursor to non–rapid eye movement (NREM) sleep. While older children and adults typically enter sleep and immediately transition into NREM sleep, the young infant enters AS immediately and spends a larger proportion of the sleep period in AS (Anders and Keener, 1985; Burnham *et al.*, 2002). In addition to these differences, the process of sleep consolidation also occurs during infancy and continues throughout early childhood. The newborn infant spends an approximately equal proportion of time asleep during day and night. However, as the circadian system develops and becomes in tune with the environmental light–dark cycle, infant sleep gradually becomes consolidated to the nighttime hours. While day-time napping continues into early childhood, the majority of sleep per 24 hours occurs at night by about 3 months of age (Kleitman and Engelmann, 1953).

The large amount of AS in the period of infancy has intrigued scientists since its discovery, yet, almost 50 years later, evidence providing an explanation for this phenomenon continues to elude the field of developmental sleep research. One of the most captivating hypotheses suggests that AS serves to aid brain development at a time when neural connections are forming at an overabundant rate (e.g., Roffwarg et al., 1964, 1966). In other words, brain activity during AS may serve as an endogenous method of stimulating synaptogenesis, axon growth and targeting, and the like, at a time when environmental stimulation is limited. As environmental stimulation increases and the need for endogenous stimulation decreases, the large amount of active/REM sleep is no longer necessary. This hypothesis has received some support from animal studies of brain development, which show that endogenous stimuli are essential for normal brain development and plasticity, and that REM sleep is the only form of such stimulation (see Graven, 2006 for a review). Depriving developing animals of REM sleep during critical periods has been shown to impact typical development of the visual, auditory, somatesthetic, olfactory, and limbic systems as well as cortical connec-tions in the hippocampus associated with learning and memory (Graven, 2006). Thus, it appears clear that active/REM sleep does play an important role in brain development.

What is not as clear is whether or not dreams, as we know them, occur during these periods of AS in infancy. Those who believe in continuity during develop-ment may purport that dreams must occur during infancy because they co-occur with REM sleep in later development. This is an erroneous conclusion, however, because REM sleep and dreaming have been shown to be doubly dissociable states (Solms, 2000). That is, although the two are correlated, recent research has shed light on two separate mechanisms controlling the expression of REM sleep and dreaming. The presence of REM sleep is not sufficient evidence to confirm the existence of dreaming. Thus, it is likely that dreaming may be discontinuous across development. That is, dreaming may not be present until certain other cognitive, social, and emotional processes become apparent. Supporting this view is brain science indicating forebrain involvement in dream generation (Solms, 2000). Given what we know about the infant brain and its early competencies and limitations, it is likely that, if infants do dream, the dreams are not characterized by a rich narrative context or the presence of self as an active agent. Indeed, if we accept the definition of dreaming offered by Foulkes (1999), as a "conscious experience ... the awareness of being in an imagined world in which things happen," developmentalists must rethink the notion that infant dreams exist in any form similar to our own (p 9). The development of a clear sense of self as separate from others is not apparent until at least 16–18 months of age

(Lewis, 1989). And, the ability to understand and manipulate symbols does not develop until sometime during the second year (Piaget, 1952). Given evidence from developmental and brain science, then, it appears unlikely that AS in infancy is associated with the same sort of dreaming that is characteristic of older children and adults.

IV. Sleep and Dreaming in Childhood

In contrast to the period of infancy, where one must speculate on the existence of dreams, it is clear that children dream. Most parents can attest to this fact, as children often spontaneously share their dreams in the morning or awaken parents in the night to describe a particularly vivid nightmare and seek comfort. Researchers have confirmed the presence of dreams, starting at least by the age of 3 (e.g., Foulkes, 1982). The sleep patterns of children are also remarkably different than those of infants. By the age of 3, most young children are taking just one afternoon nap and tend to sleep for 10–12 consolidated hours during the night (Iglowstein *et al.*, 2003). By middle childhood, the daytime nap is dropped and children sleep for 9–10 hours during the night (Iglowstein *et al.*, 2003). Children also enter sleep and transition immediately into NREM. Despite these adult-like characteristics, however, the 90 min sleep cycle that is typical of adult sleep does not appear until sometime during middle childhood (Kahn *et al.*, 1996).

Studies of dreaming in early childhood have principally come from cognitive psychologists, and thus most have taken a largely cognitive focus. Indeed, Piaget himself studied dreams in childhood and explained their development in terms of children's growing competence with regard to symbolic thought (Piaget, 1962). A more recent example comes from the work of Foulkes. In his seminal longitudinal and cross-sectional investigations of dreaming, Foulkes (1982) characterized the dreams of 3–5-year-olds as "preoperational dreams" in that they tended to reflect the characteristics of thought in Piaget's preoperational stage of development. Not only did Foulkes find that dreams were reported in only 27% of REM awakenings in the laboratory for this age group, but he also found that dream reports were typically short, straightforward, and simple. These findings illustrate an important distinction between dream reports taken in home settings versus those taken in a laboratory upon direct awakenings from REM sleep. Preschool-aged children may remember and report highly vivid, imaginative dreams at home, but laboratory studies show that these dreams are likely not typical experiences. When Foulkes studied the same children at ages 5–7, they tended to report more elaborate dream content, although the percentage of REM awakenings in which dreams were reported remained relatively stable, at 31%. By the age of 7–9, children reported dreams from REM awakenings significantly more frequently (48% of the time) and the dream reports were still more elaborate. Notably, by ages 7–9, children's dreams started to include a "self-character" and an affective nature that was rarely included in prior dream reports. Foulkes (1982) concludes that children's dreaming is "a natural and inevitable concomitant of children's waking cognitive maturation" (p 295).

Meyer and Shore (2001) expanded on the work of Woolley and Wellman (1992) and Woolley (1995) in their examination of children's dream reports in terms of the development of "theory of mind." Children's understanding of the mind involves several interrelated concepts, including that appearance and reality do not always coincide, that two people can experience the same object or event differently, and that others can have thoughts different from one's own (Flavell, 1988). Meyer and Shore (2001) hypothesize that theory of mind development is necessary for children's basic understanding of dreams. Indeed, these authors found that children's understanding of dreams as unreal and private occurrences correlated highly with both appearance-reality and perspective taking tasks.

Thus, previous work has found a strong correlation between children's cognitive development and their dream reports, which is consistent with a developmental perspective on dreaming. That is, dreams and children's understanding of dreams appear to change as developments in other domains arise. A case example will illustrate how a dream in early childhood can be used to develop a better understanding of the individual child, from a clinician's perspective.

The first case example is from a 4 year-old girl who appeared to be typical, both physically and psychologically, and who lived as an only child with both parents at the time of the dream. Her dream, consistent with 40 percent of the children in Foulkes' (1982) research, included animals. Unlike the average 3–5-year-old dreamers in Foulkes' research, however, the dream reported was emotionally charged and the dream self was active rather than the passive, something Foulkes' research did not uncover until the age of 7–9. In this case study, the dream is analyzed under the assumption that all dreams are teleological (Jung, 1974), and that childhood dreams can be anticipatory (Jung and Meyer-Grass, 2008).

A. CASE EXAMPLE 1: CHILDHOOD DREAM

A 4-year-old girl reported the following dream with vibrant and contented positive emotion. "I dreamed that I was dancing with dinosaurs. They were very friendly to me, but there were still some bad dinosaurs in the dark, but I wasn't afraid of them because I was friends with all the good dinosaurs."

To translate this dream as an anticipation of what is to come for this young child, we need to make broad, sweeping statements about the symbolism presented. Knowing nothing else about the girl except for what was presented above, we could, for instance, look to the common mythological symbolism that permeates most cultures. In this case, three major themes are readily apparent: the monster archetype, the concepts of good versus evil, and the archetypal independence of spirit. The odds are in favor that following this type of interpretation is more about the interpreter than the dreamer. After all, the girl reported loving dinosaurs, so her affinity for them can understandably amount to a quick write-off for the dream having any significant deeper meaning. Though anticipatory interpretations are not practical in nature, they do offer a glimpse into the style of analysis that can take place with older dreamers.

First, the monster archetype appearing in the form of dinosaurs is readily apparent in the dream. The dreamer interacts with the dinosaurs by dancing freely among them. She believes herself to be accepted by them. Simultaneously, the dreamer is aware of "bad dinosaurs in the dark," a reference almost directly to the definition of the shadow. According to Jung and colleagues (1964), the shadow holds the parts of our selves we do not wish to face, but it by no means has to be negative (Johnson, 1991). For a healthy 4-year-old who has not suffered abuse or trauma in any way, the monster motif could stem from the collective psyche. The battle between people and monsters is a primordial theme. At 4 years old, this dreamer could possibly be experiencing a connection to the epic battles against real and psychological monsters she will face for the rest of her life.

If the monsters in the dark are a representation of her developing shadow, then it was interesting that the dreamer had awareness but no fear of them. It seems to be only later in life when we ardently protest the shadow (Johnson, 1991). The monsters with whom she was dancing could have represented the adults in her life. As an only child, she spent more time around "giants" than people her own size. The anticipatory aspect of the monster motif is that the dinosaurs could also have represented the challenges that this young girl will face in life. That the dreamer danced among them could indicate that she will handle well the challenges that she will face in life, that she may let go to her problems and dance with what comes her way, rather than focus on the things she cannot control (i.e., monsters in the dark).

The concept of good versus evil inherent in all cultures (what [Buber, 1950] described as the "us versus them" mentality) is also fairly apparent in the dream. For this child's developing mind, perhaps the dream is indicating that these are challenges she will face. It would be interesting to find out if there would have been more of an even number of dinosaurs in the light and dark at a younger age for the dreamer. To know so might lend support to the concept that the psyche begins in a pure state, and the conscious and unconscious begin to divide evenly until the conscious mind becomes overwhelmed with ambiguity; at that point the ego develops defense mechanisms to protect itself from what Jung called our base nature (Campbell, 1971).

Finally, the archetypal independence of spirit is a theme that emerges from the first word of the dream report: "I." The identification of self exists in conjunction with the separation of self (Campbell and Moyers, 1988). The Hindu concept of Atman is that we are all one, yet we are all individuals exhibiting this oneness. From the collective narrative, the dreamer's psyche could be anticipating the eternal quest: to individually return to that from which she came. This independence is recognized with "I," but developed through the dreamer's report that she was among the dinosaurs: both with them and separate from them. This constitutes the dichotomy of opposites inherently present in all.

Viewing the dream as anticipatory is consistent with the assumption that all dreams are teleological. The interpretation presented may be no more than the Barnum effect. After all, who would expect to live a life and not face challenges that can or cannot be viewed as "monsters"? Perhaps that is exactly the point that research on childhood dreams has uncovered: there are a limited number of themes that can come up for dreamers (Foulkes, 1982; Strauch and Meier, 1996). Whether they stem from the collective psyche that holds the collective narrative, or simply the limited number of experiences humans can have, the major subjects of dreams are the major themes of being alive.

As we grow, our dreams change. Foulkes (1982) noticed that childhood dreams change throughout development. Jung (1974) saw all dreams as teleological; however, in regard to the lifespan, he believed the earliest dreams had a significance that was not regained until adolescence. Dreams have impact for people in their mid-twenties, but the dreams of people in mid-life and older can be the most profound because they represent closure to an individual life (Jung). Though in general, certain periods of life may produce more poignant dreams than others, any dream at any time can be used to help dreamers gain awareness in their lives.

V. Sleep and Dreaming in Adolescence

Adolescence marks a transition in both development and in sleep. Adolescence is functionally defined as a transitional time between childhood and adulthood, generally from the beginning of the pubertal transition to the attainment of adult roles and responsibilities (Dahl, 2004). Significant shifts occur within the circadian timing system during adolescence, and these shifts are correlated with pubertal status (Crowley et al., 2007. The adolescent's tendency to go to bed later is not only influenced by psychosocial factors such as social and school obligations; laboratory investigations have demonstrated that this shift is biologically regulated (Crowley *et al.*, 2007). Coupled with these biological shifts is the trend for schools to start earlier as children get older, thus creating a recipe for less than adequate amounts of sleep in adolescence, at least during the school week. The typical adolescent, then, experiences significant shifts in the timing and amount of sleep, although other sleep characteristics are relatively adult-like during this age period (e.g., NREM at sleep onset; percentage of REM sleep).

Dream reports of adolescents were systematically examined in Foulkes' (1982) investigation of dreaming. Compared to the lower percentage of dream reports obtained from younger children, preadolescents (aged 9–11) reported dreams on 79% of REM awakenings. The percentage remained stable when these children were 11–13 and 13–15 years of age. Foulkes also found that dream content appears to become more abstract in early adolescence, again tying the development of dreaming to cognitive developmental skills. The link between cognitive development skills and dreaming is also supported by brain science showing that dreaming is generated by forebrain mechanisms (Solms, 2000). Suggesting that dreams are related to cognitive development does not, of course, preclude their being related to other aspects of development. Indeed, all aspects of development are interrelated, and particular skills that develop in one domain inevitably impact those in another. In adolescence, understanding of oneself and the intricacies of understanding all aspects of the self are prominent. Thus, one might predict that as abstract thought develops, abstractions in dreams also may develop. Both of these may inform one's understanding of self. The clinical case study presented below illustrates this possibility.

A. CASE EXAMPLE 2: ADOLESCENT

A 17-year-old female came to counseling regularly to work on issues she was having with her family. One day, she came in and reported that she was furious with her boyfriend, and that as soon as she left our session she was going to call him and "let him have it." Naturally, I asked what had happened. "Oh he didn't do anything, but last night I had a dream where he was flirting with all these different girls. I'm so mad at him! I've been furious with him since I woke up."

I told her I was glad she didn't "go off" on him yet because I might have a view on her dream that was slightly different than how she processed it. She was curious, so I invited her to consider that some people believe that every person and object in our dreams represents a piece of our selves. I briefly explained to her the concept of her animus (the masculine aspect of her psyche), and then I invited her to consider that her boyfriend represented her animus, and that the girls in her dream may have been representations of her anima (or feminine aspect of her psyche). I invited her to explore, even momentarily, that her dream might have been all about her, meaning that instead of her boyfriend wanting to date other girls, she really had desired to date other guys.

Her face was flush and she struggled to hold in a smile. I asked her what was going on and she said, "I know my boyfriend would never cheat on me because he's whipped on me. I went to visit a community college where I want to go in the fall, and when I was there I saw like 100 cute guys, and I couldn't stop thinking how I don't want to be with my boyfriend anymore!" She went on to say that she could never see herself as someone who cheated on her boyfriend, but secretly she had been thinking about what it would be like. Perhaps her dream gave her a safe way to explore the idea, an idea that her introjections on fidelity prohibited her from allowing herself to openly consider.

For this client, viewing the people in her dream as aspects of herself brought her to an insight that may have taken much longer to get to, had she only considered her relationship consciously. The client's level of cognitive functioning allowed her to view a broader perspective of her developing identity. By challenging her introjections about gender roles and fidelity, she was able to integrate a part of her self that went largely unnoticed. When her perception changed, her reality changed. Within a week of this dream she broke up with her boyfriend of two years and let her parents know that she would attend the community college. Whether there is falsification for the process by which she gained this insight is clinically less relevant than the fact that she got the insight and was able to gain a broader understanding of her own development.

Adolescents appear to be interested in the mysteriousness of their dreams, and clinicians can benefit from connecting with adolescents through the discussion of their dreams. Developmentally, dreams for adolescents center on identity. The crisis that adolescents face is finding out who they are and what their place is among others (Erikson, 1959). Taking a clinical approach to adolescent dreams as teleological and supportive of creating or unfolding their identity can provide pragmatic results to the clinician.

VI. Sleep and Dreaming in Adulthood

The sleep of adults has been characterized in great detail (see Carskadon and Dement, 2000). The adult typically enters sleep in NREM, and then progresses into deeper NREM stages before entering the first REM episode (after 80–100 min).

Thereafter, REM and NREM sleep oscillate with a cycle period of ~90 min. REM sleep episodes become lengthened across the night, and REM sleep makes up approximately 20–25% of total sleep (Carskadon and Dement, 2000). Although it was once thought that dreams only occurred during REM sleep, it is now recognized that some mental activity occurs during NREM, and people do report "dreams" (albeit at a lower rate) when awakened directly from NREM sleep (e.g., Pivik, 2000).

The dreams of adults contain several defining characteristics that separate them from waking consciousness. According to Hobson and colleagues (1998), REM sleep dreams in adulthood include intense and vivid visual hallucinations, intense emotions, a delusional belief that one is awake, lack of self-reflective awareness, incongruity, narrative story lines, and memory deficits (i.e., most dreams are quickly forgotten). These authors purport that each of these characteristics can be mapped onto neurobiological processes underlying the generation of dreams. For example, the lack of self-reflection in dreams may be due to the deactivation of the dorsolateral prefrontal cortex during REM sleep (Maquet *et al.*, 1996;Muzur *et al.*, 2002). Hobson and colleagues' account explains both the bizarreness of dreams and their emotionally salient content as due to the brain structures involved in their generation. Dreams in adulthood may also inform the clinician interested in understanding more about the individual, given dreams' tendency to be linked to current personal concerns (e.g., Domhoff, 2001). The next cases exemplify this point.

This case example involves a woman in her mid-twenties. According to Erikson (1959), young adults typically struggle with intimacy versus isolation. The woman described here, like so many of her peers, was striving to find the balance between being differentiated and connected at the same time. To accomplish this, people in this stage of development often naturally question the meaning of their existence (May, 1991). Finding meaning in existence is the *eternus itinerius*. Because answers do not come in strictly conscious forms, relying on dream analysis can be another helpful method for supporting people in this stage of development to find healthy intimacy.

A. Case Example 2: Mid-Twenties

It seems unscientific not to be open to the possibilities of exploring dreams from different perspectives; hence, challenging the model for dream interpretation presented above is important. Perhaps sometimes dream figures represent aspects of our own psyches, and perhaps sometimes they represent the people actually depicted. Maybe exploring dreams in different ways in different times is not as comfortably systematic as many would like, but doing so may very well be the way to help others benefit from what messages, if anything, their dreams may be attempting to deliver to them.

A 25-year-old woman dreamed that her husband was downstairs in the home in which she grew up, and her father was sitting on her bed in her childhood bedroom. She reported that in the dream she wanted to go downstairs to see her husband, but she kept going back to her room to find out if it was okay with her father. She would leave her father to go down to her husband, but she found that she had to check with her husband to see if she was allowed to go outside. In turn, she went back up to see her father and ask him what he thought, and on and on it went until she woke up frustrated and confused.

After we discussed what she thought her dream meant, and whether or not she believed she should be more loyal to her father or her husband, I invited her to explore the dream symbolically. "What if the house in your dream represented your ego?" I inquired. "What if you were to explore your dreams in terms of the levels of the house?" She asked me to say more, so I went on, "Imagine that your ego is only the center of your consciousness, not the center of who you are. If that were the case, is it possible that both men in your life represent people who have controlled you?"

"Yes," she said, "I met my husband in college and went from my dad's control to my husband's." We explored her use of the word "control" so freely, and then I continued to explore with her what she thought her dream might have meant. "What if an aspect of you, your true self, say, wanted to be free of the house, of the rooms that were boxing you in, of the house (ego) that was boxing you in?" I asked. "Do you believe there is a part of you that wants to be free of what the men in your life tell you to do and how to think?" She ardently agreed. She wanted to break free from both her husband and her father, but had no idea how she could do either. I continued, "I wonder if what is outside your childhood home is a representation of the unknown world, or more importantly, the unknown you. I wonder if your dream somehow had the purpose of making your asking both men if you could go outside ridiculous enough that you would see the clear message that you will always be trapped in the 'little-girl-you' as long as you continue to ask permission from them to be who you really are?"

For this client, that unfolding interpretation of her dream hit her very deeply. Her emotional release was significant as she realized that she allowed both her husband and her father to influence not only what she did but also how she thought and who she is. The ideas behind this dream are relatively common among people in their twenties: namely, how can they hold onto what they liked from what they learned as children, and how can they break free, or let go of what they learned as children? Developmentally, the mid-twenties is the time for people to explore the differentiation of self (Gilbert, 2006). It follows that many dream motifs for people in their twenties and early thirties center on the process of differentiation.

B. Case Example 3: Middle-Aged Adult

Denny was 48 years old when he reported a dream that frustrated him because, as he said, "I believe the dream is significant, but I don't know why." He dreamed that he was standing with a backpack on, down by a river. He believed he had a picture of Maharishi on which he mediated in the backpack. He put the backpack down and walked in the water; he was halfway in the water when he turned to look at the backpack, and then descended into the water. He recalled that the water was relatively difficult to see into, and that he could only see just below the surface, so he could not see his feet.

We explored his dream from the possibility of viewing the water into which he walked as a symbol of the unconscious. Water is frequently, though not always, viewed as a symbol of the unconscious because it is the primordial source from which all life on this planet began. I remarked that it was interesting that he set the backpack down on the land (consciousness) and his dream self willingly walked into the water (unconscious). If these thoughts on the symbolism fit with the dreamer (and in this case they appeared to), then it fits with the *penultimate calx* of dreams: moving from the conscious world to the unconscious world. In any psychodynamic exploration, understanding the message of dreams is the ultimate goal.

In regard to his stopping to look back at his backpack that held the picture of Maharishi before turning and entering the water, we explored what Maharishi meant to him. He replied that he had meditated in front of a picture of Maharishi daily for the past 20 years. For the dreamer, Maharishi represented the guru figure or archetypal wise man. The wise man or woman for many people is necessary as they begin their exploration of themselves, hence the beauty of religions and spiritual teachers offering guidance. At some point in our development, however, we all must "kill the Buddha" as the Zen Buddhists say. As Heinrich Zimmer noted, "Our highest god is our highest obstruction to God" (Campbell, 2002). In the case of this dreamer, he had held Maharishi before his own Self.

When I asked Denny what he thought of the possibility of his dream saying to him that it was time to leave behind Maharishi and transcendental meditation to begin to rely solely on himself, he reported believing those thoughts to fit very deeply with what he believed was "significant" about the dream. I inquired whether or not he believed that his putting the backpack down was his dream's way of saying to him to become less attached to his guru, and his entering the water could be representative of the necessity of his delving deeper into his psyche by plunging into his unconscious. Conte (2009) noted that the subconscious is like the clear water just below the surface—it holds information that, though not in the foreground of the mind, is readily available to us—whereas the unconscious is like the darker water that we can no longer see into: it is constituted of information of which we are not aware. The unconscious is dark and full of mystery, so it makes sense that the water is usually darker in dream images representing the unconscious the deeper we go into it. That Denny began to walk deeper into the water could have signified that he was moving beyond his subconscious and into his unconscious.

He then reported that he had been considering moving away from his attachment to transcendental meditation. For him, this way of exploring his dream fit with what he had been actualizing toward: "self"-reliance. Buddhists often remind people about the "soap of the teachings." Whereas soap is needed to clean, a garment is not considered completely clean until even the soap is rinsed away. And so it is with the idea of non-attachment. Until we let go of even the teachings themselves, we cannot be fully clean. The dreamer found peace in the idea that his psyche may have been trying to tell him to let go of this "final attachment." Denny reported feeling a profound connection to the idea that Maharishi had been his "highest god" and hence his highest obstruction to God. From the work with this dream, this middle-aged client was able to tap into the generator within, which seemed to move him far away from being stagnant.

VII. Sleep and Dreaming in Older Adulthood

During later life (defined roughly as age 65 until death), sleep takes on different qualities and dreams change as well. The most marked changes in the sleep of older adults are in the decrease in proportion of the night spent in deep stages of NREM sleep (i.e., slow-wave sleep) and in the increasing proportion of waking time after sleep onset (Carskadon and Dement, 2000;Feinberg et al., 1967). Remarkably, despite these changes, REM sleep as a percentage of total sleep time remains constant into healthy old age, although changes in REM percent are seen when individuals have organic brain dysfunctions (Carskadon and Dement, 2000). Although these are typical patterns of sleep development in later life, it has also been noted that individual variability in sleep markedly increases during this time period.

The dreams of older adults also appear to change, possibly due to typical changes in brain function seen with aging and to the general tendency for more introspection during this age period. In a study of memory and dreaming, Grenier and colleagues (2005) found that older adults tended to report fewer dreams and in less detail compared with younger adults in the laboratory, but not when dreams were collected in the home. In addition, these researchers found that autobiographical dream content followed the same pattern of development as autobiographical memory does in waking cognition. That is, most dreams referred back to adolescence and early adulthood; the same pattern is seen in waking autobiographical memory. This pattern indicates that dreams continue to parallel typical waking cognitive developments into later adulthood. Other research has found that older adults report fewer nightmares than younger adults, their dreams tend to have a more positive affective tone, and their dreams tend to contain strikingly less visual imagery than the dreams of younger adults (Fein et al., 1985; Funkhouser et al., 1999; St-Onge et al., 2005). This does not, however, imply that dreams cannot inform clinical work with older adults. On the contrary, dreams have been shown to play an important role in understanding the individual throughout the lifespan, including in later life (e.g., Funkhouser et al., 1999). The following clinical case example supports this consideration.

A. Case Example 3: Older Adult

At 66 years old, Barb entered therapy to "just talk." The first couple of sessions centered on finding ways to help her be assertive with her twin sister. The two of them and a friend ate breakfast at the local restaurant twice a week for years. Her twin sister refused to let Barb sit in the booth seat facing the restaurant, and so Barb decided that she wanted to be able to "stand up to her sister some time", but she would only say that with a smile on her face.

The setting is very relevant to this case. My office was one of four in an old Victorian house that was turned into office space. In my office was a door to the attic where we kept various resources. As Barb was walking into my office, a colleague asked if he could please get something he needed out of the attic. Barb sat down on the couch, and my colleague came back down, turned, closed the door behind him, and slid the lock across the door. He turned toward me, motioned he was sorry for interrupting, and closed the door on the way out. When I turned back to Barb, she had her head in her hands and was experiencing a panic attack.

I asked her to take some deep breaths and inquired as to what had happened. It took her some time, but finally, in a regressed voice, she said that the moment my colleague locked the door, the sound took her back to her childhood when her father and his brother would lock her in the closet and slide the very same kind of lock shut. She had not heard the sound since she was a child. Her body shook and she experienced strong emotional reactions as she described how she would be locked in the closet until she would come out naked, and her father and uncle would laugh at her. From that moment, she was haunted with a "something else" that she could not identify. Something else was bothering her, but she had a difficult time putting it into words.

In subsequent sessions, Barb uncovered that her uncle had fondled her, but more disturbingly, she remembered her father frequently taking her to an old, abandoned movie theater. As much as she tried, she could not consciously recall what happened there, but she recognized that whatever did happen was significant in her life. We had reached an impasse in therapy. Drawing on Jung's suggestion that the impasse is one of the most effective times to turn to dreams, I asked her to pay attention to her dreams in the upcoming week.

The following week Barb came into therapy anxious to see me because she had a dream about the theater two nights after we met the previous week. As would be expected of someone who was struggling with assertiveness, it never occurred to her to contact me before her scheduled appointment to let me know that something monumental had happened. The details of her dream revealed that the carpet in the theater was red, and that her father took her down to the bottom floor of the theater to meet with an older man. Her father left; she did not want him to leave her with the other man. She turned to talk to the man, then woke up.

The dream was powerful for her, and she simultaneously now wanted to know and remain unaware of what happened there. Whatever happened occurred when she was 7 years old; she was now 66 years old. She knew that something changed for her when she turned 7, but she did not know what. We tried to talk through what she could remember, including doing guided imagery to the point that she described in the dream; however, her conscious mind had genuinely repressed whatever had occurred. Again, I asked her to pay attention to her dreams.

She experienced two dreams in the subsequent weeks: in the first dream, she had an identical dream to her first one, except that this time, she recalled more information about a tree outside of the theater, and more details about what happened after her father left her alone with the man. She saw the man touching her above her clothes. In the second dream, she again experienced the same dream, except this time, she recalled having been made to lie down on a bed. In therapy, she was torn between her thoughts that she uncovered what was missing and her feelings that indicated more went on. We worked through what was conscious, and I attempted to implant a subconscious message when I told her before she left, "I wouldn't be surprised if you had a dream this week that helped you find what you are looking for."

The following week she called to ask if she could come in before her scheduled appointment. During this session, she provided more detail about the dream and the experience, and shared information that empirically validated this apparent repressed memory. Barb reported having the same dream, but this time she described it without the smiles that she had typically used to keep the listener believing everything is okay. In the dream, her father took her and her twin sister to the theater; her sister was told to play under the tree, while her father took her inside. She walked the long walk down the theater with her father.

The man greeted her father at the door and handed him an envelope. Her father turned and left. The man had her go to the bed; he raped her. He told her she deserved it, then sent her back to her father. When she got outside, she saw her father playing with her sister under the tree. She woke up.

She called her older sister to tell her about the dream. Her sister sounded a blood-curling scream through the phone and broke down crying. She told Barb that she knew about her father prostituting her out to that man, but her mother always threatened her not to tell Barb; then, on her deathbed, their mother made her older sister promise to never tell Barb unless she could discover it for herself. Barb broke down as she told me the story, but her energy was different in this breakdown. She experienced a completion to that "something else" which had unconsciously been bothering her.

Barb's journey through therapy ended a few sessions after her discovery. She was able to be assertive with her twin sister; a sister, she learned, who was to no fault of her own, always favored. Regardless of where fault or blame could fall, the Gestalt that Barb achieved through dream work helped her gain the voice of her Self that was buried 59 years prior. Because she continues to write to me from time to time, I am aware that the changes she made have stayed with her to date.

Repressed for over 50 years, Barb gained through the exploration of her dreams a reality that was confirmed through a third party. With the awareness she encountered from her dreams, she was able to change her energy and the direction of her life. Though the approach to the dream was significantly different from that of the dream work with the adolescent, the result was the same: an impact of awareness on a person's life because she listened to her dream.

Dreams in later life are often the most significant (Jung, 1974) because they represent the culmination of the life of a conscious being. For some, dreams in later life can be viewed as mystical, nonsensical, psychological (Funkhouser *et al.*, 1999), predictive (Jung and Meyer-Grass, 2008), or bringing closure to unfinished business that has perplexed or otherwise disturbed the psyche (Polster and Polster, 1973). Investigating dreams for older adults as psychological data and facilitating older adults exploring completion for apparent unfinished business appears to be the most clinically pragmatic use for dreams in therapy.

VIII. Conclusion

Dreaming appears to undergo changes across the lifespan. Existing literature suggests that dreams and their content map onto cognitive developmental changes in waking life, and some research suggests that social and emotional developments also may play a role. Dreams can be useful tools for the clinician interested in understanding individuals, if not through their direct content or symbolism, then certainly through the dreamer's interpretation of the dream. Whether one accepts a neo-Jungian or a more pragmatic approach to dreaming, the possibility exists for dreams to inform practice. Even if dreams are simply by-products of an active neural network during sleep, or an epiphenomenon of REM sleep (Hobson *et al.*, 1998), the literature and clinical cases reviewed here show that, at the very least, this epiphenomenon changes with development and has proven useful in understanding clients.

References

Anders, T. F., and Keener, M. (1985). Developmental course of nighttime sleep–wake patterns in full-term and premature infants during the first year of life: I. *Sleep*, **8**(3), 173–192.

Anders, T. F., and Sostek, A. M. (1976). The use of time lapse video recording of sleep-wake behavior in human infants. *Psychophysiology* **13**(2), 155–158.

Buber, M. (1950). I and Thou. Morrison & Gibb Limited, Edinburgh.

Burnham, M. M., Goodlin-Jones, B. L., Gaylor, E. E., and Anders, T. F. (2002). Nighttime sleep–wake patterns and self-soothing from birth to one year of age: a longitudinal intervention study. *J. Child Psychol. Psychiatry*. **43**, 713–725.

Campbell, J. ed. (1971). The Portable Jung. Penguin Books, New York.

Campbell, J., and Moyers, B. (1988). *The power of myth*. Double Day, New York.

Campbell, J. (2002). Mythos: The Shaping of Our Mythic Tradition (Narrated by Susan Sarandon). DVD. Wellspring Media, New York.

Carskadon, M. A., and Dement, W. C. (2000). Normal human sleep: an overview. In: Principles and Practice of Sleep Medicine, 3rd ed. (M. H. Kryger, T. Roth, and W. C. Dement, eds.), W.B. Saunders, Philadelphia, PA, pp. 15–25.

Conte, C. (2009). Advanced Techniques for Counseling and Psychotherapy. Springer Publishing, New York.

Crowley, S. J., Acebo, C., and Carskadon, M. A. (2007). Sleep, circadian rhythms, and delayed phase in adolescence. *Sleep Med*. **8**, 602–612.

Dahl, R. E. (2004). Adolescent brain development: a period of vulnerabilities and opportunities. *Ann. N. Y. Acad. Sci*. **1021**, 1–22.

Domhoff, G. W. (2001). A new neurocognitive theory of dreams. *Dreaming* **11**(1), 13–33.

Erikson, E. H. (1959). Identity and the Lifecycle: Selected Papers. International Universities Press, Oxford.

Fein, G., Feinberg, I., Insel, T. R., Antrobus, J. S., Price, L. J., Floyd, T. C., and Nelson, M. A. (1985). Sleep mentation in the elderly. *Psychophysiology* **22**(2), 218–225.

Feinberg, I., Koresko, R. L., and Heller, N. (1967). EEG sleep patterns as a function of normal and pathological aging in man. *J. Psychiatr. Res*. **5**, 107–144.

Flavell, J. H. (1988). The development of children's knowledge about the mind: from cognitive connections to mental representations. In: Developing Theories of Mind (J. W. Astington, P. L. Harris, and Dr. R. Olson, eds.), Cambridge University Press, Cambridge.

Foulkes, D. (1982). Children's Dreams: Longitudinal Studies. John Wiley & Sons, New York.

Foulkes, D. (1999). Children's Dreaming and the Development of Consciousness. Harvard University Press, Cambridge, MA.

Frank, M. G., and Heller, C. (2003). The ontogeny of mammalian sleep: a reappraisal of alternative hypotheses. *J. Sleep Res.* **12**, 25–34.

Freud, S. (1900). The Interpretation of Dreams. Macmillan, New York.

Funkhouser, A. T., Hirsbrunner, H. -P., Cornu, C., and Bahro, M. (1999). Dreams and dreaming among the elderly: an overview. *Aging. Ment. Health* **3**(1), 10–20.

Gilbert, R. M. (2006). The Eight Concepts of Bowen Therapy: A New Way of Thinking about the Individual and the Group. Leading Systems Press, Falls Church & Basye, VA.

Graven, S. (2006). Sleep and brain development. *Clin. Perinatol.* **33**, 693–706.

Grenier, J., Cappeliez, P., St-Onge, M., Vachon, J., Vinette, S., Roussy, F., Mercier, P., Lortie-Lussier, M., and De Koninck, J. (2005). Temporal references in dreams and autobiographical memory. *Mem. Cognit.* **33**(2), 280–288.

Hobson, J. A., Stickgold, R., and Pace-Schott, E. F. (1998). The neuropsychology of REM sleep dreaming. *NeuroReport* **9**, R1–R14.

Johnson, R. (1991). Owning Your Own Shadow. Harper Collins, New York.

Jung, C. G. (1950). Modern Man in Search of a Soul. (W. S. Dell and C. F. Baynes, Trans.) Harcourt Brace Jovanovich, New York. (Original work published in 1933).

Jung, C. G. (1961). Memories, Dreams, Reflections. Random House, New York.

Jung, C. G. ed. (1964). Man and His Symbols. Laurel, New York.

Jung, C. G. (1974). Dreams. Princeton University Press, Princeton, NJ.

Jung, L., and Meyer-Grass, M. eds. (2008). Children's Dreams: Notes from the Seminar Given in 1936–1940 by C.G. Jung. (E. Falzeder and T. Woolfson, Trans.) Princeton University Press, Princeton, NJ. (Original work published in 1987).

Iglowstein, I., Jenni, O. G., Molinari, L., and Largo, R. H. (2003). Sleep duration from infancy to adolescence: Reference values and generational trends. *Pediatrics,* **111**(2), 302–307.

Kahn, A., Dan, B., Groswasser, J., Franco, P., and Sottiaux, M. (1996). Normal sleep architecture in infants and children. *J. Clin. Neurophysiol.* **13**(3), 184–197.

Kleitman, N., and Engelmann, T. G. (1953). Sleep characteristics of infants. *J. Appl. Physiol.* **6**, 269–282.

Lewis, M. (1989). Emotional development in the preschool child. *Pediatr. Ann.* **18**, 316–327.

Maquet, P., Péters, J. -M., Aerts, J., Delfiore, G., Degueldre, C., Luxen, A., and Franck, G. (1996). Functional neuroanatomy of human rapid-eye-movement sleep and dreaming. *Nature* **383**, 163–166.

Mattoon, M. A. (2006). Dreams. In: The Handbook of Jungian Psychology: Theory, Practice and Applications (R. K. Papadopoulos, ed.), Routledge, New York.

May, R. (1991). The Cry for Myth. Dell Publishing, New York.

Meyer, S., and Shore, C. (2001). Children's understanding of dreams as mental states. *Dreaming* **11**(4), 179–194.

Montgomery-Downs, H. E. (2008). Normal sleep development in infants and toddlers. In: Sleep and Psychiatric Disorders in Children and Adolescents (A. Ivanenko, ed.), Informa, New York, pp. 11–21.

Muzur, A., Pace-Schott, E. F., and Hobson, J. A. (2002). The prefrontal cortex in sleep. *Trends. cogn. sci.* **6**(11), 475–481.

Piaget, J. (1952). The Origins of Intelligence in Children. International Universities Press, New York.

Piaget, J. (1962). Play, Dreams, and Imitation in Childhood. W. W. Norton & Company, New York.

Pivik, R. T. (2000). Sleep and dreaming. In: Handbook of Psychophysiology, 2nd ed. (J. Cacioppo, L. G. Tassinary, and G. G. Berntson, eds.), Cambridge University Press, New York, pp. 687–716.

Polster, E., and Polster, M. (1973). Gestalt Therapy Integrated: Contours of Theory and Practice. Vintage Books, New York.

Roffwarg, H. P., Dement, W. C., and Fisher, C. (1964). Preliminary observations of the sleep–dream pattern in neonates, infants, children, and adults. In: Problems of Sleep and Dream in Children (E. Harms, ed.), Macmillan Company, New York.

Roffwarg, H. P., Muzio, J. N., and Dement, W. C. (1966). Ontogenetic development of the human sleep–dream cycle. *Science* **152**, 604–619.

Solms, M. (2000). Dreaming and REM sleep are controlled by different brain mechanisms. *Behav. Brain Sci.* **23**, 843–850.

St-Onge, M., Lortie-Lussier, M., Mercier, P., Grenier, J., and De Koninck, J. (2005). Emotions in the diary and REM dreams of young and late adulthood women and their relation to life satisfaction. *Dreaming* **15**(2), 116–128.

Strauch, I., and Meier, B. (1996). In Search of Dreams: Results of Experimental Dream Research. State University of New York Press, Albany, NY.

von Franz, M. L. (1964). The process of individuation. In: Man and His Symbols (C. G. Jung, ed.) Laurel, New York.

Woolley, J. D. (1995). The fictional mind: young children's understanding of imagination, pretense, and dreams. *Dev. Rev.* **15**, 172–211.

Woolley, J. D., and Wellman, H. M. (1992). Children's conceptions of dreams. *Cogn. Dev.* **7**, 365–380.

REM AND NREM SLEEP MENTATION

Patrick McNamara*, Patricia Johnson*, Deirdre McLaren*, Erica Harris*,
Catherine Beauharnais*, and Sanford Auerbach[†]

*Department of Neurology, Boston University School of Medicine and VA New England
Healthcare System, Boston, MA, USA
[†]Sleep Disorders Center, Boston Medical Center, Boston University School of Medicine, Boston,
MA, USA

We review the literature on the neurobiology of rapid eye movement (REM) and non–rapid eye movement (NREM) sleep states and associated dreams. REM is associated with enhanced activation of limbic and amygdalar networks and decreased activation in dorsal prefrontal regions while stage II NREM is associated with greater cortical activation than REM. Not surprisingly, these disparate brain activation patterns tend to be associated with dramatically different dream phenomenologies and dream content. We present two recent studies which content-analyzed hundreds of dream reports from REM and NREM sleep states. These studies demonstrated that dreamer-initiated aggressive social interactions were more characteristic of REM than NREM, and dreamer-initiated friendliness was more characteristic of NREM than REM reports. Both REM and NREM dreams therefore may function to simulate opposing types of social interactions, with the REM state specializing in simulation of

INTERNATIONAL REVIEW OF
NEUROBIOLOGY, VOL. 92
DOI: 10.1016/S0074-7742(10)92004-7

69

aggressive interactions and the NREM state specializing in simulation of friendly interactions. We close our review with a summary of evidence that dream content variables significantly predict daytime mood and social interactions.

I. Introduction

In humans, sleep is composed of rapid eye movement (REM or "R") sleep and non–rapid eye movement (NREM or "N") sleep, with NREM further subdivided into three or four "descending" stages. The terminology differs a bit according to the system used, and the current system endorsed by the American Academy of Sleep (AASM, 2007) uses the "signs" of R and N to represent REM sleep and NREM sleep, respectively. NREM sleep is further subdivided into N1, N2, N3, with N3 designating slow-wave sleep (SWS). As one progresses from N1 to N2 to N3, there is progressive slowing of the electroencephalographic (EEG) background with a gradual drop in muscle tone and a loss of eye movements. In the past, N3 was further subdivided into stage III and stage IV of NREM sleep, but general use consolidates these stages into SWS. N2, or stage II, of NREM sleep is further distinguished by the presence of spindles (short bursts of rhythmic high-frequency waves) and K-complexes (high amplitude, sharp activity, followed by a slower, positive wave.) These sleep stages do not appear randomly. They display an "ultradian" rhythm through the night. Each cycle consists of a period of NREM sleep followed by a period of REM sleep and lasts for about 90–120 min. Typically, one enters NREM sleep and experiences a sequence of "descending" stages and then an "ascend" (indicating greater brain activation levels) with an entry into REM sleep. The cycle repeats through the night, but with each cycle, the duration of REM sleep increases and the duration of SWS decreases. The cycles repeat every 90–120 min. As a consequence of the shifting durations, REM sleep is typically skewed toward the end of the night and SWS, or N3, is skewed toward the beginning of the night. The various nomenclature systems are quite similar, and for the purpose of this chapter, we will use the terms REM, NREM, and SWS because they are most commonly seen in the relevant literature on this topic.

II. Dreams in REM and NREM Sleep

When subjects are awakened from REM sleep, they generally report a "dream"—a narrative involving the dreamer who interacts with others in ordinary or extraordinary ways in both familiar and strange settings. The emotions in the dream are often unpleasant and the events recounted occasionally involve bizarre and improbable elements (Domhoff, 2003; Foulkes, 1962; Hobson and Pace-Schott, 2002; Nielsen *et al.*, 2001; Rechtschaffen *et al.*, 1963; Snyder *et al.*, 1968; Strauch and Meier, 1996). When someone is awakened from stage II NREM sleep, however, reporting bizarre and improbable elements and unpleasant emotions is less likely and the settings are more often familiar. Such is the typical, though by no means consensus, view of the content differences in dreams associated with REM and NREM sleep states. Many investigators dispute the claim that there are such clear-cut content differences between the two sleep states, or, if differences are admitted, they are dismissed as one of degree or intensity. For these latter investigators, there is but one "dream generator" in the brain and that is REM sleep—all sleep mentation derives its fuel and phenomenology from that generator. Our own position is that sleep mentation is colored by the tone of the prevailing sleep stage—it varies over the course of a sleep episode as a function of brain activation patterns. Those brain activation patterns are indexed traditionally via EEG sleep staging but have more recently been studied via neuroimaging methods as well. As sleep cycles into an REM sleep episode, mentation reports (what people report they were thinking or imagining or "dreaming" while asleep) become more dream-like and bizarre, and as sleep cycles into an NREM sleep episode, mentation reports become more thought-like and less bizarre. In short, there is a continuum of mental activity that varies along several dimensions and in tandem with brain activation patterns that fluctuate in a relatively predictable manner across the entire sleep episode. None of this is surprising if one holds that mind or content is a reflection of or in some sense reducible to brain activity. What is surprising, however, is that when one looks at content differences taken from reports emerging from REM versus NREM sleep, one finds very dramatic differences for a small but extremely significant set of content indicators—namely, social interactions. It appears then for this small set of content indicators that there may be two dream generators: one for REM and one for NREM sleep, each with specialized functions.

III. Neurobiologic Correlates of REM and NREM Sleep That Are Consistent with REM and NREM Sleep Processing Specializations

Brain activation patterns are significantly different for REM and NREM sleep, with REM sleep demonstrating high activation levels in limbic/amygdaloid sites and deactivation of dorsolateral prefrontal cortex sites (Braun *et al.*, 1997; Hobson *et al.*, 1998; Maquet and Franck, 1997; Maquet *et al.*, 1996; Nofzinger *et al.*, 1997), and NREM/SWS associated with deactivation of thalamic functions and emergence of synchronized wave activity throughout neocortical sites. Stage

II NREM sleep, however, involves higher activation levels in cortex than SWS sleep, and it is stage II NREM sleep where we will look for evidence of NREM sleep processing specializations. Hobson *et al.* (2002, for a recent review), Steriade and McCarley (1990), and others (e.g. Datta and Maclean, 2007) have presented a great deal of experimental evidence which suggests that regulation of the REM/NREM sleep cycle is governed by two major neural ensembles that act in an antagonistic fashion to turn on or turn off REM sleep. Briefly, REM sleep is generated by cholinergic neurons originating within the peribrachial regions known as the laterodorsal tegmental and pedunculopontine tegmental (LDT/PPT) nuclei, and is inhibited by noradrenergic and serotonergic neurons in the locus coeruleus and dorsal raphe, respectively. Activation of cholinergic REM is due to removal of inhibition of cholinergic cells in the LDT/PPT normally sustained by aminergic efferents. In sum, expression of the REM–NREM sleep cycle is regulated, in part, by antagonistic cellular groups within the brainstem with aminergic cell groups inhibiting expression of REM sleep (with emergence of NREM sleep) and cholinergic groups promoting expression of REM sleep. When cholinergic REM-on cells are activated, aminergic REM-off cell groups are inhibited, and vice versa. REM and NREM sleep may also be regulated by separate sets of genes. Tafti and Franken (2002) have shown that inbred mice strains C57BL and C57BR are associated with increased REM and short SWS episodes, while the BALB/c strain is associated with short REM and long NREM episodes, indicating separate genetic influences on REM and NREM sleep amounts. Electrophysiologic measures of sleep in humans are also influenced by genes. Studies of sleep EEG in twins reared apart show separable genetic effects on REM sleep versus delta waves of NREM sleep as well as stronger correlations of EEG indices of sleep among monozygotic twin pairs as compared to dizygotic twin pairs (see reviews in Linkowski, 1999; Tafti and Franken, 2002; Taheri and Mignot, 2002).

A. Study I: REM versus NREM Sleep Specializations in Simulation of Social Interactions

Noting that REM sleep is associated with very high activation levels in the amygdala and NREM sleep is associated with deactivation in some limbic sites, we (McNamara *et al.*, 2005) hypothesized that representations of social interactions in REM and NREM sleep would reflect differing regional brain activation patterns associated with the two sleep states, with higher levels of aggressive interactions in REM versus NREM sleep. One hundred REM sleep, 100 NREM sleep, and 100 wake reports (equated in length in a study conducted by Stickgold (2001)) were collected in the home from eight men and seven women using the "Nightcap" sleep/wake mentation monitoring system, and scored for number and variety of social interactions. The Nightcap (Mamelak and Hobson, 1989) consists of a 25–8-mm piezoelectric eyelid movement sensor and a cylindrical, multipole mercury switch that detects head movements (Ajilore et al., 1995). The Nightcap counts eyelid and head movements in intervals of 250 ms, identifying an eyelid movement interval whenever a voltage in excess of 10 mV is detected within an interval. The sensor and associated circuitry are sensitive to REMs and twitches of the levator palpebrae and orbicularis oculi (eyelid muscles), but not to the slow eye movements characteristic of other sleep states. After scoring the reports with standardized techniques for scoring dream content, we found dramatic differences in the types of social interactions depicted in the two dream states (see Table IV and McNamara et al., 2005).

We found the following:

- Social interactions were more likely to be depicted in dream than in wake reports (Fig. 1). There were 56 social interactions in the 100 REM reports, 34 in the 100 NREM reports, and 26 in the 100 wake reports. The REM versus NREM and the REM versus wake differences were statistically reliable (both $p < 0.001$).
- Aggressive social interactions were more characteristic of REM sleep than NREM sleep or wake reports. Sixty-five percent of the REM reports contained an aggressive social interaction while one-third of the NREM reports and 23% of wake reports contained an aggressive interaction.

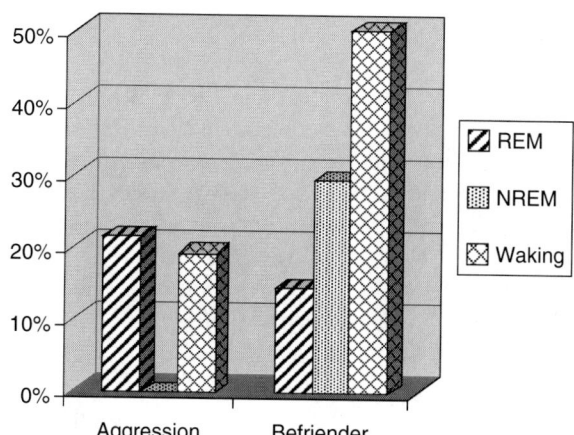

FIG. 1. Dreamer-initiated aggression befriending percents out of total number of social interactions in REM, NREM, and waking reports.

The REM versus NREM and REM versus wake differences were statistically significant ($p < 0.01$). Twenty-five percent of the aggressions in REM dreams were physical aggressions while there were no physical aggressions recounted in wake reports. Eighteen percent of NREM reports contained physical aggressions. The dreamer was the aggressor in 52% of REM reports (with aggressions) and 0% of NREM reports with aggressions and 100% of the few wake reports that contained an aggression.

- Dreamer-initiated friendliness was more characteristic of NREM than of REM sleep. It is important to note that dreamer-initiated aggressive interactions were *reduced to zero* in NREM sleep dreams, while dreamer-initiated friendly interactions were twice as common in NREM as in REM sleep (90% vs 54%, respectively; $p < 0.05$).

Note that these data were gathered using Nightcap technology. The Nightcap can only identify REM and NREM sleep states and it cannot differentiate sleep states within NREM sleep. So, for example, we do not know if all dreams in the NREM sleep category came from stage II or some other NREM sleep stage. One needs standard EEG technology to identify sleep states within NREM sleep. We, therefore, conducted a second set of studies in order to replicate this set of findings using standard EEG technology.

B. Study 2: Replication of Social Interaction Content Differences in REM versus NREM Sleep

1. *Participants*

A convenience sample of 64 healthy participants (28 males, 36 females; mean age = 20.89, SD 2.56 years) between the ages of 18 and 32 years was recruited from the Boston area and local universities. Participants averaged 14 years of education (SD = 1.85) and had an average scaled Wechsler Test of Adult Reading (WTAR; Wechsler, 2001) score, a measure of verbal intelligence, of 119.80 (SD = 8.04) and an average Depression Anxiety Stress Scale (DASS; Lovibond and Lovibond, 1995) score, a measure of stress and other mood states, of 9.80 (6.38). Males had an average total DASS score of 9.07 (5.55) and a scaled WTAR score of 118.25 (8.29); females had an average total DASS score of 10.36 (6.98) and a scaled WTAR score of 121.00 (7.74). Participants were required not to have a prior history of any neurological, psychiatric, sleep (particularly, sleep apnea), drug or alcohol problems, or head injury. In addition, they were asked to refrain from using alcohol throughout the study period. Informed consent was obtained from all participants prior to entering the sleep lab or completing any measures. In the following summary of analyses of these participants' sleep and dreams, N will vary because of random data-recording errors.

2. *Procedures*

a. Pre-Sleep Lab Assessments

Before coming into the sleep lab, each participant completed a battery of tests measuring personality, mood, sleep, and behavior. Participants began a sleep diary and an activity log 4 days before coming into the lab (data not reported here), which they continued until after the experimental night. Data from sleep diaries identified no participants having abnormal sleep cycles during the 4 days before entering the study.

b. Habituation Night in Sleep Lab

The first night in the sleep lab was the habituation night during which participants were instrumented for polysomnography (PSG) and slept in the sleep lab, without awakenings, as adaptation for the following experimental night. They arrived at the lab at 9:00 PM and were awoken at 6:30 AM in the morning, given breakfast, and then allowed to go about their normal waking activities.

c. Experimental Night

Participants returned to the lab at 9:00 PM the following evening and were again instrumented for PSG. Participants completed a booklet of pencil-and-paper tasks at four different times during this second visit to the sleep lab: right before going to bed, when awoken from REM sleep, when awoken from NREM sleep, and in the morning following awakening. Booklets contained short mood scales (data not reported here) that took ~15 min to complete, after which participants audiotaped any dreams they were having right before being awoken. Mood tests after awakenings revealed no differences in mood after awakenings from REM versus those from NREM sleep. The order of awakenings from REM and NREM sleep was counterbalanced as described by Walker *et al.* (2002).

d. Overnight PSG

PSG data were obtained from nine channels with six EEG (F3, F4; C3, C4; O1, O2) referenced to contralateral mastoid (A1, A2), two of electrooculogram (EOG) (both referenced to A2), and submental electromyogram (EMG). Respiratory effort during sleep was also recorded. A full complement of sleep architecture and quality measures was also obtained. These included total sleep time (TST) and sleep efficiency (SE) as indices of sleep quality, SWS percent (SWS%) as an indicator of achieved depth of NREM sleep, as well as REM percent (REM%) and REM latency (REML) as indicators of REM sleep propensity. Finally, stage II NREM% was used as our NREM sleep state.

3. *Dream Analyses*

We analyzed the dream reports with the Hall–Van de Castle procedures described above as well as with an unbiased computerized word count procedure (Linguistic Inquiry and Word Count [LIWC]; Pennebaker *et al.*, 2001). The LIWC has been used successfully in several studies of language and mental function (see www.liwc.net). For the purposes of the current study, to minimize type 1 error, only total word count (TWC) and four emotion- and personality-relevant word categories were used in analyses: positive emotion words ("LIWC positive"), negative emotion words ("LIWC negative"), words that refer to mental states (e.g., thought, felt, wanted, believed; abbreviated "cognition"), and words referring to social situations ("social") (Table I–IV).

REM sleep mentation reports had significantly greater numbers of references to an Other as compared to NREM sleep reports. In addition, social interactions were more frequent in REM sleep reports than in NREM sleep reports. There were no significant differences, however, in emotion types as a function of sleep state. But this lack of emotional differences between sleep states disappeared when we looked at dreamer-involved events in the dreams (Fig. 2).

Figure 2 represents the dreamer's role in all dreamer-involved social interactions. There were 38 REM dreams with dreamer-involved social interactions and

TABLE I

HALL–VAN DE CASTLE CONTENT RATIOS

Known and unknown characters by gender	
Number and variety of emotions	
Aggression/friendliness percent	Dreamer-involved aggression/(dreamer-involved aggression + dreamer-involved friendliness).
Befriender percent	The percentage of all dreamer involved friendly interactions in which the dreamer befriends some other character. Dreamer as befriender/(dreamer as befriender + dreamer as befriended)
Aggressor percent	The percentage of all dreamer involved aggressions in which the dreamer is the aggressor. Dreamer as aggressor/ (Dreamer as aggressor + Dreamer as victim)
Physical aggression percent	The percentage of all aggressions appearing in reports whether witnessed or dreamer involved that are physical in nature. Physical aggressions/all aggressions
Aggression (A/C) index	Frequencies of aggressions per character (all aggressions/all characters).
Friendliness (F/C) index	Frequencies of friendly interactions per character (all friendliness/all characters).
Sexuality (S/C) index	Frequencies of all sexual encounters per character (sexual encounters/all characters).

TABLE II

FREQUENCY OF SOCIAL INTERACTIONS BY STATE

	REM	NREM	WAKE	REM vs. NREM	REM vs. WAKE	NREM vs. WAKE
Total social interactions	56	34	26	0.002**	0.0001**	0.21
Social interactions/social reports	1.4	1.55	1.13	0.18	.015*	.001**
Reports with at least one social interaction (of any type)	40	22	23	.005**	.009**	0.86
Reports with at least one Aggressive interaction	24	12	8	0.025*	0.001**	0.34
Reports with at least one friendly interaction	17	15	17	0.70	1.00	0.70
Reports with at least one sexual interaction	2%	0%	0	0.045*	0.045*	1.00

Note: *p<.05; **p<.01

TABLE III

HALL/VAN DE CASTLE SOCIAL INTERACTION PERCENTS

	REM (%)	NREM (%)	WAKE (%)	REM vs. NREM	REM vs. WAKE	NREM vs. WAKE
Aggression/ Friendliness (%)	65	33	23	0.026*	0.001**	0.456
Befriender (%)	54	90	76	0.043*	0.192	0.354
Aggressor (%)	52	0	100	0.0001**	0.014*	0.0001**
Physical aggression (%)	25	18	0	0.540	0.007**	0.043*

Note: *p <.05; **P <.01   Aggression/friendliness % = Dreamer-involved aggression / (Dreamer-involved aggression + Dreamer-involved friendliness); Befriender % = Dreamer as befriender / Dreamer as befriender + Dreamer as befriended); Aggressor % = Dreamer as aggressor / (Dreamer as aggressor + Dreamer as victim); Physical aggression % = Physical aggressions / All aggressions.

TABLE IV

PERCENTAGE OF EACH TYPE OF WORD USING LINGUISTIC ANALYSIS FOR REM AND NREM DREAM

	REM Dreams	NREM Dreams	p-value
Word count (number of words)	104.11 (95.02)	110.16 (150.87)	n.s.
Self	10.61 (3.58)	11.25 (3.44)	0.318
Other	2.36 (2.71)	1.34 (2.16)	0.023*
Negative emotion	0.54 (0.84)	0.77 (1.37)	0.262
Positive emotion	1.09 (1.69)	1.25 (1.64)	0.609
Social	8.21 (5.68)	6.15 (4.31)	0.026*

Note: *p < 0.05.

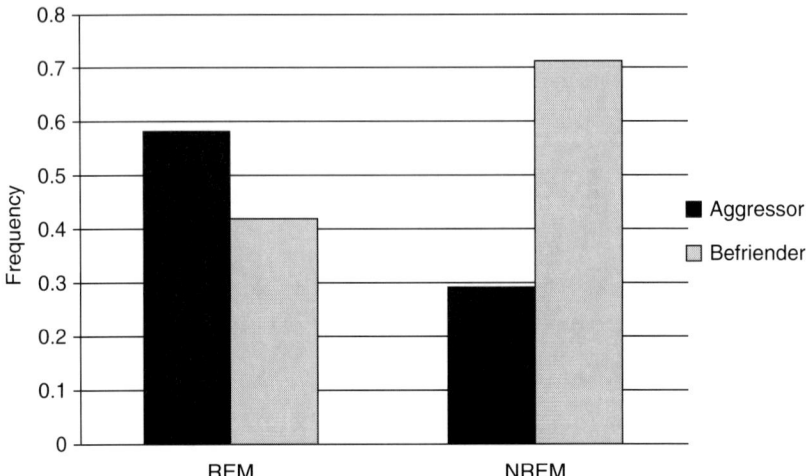

Fig. 2. Frequency of dreamer's role in social interactions for REM and NREM dreams.

37 NREM dreams. This means that in 71% of the dreamers' social interactions in NREM dreams, the dreamer was a befriender.

Figure 2 shows that when we look only at those dreams where the dreamer was directly involved in (e.g., initiating) a social interaction, then clear differences emerge. As in our previous study with the Nightcap technology, we find (with standard EEG methods) that in dreamer-involved friendly interactions, the dreamer was the befriender in only 42% in REM sleep dreams ($N = 24$) but was the befriender in 71% in NREM sleep dreams ($n = 14$; $p = 0.070$). In dreamer-involved aggressive interactions, the dreamer was the aggressor in 58% in REM sleep dreams ($N = 12$) and in only 29% in NREM sleep dreams ($N = 17$).

It is important to note that these REM–NREM differences emerge only when you look at *dreamer-involved* social interactions. One has apparently to take the point of view of the dream ego to understand what is happening in dreams. When we analyzed the entire REM–NREM dream set using the standard Hall–Van de Castle categories mentioned above, we found that among the social interaction scales, REM sleep dreams were *less aggressive* and more friendly than NREM sleep dreams. Of aggressive and friendly social interactions in REM sleep dreams, only 37% ($N = 38$) were aggressive compared to 57% of interactions in NREM sleep dreams ($N = 37$; $p = 0.082$). Fifty percent of aggressive interactions in REM sleep dreams ($N = 24$) were physically aggressive (as opposed to verbal aggression or other types) as opposed to 43% of aggressive interactions in NREM sleep dreams ($N = 23$; $p = 0.654$). *However*, as depicted in Fig. 2, looking at just dreamer-involved social interactions, the dreamer tended to be the aggressor in REM sleep dream interactions and the befriender in NREM sleep dream interactions.

Characters in both REM and NREM sleep dreams were 48% male and 52% female ($p = 0.967$). Characters tended to be familiar in both REM sleep dreams (60% of characters were familiar; $N = 132$) and NREM sleep dreams (68% of characters were familiar; $N = 117$; $p = 0.207$). The friends percent (number of friends/number of all human characters) was 43% in REM sleep dreams ($N = 132$) and 45% in NREM sleep dreams ($N = 117$; $p = 0.737$). The family percent (number of family member/number of all human characters) was 16% in REM ($N = 132$) and 21% in NREM ($N = 117$; $p = 0.347$). The dead and imaginary percent (number of dead and imaginary characters/all characters) was 1% in both REM and NREM sleep dreams. The animal percent (number of animals/all human/animal characters) was 6% in REM sleep dreams ($N = 142$) and 7% in NREM sleep dreams ($N = 126$; $p = 0.614$).

IV. Specializations in Emotional Processing

Given the now replicated findings of specialization differences in simulation of social interactions between REM sleep and stage II NREM sleep, it follows that other types of dream content would be related to the functional differences in social interaction. In our studies, we did not find any emotional differences between sleep states unless we looked at emotions of the dream ego versus emotions attributed to other characters or the emotional tinge of the overall dream itself. If aggressive interactions are more frequent in REM sleep dreams and the dreamer is the aggressor, then we would expect more aggression-related emotions in REM versus NREM sleep dreams and so forth. Spontaneously recalled dreams are probably REM sleep dreams and are typically filled with emotion. Between 75% and 95% of spontaneously recalled dreams (likely REM dreams) contain at least one emotion (Domhoff, 1996; Merritt et al., 1994). Merritt et al. (1994) asked their subjects to specifically indicate, on a line-by-line basis, the presence of one or more specific emotions associated with the content of their spontaneously recalled dreams. Using this method, Merritt et al. found that 95% of all their dream reports ($N = 200$ reports) contained emotions and that each dream report averaged about 3.6 emotions per dream. Interestingly, many emotional dreams were not unpleasant from the point of view of the dreamer. Smith et al. (2004) scored both REM and NREM dream reports for emotion content from 25 dreamers. They identified eight emotions that they then divided into positive and negative categories. There was no clear demarcation between REM and NREM sleep dreams in terms of positive versus negative emotionality though REM sleep dreams were more intense. Aggression is not necessarily unpleasant unless you are on the receiving end, it seems.

V. REM–NREM Interactions in Processing of Memories across a Single Night

Many authors have suggested various ways in which NREM sleep interacts with REM sleep in processing of memories. Giuditta *et al.* (1995) suggested that NREM sleep selects memories that will be consolidated, and then consolidation occurs under REM sleep. Stickgold and his associates have drawn on the work of Buzsaki (1996) and their own empirical findings in analyses of dream content (Fosse *et al.*, 2003) to suggest that one can "see" aspects of the process of memory consolidation as a function of REM–NREM sleep interactions *by looking at dream content*. It was hypothesized that during slow-wave NREM sleep, memories of life episodes and events are consolidated into episodic memories and transferred to the cortex for storage. During REM sleep, hippocampal outflow to the cortex is blocked. Instead, the hippocampus receives information from cortical networks, thus preventing transfer of newly consolidated episodic memories to the cortex. Thus, one should not see instances of fully formed episodic memories in REM sleep dreams, and that is what Fosse *et al.* in fact found. Instead, semantic and procedural aspects of memories are thought to be processed during REM sleep. Consistent with the Stickgold *et al.* model, Cavallero *et al.* (1988) elicited dreams from the onset phase, from NREM and REM sleep, and asked their subjects to freely associate to individual dream segments. Associations to REM sleep dreams were more often related to general knowledge, whereas dream onset and NREM sleep dreams were more closely related to memories of life episodes.

VI. Which (If Any) Elements of REM or NREM Sleep Dream Content Are Correlated with Daytime Mood and Behavioral Variables?

Kramer (1993) reported that selected emotional content variables showed statistically significant change across the night's REM sleep dreams and were predictive of mood improvement in the morning. These content variables involved change in the *number and variety of characters* in dreams across a single night of dreaming. Self-reported mood upon awakening in the morning was related to the increase in number of characters in dreams across the night. The latter increase may reflect an increase in REM sleep density measures across the night. The typical REM sleep dream contains between two and three characters in addition to the dreamer, and these characters typically interact with the dreamer. According to the Hall–Van de Castle norms, about 50% of characters in dreams are *not* familiar to the dreamer. In some dream series, *up to 80% of characters are unknown to the dreamer.* Kavanau (2002) suggested that the appearance of strangers in dreams were epiphenomenal expressions of faulty or "incompetent" memory editing or synaptic circuits. Payne and Nadel (2004) similarly suggested that dream elements may reflect ongoing memory editing procedures. Empirical analyses of the properties and relationships of "unknown characters" in dreams reveals that they appear in rule-governed ways, thus supporting claims that dream elements reflect ongoing memory editing procedures. For example, analyses of the appearance of strangers in dreams demonstrate that they most often appear as male, emotionally threatening, and aggressive! In an early study of over thousand dreams, Hall (1963) reported that (1) strangers in dreams were most often male, (2) aggressive encounters were more likely to occur in interaction with an unknown male than with an unknown female or a familiar male or female, and (3) unknown males appeared more frequently in dreams of males than of females. Using the Hall–Van de Castle system, Domhoff (1996) looked at the role of "enemies" in dreams. Enemies were defined as those dream characters who typically interacted (greater than 60% of the cases) with the dreamer in an aggressive manner. Those enemies turned out to be male strangers and animals. Interactions with female strangers are predominantly friendly in the dreams of both males and females. Schredl (2000) reports that almost all murderers and soldiers in dreams are male. Domhoff (2003) more recently has shown that when male strangers appear in a dream, the likelihood that physical aggression will occur in that dream far exceeds what would be expected on the basis of chance. In short, *male strangers signal physical aggression.* In summary, it appears that there is some evidence suggesting that the number and type of dream characters both affect the affective content of the dream itself and may also predict mood of the dreamer during the waking period.

VII. Summary

From the above-reviewed series of dream content studies, it is reasonable to conclude that (1) sleep mentation may exhibit striking processing specializations, (2) these specializations include a number and variety of social interactions, and (3) specific dream content variables (including number and variety of dream characters) may predict mood states during the following morning.

VIII. Significance

Identification of reliable REM/NREM dream content and processing spe-cializations, around emotion, memory, personality, and social cognition, will carry significant implications for clinical research and treatment of disorders involving these basic functions. The finding that REM sleep dreams "specialize" in simulations of aggressive and unpleasant social interactions raises several important questions. How does this content indicator vary across mood and psychiatric disorders? What is its level in various disorders of mood and personality? Do depressives, or criminal offenders, or personality-disordered individuals exhibit especially high levels of aggression in REM versus NREM sleep dreams? Is severity of the disorder linked to frequency of the relevant content indicators? Is the subjective quality of sleep correlated with dream content? If dream content influences sleep quality, might it also influence daytime functioning? Can manipulation of the relevant content indicators ameliorate or treat symptoms of both sleep and other types of sleep-related disorders? To underline the potential importance of this question, consider recent reports citing REM sleep dreams and nightmare content indicators as significant predictors of suicidal ideation in depressed individuals (Agargun and Cartwright, 2003; Agargun *et al.*, 1998). If a clinician monitored such dream content indicators in individuals at risk for suicide attempts, he or she could potentially identify early warning signs of new ideation around suicide and could therefore act to prevent a new suicide attempt.

The potential clinical importance of establishing reliable REM and NREM sleep processing specializations is underlined by the wealth of data linking dream content variables to specific clinical outcomes, such as depression, nightmares, recurrent dreams, and unpleasant everyday dreams (e.g., Blagrove *et al.*, 2004; Brown and Donderi, 1986; Kramer, 2000; Schredl and Engelhardt, 2001; Zadra and Donderi, 2000). It is also fairly well established that specific content indicators (e.g., fearful or unpleasant emotional imagery) of REM sleep-related mentation of persons with posttraumatic stress disorder (PTSD) predicts the severity of PTSD (Germain and Nielsen, 2003). Indeed incorporation of trauma-related memories into dreams is one of the *DSM-IV* criteria for the disorder. REM and NREM dream content variables are also strongly correlated with selected personality dispositions, such as attachment status (McNamara *et al.*, 2001), extraversion (Bernstein and Roberts, 1995; Samson and De Koninck, 1986), neuroticism (Schredl *et al.*, 2003), and psychological boundaries (Hartmann *et al.*, 1991; Schredl *et al.*, 1999).

Study of potential links between dream content and personality dimensions may lead to new perspectives on the ways in which personality dispositions influence health and well-being. Zadra *et al.* (2006), for example, recently reported that increased numbers of aggressive social interactions in dreams predict *poorer* health and psychological well-being. When linked with our finding that REM sleep "specializes" in simulations of aggressive interactions, it raises the clinical question of whether persons with high amounts of REM sleep may be at risk for increased hostility or aggression. Independent studies of links between sleep architecture and health have, in fact, shown that high amounts of REM sleep predict poor health and decreased longevity (Brabbins *et al.*, 1993;

Dew *et al.*, 2003; Kripke, 2003). It may be that REM sleep dream content plays a role in the health-related risks of excessive REM sleep. Several recent reviews of REM sleep dream content revealed that REM sleep dreams are typically characterized by high levels of negative affect (Domhoff, 2003; Revonsuo, 2000). With respect to the impact of REM/NREM dream content/processing specializations on memory, once again, we find evidence that specific content indicators are correlated with clinically relevant outcomes such as memory consolidation. For example, Nielsen *et al.* (2004) have noted a so-called "dream-lag effect" in the memory consolidation process whereby an image associated with a significant daytime event makes an appearance in dreams that same night but not necessarily on following nights, where the probability of appearance drops precipitously. In about 1 week, however, the probability rises again for a replay of the image in a dream. The dream-lag effect suggests that significant amounts of memory processing are reflected in dreams. If so, studies of dream content may give us a unique investigative window onto specific aspects of memory processing and breakdowns in memory processing.

Acknowledgments

This research was supported by NIMH grant 5R21MH076916-02 (P.M.). In addition, the project described was supported by the CTSA grant 1UL1RR025771 from the National Center for Research Resources (NCRR), a component of the National Institutes of Health (NIH). Its contents are solely the responsibility of the authors and do not necessarily represent the official view of NCRR or NIH.

Disclosures: None of the authors have any potential conflicts of interest, financial or otherwise.

References

Agargun, M. Y., and Cartwright, R. (2003). REM sleep, dream variables and suicidality in depressed patients. *Psychiatry Res.* **119**(1–2), 33–39.

Agargun, M. Y., Cilli, A. S., Kara, H., Tarhan, N., Kincir, F., and Oz, H. (1998). Repetitive frightening dreams and suicidal behavior in patients with major depression. *Compr. Psychiatry* **39**, 198–202.

Ajilore, O. A., Stickgold, R., Rittenhouse, C., and Hobson, J. A. (1995). Nightcap: laboratory and home-based evaluation of a portable sleep monitor. *Psychophysiology* **32**, 92–98.

Bernstein, D. M., and Roberts, B. (1995). Assessing dreams through self-report questionnaires: relations with past research and personality. *Dreaming* **5**, 13–27.

Blagrove, M., Farmer, L., and Williams, E. (2004). The relationship of nightmare frequency and nightmare distress to well being. *J. Sleep Res.* **13**(2), 129–136.

Brabbins, C. J., Dewey, M. E., Copeland, J.R.M., Davidson, I. A., McWilliam, C., Saunders, P., Sharma, V. K., and Sullivan, C. (1993). Insomnia in the elderly: prevalence, gender differences and relationships with morbidity and mortality. *Int. J. Geriatr. Psychiatry* **8**, 473–480.

Braun, A. R., Balkin, T. J., Wesenstein, N. J., Varga, M., Baldwin, P., Selbie, S., Belenky, G., and Herscovitch, P. (1997). Regional cerebral blood flow throughout the sleep–wake cycle. *Brain* **120**, 1173–1197.

Brown, R. J., and Donderi, D. C. (1986). Dream content and self-reported well-being among recurrent dreamers, past-recurrent dreamers, and nonrecurrent dreamers. *J. Pers. Soc. Psychol.* **50**, 612–623.

Buzsaki, G. (1996). The hippocampo-neocortical dialogue. *Cereb. Cortex* **6**(2), 81–92.

Cavallero, C., Cicogna, P., and Bosinelli, M. (1988). Mnemonic activation in dream production. In: Sleep '86 (W. P. Koella, F. Obal, H. Schulz, and P. Visser Eds.eds), Gustav Fischer, Stuttgart, pp. 91–94.

Datta, S., and Maclean, R. R. (2007). Neurobiological mechanisms for the regulation of mammalian sleep–wake behavior: reinterpretation of historical evidence and inclusion of contemporary cellular and molecular evidence. *Neurosci. Biobehav. Rev.* **31**(5), 775–824.

Dew, M. A., Hoch, C. C., Buysse, D. J., Monk, T. H., Begley, A. E., Houck, P. R., Hall, M., Kupfer, D. J., and Reynolds, C. F. III (2003). Healthy older adults' sleep predicts all-cause mortality at 4 to 19 years of follow-up. *Psychoso. Med.* **65**(1), 63–73.

Domhoff, G. W. (1996). Finding Meaning in Dreams: A Quantitative Approach. Plenum Press, New York.

Domhoff, G. W. (2003). The Scientific Study of Dreams: Neural Networks, Cognitive Development, and Content Analysis.American Psychological Association, Washington, DC.

Fosse, M. J., Fosse, R., Hobson, J. A., and Stickgold, R. (2003). Dreaming and episodic memory: a functional dissociation? *J. Cogn. Neurosci.* **15**, 1–9.

Foulkes, W. D. (1962). Dream reports from different stages of sleep. *J. Abnorm. Soc. Psychol.* **65**, 14–25.

Germain, A., and Nielsen, T. (2003). Impact of imagery rehearsal treatment on distressing dreams, psychological distress and sleep parameters on nightmare patients. *Behav. Sleep Med.* **1**(3), 140–154.

Giuditta, A., Ambrosini, M. V., Montagnese, P., Mandile, P., Cotugno, M., Grassi Zucconi, G., and Vescia, S. (1995). The sequential hypothesis of the function of sleep. *Behav. Brain Res.* **69**(1-2), 157–166.

Hall, C. (1963). Strangers in dreams: an empirical confirmation of the Oedipus complex. *J. Pers.* **31**, 336–345.

Hartmann, E., Elkin, R., and Garg, M. (1991). Personality and dreaming: the dreams of people with very thick and very thin boundaries. *Dreaming* **1**, 311–324.

Hobson, J. A., and Pace-Schott, E. F. (2002). The cognitive neuroscience of sleep: neuronal systems, consciousness and learning. *Nat. Rev. Neurosci.* **3**(9), 679–693.

Hobson, J. A., Stickgold, R., and Pace-Schott, E. F. (1998). The neuropsychology of REM sleep dreaming. *NeuroReport* **9**(3), R1–R14.

Kavanau, J. L. (2002). Dream contents and failing memories. *Arch. Biol. (Liege)*, **140**(2), 109–127.

Kramer, M. (1993). The selective mood regulatory function of dreaming: an update and revision. In: The Functions of Dreaming (A. Moffit, M. Kramer, and R. Hoffman , eds.), State University of New York Press, Albany, pp. 139–145.

Kramer, M.(2000). Dreams and psychopathology. In: Principles and Practice of Sleep Medicine, 3rd ed. (M. H. Kryger, T. Roth, and W. C. Demet eds.), Saunders, Philadelphia, pp. 511–519.

Kripke, D. F. (2003). Sleep and mortality. *Psychosom. Med.* **65**(*1*), 74.

Linkowski, P. (1999). EEG sleep patterns in twins. *J. Sleep Res.* **8**(Suppl. 1), 11–13.

Lovibond, S. H., and Lovibond, P. F. (1995). Manual for the Depression Anxiety Stress Scales, The Psychology Foundation of Australia, Inc, Sydney.

Mamelak, A. N., and Hobson, J. A. (1989). Nightcap: a home-based sleep monitoring system. *Sleep* **12**, 157–166.

Maquet, P., and Franck, G. (1997). REM sleep and amygdala. *Mol. Psychiatry* **2**(3), 195–196.

Maquet, P., Peters, J. M., Aerts, J., Delfiore, G., Degueldre, C., Luxen, A., and Franck, G. (1996). Functional neuroanatomy of human rapid-eye-movement sleep and dreaming. *Nature* **383**, 163–166.

McNamara, P., Andresen, J., Clark, J., Zborowski, M., and Duffy, C. (2001). Impact of attachment styles on sleep and dreams: a test of the attachment hypothesis of REM sleep. *J. Sleep Res.* **10**, 117–127.

McNamara, P., McLaren, D., Smith, D., Brown, A., and Stickgold, R. (2005). A "Jekyll and Hyde" within: aggressive versus friendly social interactions in REM and NREM dreams. *Psychol. Sci.* **16**(2), 130–136.

Merritt, J. M., Stickgold, R., Pace-Schott, E. F., Williams, J., and Hobson, J. A. (1994). Emotion profiles in the dreams of men and women. *Conscious. Cogn.* **3**, 46–60.

Nielsen, T. A., Kuiken, D., Alain, G., Stenstrom, P., and Powell, R. A. (2004). Immediate and delayed incorporations of events into dreams: further replication and implications for dream function. *J. Sleep Res.* **13**(4), 327–336.

Nielsen, T. A., Kuiken, D., Hoffmann, R., and Moffitt, A. (2001). REM and NREM sleep mentation differences: a question of story structure? *Sleep Hypnosis* **3**(1), 9–17.

Nofzinger, E. A., Mintun, M. A., Wiseman, M. B., Kupfer, D. J., and Moore, R. Y. (1997). Forebrain activation in REM sleep: an FDG PET study. *Brain Res.* **770**, 192–201.

Payne, J. D., and Nadel, L. (2004). Sleep, dreams, and memory consolidation: the role of the stress hormone cortisol. *Learn. Mem.* **11**(6), 671–678.

Pennebaker, J. W., Francis, M. E., and Booth, R. J. (2001). Linguistic Inquiry and Word Count. Erlbaum Publishers, Mahwah, NJ.

Rechtschaffen, A., Verdonne, P., and Wheaton, J. (1963). Reports of mental activity during sleep. *Can. Psychiatr. Assoc. J.* **257**, 409–414.

Revonsuo, A. (2000). The reinterpretation of dreams: an evolutionary hypothesis of the function of dreaming. *Behav. Brain Sci.* **23**(6), 877–901 (discussion 904–1121).

Samson, H., and De Koninck, J. (1986). Continuity or compensation between waking and dreaming: an exploration using the Eysenck Personality Inventory. *Psychol. Rep.* **58**, 871–874.

Schredl, M. (2000). Dream research: integration of physiological and psychological models. *Behav. Brain Sci.* **23**(6), 1001–1003.

Schredl, M., and Engelhardt, H. (2001). Dreaming and psychopathology: dream recall and dream content of psyhiatric inpatients. *Sleep and Hypnosis* **3**, 44–54.

Schredl, M., Landgraf, C., and Zeiler, O. (2003). Nighmare frequency, nightmare distress and neuroticism. *North Am. J. Psychol.* **5**, 345–350.

Schredl, M., Schäfer, G., Hofmann, F., and Jacob, S. (1999). Dream content and personality: thick vs thin boundaries. *Dreaming* **9**, 257–263.

Smith, M. R., Antrobus, J. S., Gordon, E., Tucker, M. A., Hirota, Y., Wamsley, E. J., Ross, L., Doan, T., Chaklader, A., and Emery, R. N. (2004). Motivation and affect in REM sleep and the mentation reporting process. *Conscious. Cogn.* **13**, 501–511.

Snyder, F., Karacan, I., Tharp, V. K. Jr., and Scott, J. (1968). Phenomenology of REMS dreaming. *Psychophysiology* **4**(3), 375.

Steriade, M., and McCarley, R. (1990). Brainstem Control of Wakefulness and Sleep. Plenum, New York.

Stickgold, R., *et al.*, Scott, L., Fosse, R., and Hobson, J. A. (2001). Brain-mind states: I. Longitudinal field study of wake–sleep factors influencing mentation report length. *Sleep* **24**(2), 171–179.

Strauch, I., and Meier, B. (1996). In Search of Dreams: Results of Experimental Dream Research. State University of New York Press, Albany, NY.

Tafti, M., and Franken, P. (2002). Invited review: genetic dissection of sleep. *J. Appl. Physiol.* **92**(3), 1339–1347.

Taheri, S., and Mignot, E. (2002). The genetics of sleep disorders. *Lancet Neurol.* **1**(4), 242–250.

Walker, M. P., and Stickgold, R. (2002). Cognitive Flexibility across the sleep-wake cycle: REM-sleep enhancement of anagram problem solving. *Cogn. Brain Res.* **14**, 314–324.

Wechsler, D. (2001). Wechsler Test of Adult Reading™ (WTAR™). Pearson Assessment and Information, Upper Saddle River, NJ.

Zadra, A., Desjardins, S., and Marcotte, E. (2006). Evolutionary funcion of dreams: a test of the threat of simulation theory in recurrent dreams. *Conscious. Cogn.* **15**(2), 450–463.

Zadra, A., and Donderi, D. C. (2000). Nightmares and bad dreams: their prevalence and relationship to well-being. *J. Abnorm. Psychol.* **109**(2), 273–281.

Zadra, A., Pilon, M., and Donderi, D. C. (2006). Variety and intensity of emotions in nightmares and bad dreams. *J. Neural Eng.* **194**(4), 249–254.

NEUROIMAGING OF DREAMING: STATE OF THE ART AND LIMITATIONS

Caroline Kussé, Vincenzo Muto, Laura Mascetti, Luca Matarazzo, Ariane Foret, Anahita Shaffii-Le Bourdiec, and Pierre Maquet

Cyclotron Research Centre, University of Liège, Liège, Belgium

During the last two decades, functional neuroimaging has been used to characterize the regional brain function during sleep in humans, at the macroscopic systems level. In addition, the topography of brain activity, especially during rapid eye movement sleep, was thought to be compatible with the general features of dreams. In contrast, the neural correlates of dreams remain largely unexplored. This review examines the difficulties associated with the characterization of dream correlates.

ἓν οἶδα ὅτι οὐδὲν οἶδα

Σωκράτης

(The only thing I know is that I know nothing)

Socrates

I. Introduction

Dreams have always fascinated mankind. Their perceptual and motor vividness, their emotionality, their irrational plot, their apparent ability to find solutions to unfathomable problems or to reconstruct the past and project it into novel incongruous situations potentially explain why men thought their dreams

INTERNATIONAL REVIEW OF
NEUROBIOLOGY, VOL. 92
DOI: 10.1016/S0074-7742(10)92005-9

87

might predict the future (Artemidorus, 1992), reflect their mental health (for example, incubus Bond, 1753), access the depths of their unconscious self (Freud, 1926), or foster their creativity (Maquet and Ruby, 2004).

Yet, dreams eventually result from brain activity. They occur in the specific context of sleep [more often in rapid eye movement (REM) sleep than in non-REM (NREM) sleep Hobson *et al.* 2000], a simple but remarkable finding that links dreaming activity to the mode of neural function selectively implemented during sleep. Moreover, focal brain lesions are known to alter the amount and quality of dreams, depending on their topography (Solms, 1997). Finally, the behavioral correlates of dreaming can be objectively recorded if muscular atonia of REM sleep is hindered by brainstem lesions. This has been observed in man (Schenck *et al.*, 1986) and arguably also in other mammals (Sastre and Jouvet, 1979).

An important objective of the study of dreams is obviously to characterize their neural correlates. The advent of neuroimaging techniques that allow for the atraumatic assessment of regional brain activity in humans resulted in a number of significant advances in this direction. A number of excellent reports have already reviewed and extensively discussed these results (Hobson and Pace-Schott, 2002; Hobson *et al.*, 2000). Before summarizing our current views about the neural underpinnings of dreaming, we want to examine the difficulties inherent to this endeavor and the basic assumptions which frame the interpretation of current sleep neuroimaging studies.

II. Necessity of and Difficulties in the Assessment of Dream Reports

The collection of subjective dream reports should be an essential step in any functional imaging study of dreaming. Obviously, the content of a dream. This absence of objective assessment is not specific to dream research and is pervasive in the study of conscious processes (liminal perception Del Cul *et al.*, 2007, bistable perception Sterzer and Kleinschmidt, 2005, perceptual rivalry Tong *et al.*, 2006, etc.). In this section, we will discuss the main difficulties in obtaining reliable dream reports, thereby hindering the characterization of their neural correlates.

First, dream reports can be obtained only after a change in the state of vigilance, i.e., after awakening which might alter the quality of dream report. The consistency of dream content across subjects or within subjects with repeated sampling in the same conditions is usually felt as a partial solution to this limitation (Schwartz and Maquet, 2002). However, this strategy restricts the characterization of dreams to its statistical invariants.

Second, the ability to recall dreams at awakening considerably varies across individuals and across conditions (e.g., the laboratory environment against home setting). Selecting individuals who are trained or particularly proficient in remembering their dreams appears as a good strategy to circumvent these difficulties (Strauch and Meier, 1996).

Third, a further hurdle in dream research arises from the fact that memory is reconstructive: the original dream content can ultimately be distorted by interference with waking material, by limitations imposed by verbal reports, or by moral censorship (Hobson *et al.*, 2000). As a result, the accurate retrieval of the sequence of mental representations that originally formed a given dream is never guaranteed and cannot be objectively assessed.

Fourth, the absence of dream reports does not entail the absence of dreaming: dreams are easily and quickly forgotten. Dream amnesia potentially arises from a combination of factors including a specific brain functional organization during sleep hindering episodic retrieval (see below), the change in vigilance state at awakening which potentially modifies the accessibility to dream information, and the time lag between the actual dreaming experience and its restitution during wakefulness. In order to minimize the latter, instead of collecting dream reports in the morning, repeated awakenings during sleep should be used to probe ongoing dream experience. However, this procedure is hard to combine with functional neuroimaging.

III. Basic Assumptions

Several basic assumptions, usually overlooked, have profound consequences on the interpretation of sleep neuroimaging studies, especially those that would address the topic of the neural correlates of dreams.

During wakefulness, brain function is mainly driven by cognitive processes (perception, motor behavior, attention, memory, language, etc.). In contrast, one can argue that during sleep, regional brain function is primarily organized by specific sleep processes. For instance, during NREM sleep, a coalescence of oscillations organizes neural firing in specific temporal patterns (slow oscillation, delta rhythm, spindles). These oscillations are characterized by consistent regional brain responses (see below) and are probably related to specific functional processes (e.g., synaptic downscaling, memory replay). These processes are obviously different from those occurring during wakefulness in the same brain areas.

However, when sleep processes give rise to retrievable mental representations, e.g., dreams, it is tempting to assume that the distribution of regional brain function is related to these mental processes. In this context, the information processing during sleep occurs in a regionally specific way that matches the functional specialization prevailing during wakefulness. In other words, it is assumed that dream content informs us on how regional brain function is organized during sleep. This hypothesis is based on two basic assumptions. First, brain specialization does not change between sleep and wakefulness (Schwartz and Maquet, 2002). Second, structural brain connectivity, the main determinant of brain specialization, does not substantially change across vigilance states.

A more qualified assumption would therefore be that brain activity associated with dreams results from genuine sleep processes taking place in a regionally specialized brain. As a consequence, describing the spatial distribution of brain activity during sleep provides only a partial view on the neural correlates of dreaming. For example: if lucid dreamers were able to signal their dreams by sequences of saccades while being studied by functional neuroimaging, it would be trivial to observe a related increase in activity in the relevant brain areas, e.g., the frontal eye field.

If brain specialization does not dramatically change between wakefulness and sleep, then what makes sleep and wakefulness so behaviorally different? It is clear that the main unknown resides in functional integration, namely, how brain areas interact with each other during sleep. It is likely that interactions between functional brain units are profoundly modified by genuine sleep processes (slow oscillations, spindles, pontine waves, etc.). For instance, modifications of connectivity associated with the slow oscillation of NREM sleep were observed using transcranial magnetic stimulation (Massimini *et al.*, 2005), but the impact of these changes on dream reports has not yet been reported. In addition, the dramatic change in neuromodulation that differentiates sleep states from wakefulness is another potential factor influencing functional brain integration (Steriade and McCarley, 2005a). In the following sections, it will become clear that whereas some advances have been made in characterizing brain segregation during sleep, very little is known about functional integration during sleep.

IV. Current Data

A. Neuroimaging of Sleep Processes: Cerebral Correlates of NREM Sleep Oscillations

The slow rhythm (<1 Hz) constitutes the fundamental rhythm which characterizes NREM sleep. Originally, unit recordings in cats showed that neuronal membrane potential oscillates at low frequency, around 1 Hz (Steriade *et al.*, 1993a). This oscillation shapes neuronal activity, by alternating a depolarizing phase, associated with important neuronal firing ("up state"), and a hyperpolarizing phase, during which cortical neurons remain silent for a few hundred milliseconds ("down state") (Steriade et al., 1993a, Borbely, 2001). This so-called slow oscillation (<1 Hz) is recorded during NREM sleep in all major types of neocortical neurons (both excitatory and inhibitory) and occurs synchronously in large neuronal populations. At the population level, the activity is therefore made up of the alternation of "ON" states and "OFF" states." Because these events represent massive and synchronous changes in large neuronal populations, they can be reflected on electroencephalographic (EEG) recordings as large amplitude low-frequency waves (Steriade et al., 1993b). The slow oscillation is generated by the cortex as it can be observed after thalamic destruction (Steriade et al., 1993b), in cortical slabs isolated from thalamic influence (Timofeev et al., 2000) or in cortical slices (Sanchez-Vives and McCormick, 2000). However, two intrinsic conditional thalamic oscillators also participate in the generation of the slow oscillation (Crunelli and Hughes, 2009).

In humans, the taxonomy of slow wave sleep (SWS) waves is not always clear. A slow rhythm was initially identified on scalp EEG recordings as the recurrence of spindles (Achermann and Borbely, 1997) or their grouping by slow waves (Molle et al., 2002). More recently, high-amplitude slow waves themselves were taken as realization of the slow rhythm (Massimini et al., 2004). On the other hand, historically, the power density in the 0.75–4 Hz frequency band, usually referred to as "slow wave activity" (SWA), has proved a very useful and popular parameter because it quantifies the dissipation of homeostatic sleep pressure during NREM sleep (Borbely, 2001). The frequency bounds of SWA do not respect the dichotomy between slow (<1 Hz) and delta rhythms (1–4 Hz), which is based on differences in the respective cellular correlates of these rhythms in animals (Steriade and McCarley, 2005b). In the temporal domain, the amplitude of SWS waves is classically larger than 75 μV (Rechtschaffen and Kales, 1968) but only the largest waves (>140 μV) are taken as realizations of the slow oscillation (<1 Hz) (Massimini et al., 2004; Molle et al., 2002). This approach suggests that relatively smaller waves (amplitude between 75 and 140 μV) correspond to delta waves (1–4 Hz). These faster waves of smaller amplitude would also be an expression of the slow oscillation but arise when the synchronization in the network is less marked (Esser et al., 2007; Vyazovskiy et al., 2009). On scalp EEG recording, SWA predominates over frontal areas (Finelli et al., 2001), where indeed the largest waves are typically recorded. However, an analysis of individual waves demonstrated the spatial variability of slow waves. Each wave originates at a specific site and travels over the scalp following a particular trajectory (Massimini et al., 2004). Waves originate more frequently in frontal regions and travel backward to posterior areas. Beyond this variability, slow waves seem to systematically recruit various brain regions. Early studies based on cerebral blood flow (CBF) measurement by positron emission tomography (PET) reported decreases in CBF during NREM sleep, relative to wakefulness (Braun *et al.*, 1997; Kajimura *et al.*, 1999; Madsen *et al.*, 1991a). In addition, the power density of delta waves (1–4 Hz) during NREM sleep was negatively correlated with CBF in the ventromedial prefrontal cortex, the basal forebrain, the striatum, the anterior insula, and the precuneus (Dang-Vu *et al.*, 2005).

However, capitalizing on the better temporal resolution of simultaneous EEG and event-related functional magnetic resonance imaging (fMRI), it was possible to show that slow waves were associated with transient increases in regional brain activity (Steriade and McCarley, 2005a), in line with previously acquired data from animals (Dang-Vu *et al.*, 2008). Slow waves were associated with significant increases in activity in inferior and medial frontal cortices, precuneus, and posterior cingulate. Compared to baseline activity, the largest waves (>140 µV) were associated with significant activity in the parahippocampal gyrus, cerebellum, and brainstem, whereas delta waves were related to frontal responses. Source reconstruction of scalp high-density EEG recordings confirmed these results. Slow waves originate more frequently in the insula and cingulate gyrus, even though they mainly involve the precuneus, the posterior cingulate, ventrolateral, and medial frontal areas afterward (Murphy *et al.*, 2009). It is currently believed that these areas constitute a preferred propagation pathway because they correspond to major structural connectivity nodes in the human brain (Murphy *et al.*, 2009).

Spindles constitute the hallmark of light NREM sleep, although they persist in smaller amounts during deep NREM sleep. In humans, spindles consist of waxing-and-waning 11–16 Hz oscillations, lasting 0.5–3 seconds. At the cellular level, spindles arise from cyclic inhibition of thalamo-cortical neurons by reticular thalamic neurons. Post-inihibitory rebound spike bursts in thalamo-cortical cells entrain cortical populations in spindle oscillations (Steriade and McCarley, 2005a). In addition, two kinds of spindles are observed in humans, namely slow spindles (grossly <13 Hz) and fast spindles (>13 Hz). These two spindling activities differ in many ways: by their circadian and homeostatic regulations, pharmacological reactivity, development in infancy, evolution during aging, modulation during menstrual cycle and pregnancy (De Gennaro and Ferrara, 2003), and, intriguingly, by their association with general cognitive capabilities (Bodizs *et al.*, 2005) and memory processing (Schabus, 2009). Despite these functional differences, it is still debated whether slow and fast spindles reflect the activity of different neural networks or the differential modulation of a single generator.

Additionally, these two kinds of spindles differ in their topography. Scalp multi-channel EEG recordings showed that slow spindles (centered around 12 Hz) exhibit a variable topography, primarily over the frontal cortex (Doran, 2003). Fast spindles (centered at 14 Hz) are topographically and dynamically limited to the superior central and parietal cortex (Doran, 2003). Source reconstruction of scalp EEG recordings identified two sources: one for slow spindles in a mesial frontal region and another for fast spindles in the precuneus (Anderer et al., 2001).

Little is known about the cerebral correlates of human spindles. Early PET studies reported a negative relationship between thalamic CBF and the power spectrum in the spindle frequency band (Hofle et al., 1997). Taking advantage of the high temporal resolution of EEG/fMRI, it was later shown that human spindles were also associated with transient surge in activity in the thalami, paralimbic areas (anterior cingulate and insular cortices) and superior temporal gyri (Schabus et al., 2007). Slow spindles were further associated with increased activity in the superior frontal gyrus. In contrast, fast spindles recruited a set of cortical regions involved in sensorimotor processing, as well as the mesial frontal cortex and hippocampus. The recruitment of partially segregated cortical networks for slow and fast spindles further supports the existence of two spindle types during human NREM sleep, with potentially different functional significance.

B. Neuroimaging of Sleep Processes: Functional Neuroanatomy of REM Sleep

Cerebral neurons typically adopt a tonic firing pattern during rapid eye movement sleep (REMS) which on average is as intense as during wakefulness (Steriade and McCarley, 2005a). Accordingly, cerebral energy metabolism (Maquet et al., 1990) and blood flow (Madsen et al., 1991b) reach similar levels during REMS as during wakefulness. However, the distribution of regional brain activity considerably differs between REMS and wakefulness and, in humans, is characterized by three main features.

First, functional imaging studies in man reported a high activity in the brainstem and thalamic nuclei. In humans, the activation of mesopontine tegmentum and thalamic nuclei has been systematically reported during REMS (Braun et al., 1997; Maquet et al., 1996; Nofzinger et al., 1997). This pattern of activity is easily explained by the known neurophysiological mechanisms which generate REMS in animals (Steriade and McCarley, 2005a). More specifically, during REMS, neuronal populations in the mesopontine tegmentum are the source of a major activating input to the thalamic nuclei (Steriade and McCarley, 1990, 2005a), which in turn forward this activation to the entire cortical mantle.

Second, a high activity has been observed in limbic and paralimbic areas. This confirms earlier measurements of brain energy metabolism in animals (Lydic et al., 1991; Ramm and Frost, 1983, 1986). In the forebrain, REMS is characterized by high activity levels in the amygdala, the hippocampal formation, and the anterior cingulate, orbito-frontal, and insular cortices (Braun et al., 1997; Maquet et al., 1996; Nofzinger et al., 1997).

In addition to these limbic and paralimbic areas, temporal and occipital cortices were also shown to be very active (Braun *et al.*, 1997), although this result was less systematically reported (Maquet *et al.*, 1996). Finally, the motor and premotor cortices were also very active during REMS (Maquet *et al.*, 2000a).

Third, previous research further highlighted the contrast between this limbic activation and the relative quiescence of the associative frontal and parietal cortices during REMS, relative to wakefulness (Braun *et al.*, 1997; Maquet *et al.*, 1996). More precisely, the hypoactive areas were located bilaterally in the inferior and middle frontal gyrus as well as in the posterior part of the inferior parietal lobule (Maquet *et al.*, 2005). Interestingly, the superior frontal gyrus, the medial frontal areas, the intraparietal sulcus, and the superior parietal cortex were not less active in REMS than during wakefulness (Maquet *et al.*, 2005).

Not only the distribution of brain activity but also its functional connectivity is modified during human REMS. The functional relationship between striate and extrastriate cortices—excitatory during wakefulness—was shown inverted during REMS (Braun *et al.*, 1998). Likewise, the functional relationship between the amygdala and the temporal and occipital cortices was different during REMS than during wakefulness or NREM sleep (NREMS) (Maquet and Phillips, 1998).

The reasons which explain these peculiar functional segregation and integration remain unclear. It is usually assumed that changes in neuromodulation might contribute to a modification of forebrain activity and responsiveness during REMS because REMS is characterized by a prominent cholinergic tone and a decrease in noradrenergic and serotonergic modulation (Steriade and McCarley, 2005a). However, objective evidence supporting this hypothesis is still lacking. It was also assumed that the regional distribution during REMS is partly driven by phasic events concomitant to REMs, called pontine waves or ponto-geniculo-occipital waves. However, brain activity associated with bursts of saccades, supposedly associated with pontine waves in humans, showed a different distribution. Using PET and CBF measurements, it was shown that the activity in the right geniculate body and the primary occipital cortex increased proportionally to the density of eye movements to a larger extent during REMS than during wakefulness (Peigneux *et al.*, 2001). Similar results data were eventually reported with fMRI (Hong *et al.*, 2009; Miyauchi *et al.*, 2009; Wehrle *et al.*, 2005).

C. Neuroimaging and Dream Correlates during REM Sleep

To our knowledge, functional neuroimaging research specifically devoted to the characterization of dream correlates has been conducted only during REM sleep. Indeed, mentation during REMS is more abundant, vivid, and story-like and hence more detailed dream reports can be obtained from REM than from NREM sleep (Stickgold *et al.*, 1994). The particular pattern of cerebral activity observed during REMS (a high limbic contrasting with a low prefrontal and parietal activity) is usually assumed to correlate with, and possibly influence, the main characteristics of dreaming activity (Hobson *et al.*, 2000; Maquet and Franck, 1997; Maquet *et al.*, 1996, 2000a)(see also Fig. 1). The activation synthesis of dreams (Hobson and McCarley, 1977; McCarley and Hoffman, 1981) states that several features of dream phenomenology correspond remarkably well with the distribution of concomitant brain activity.

Vision is the most prominent dreaming sensory modality, occurring almost in all dreams. This might be related to the activation of the occipital cortex. Indeed, PET studies revealed significant increases in regional CBF in extrastriate visual cortices, particularly within the ventral visual stream (Braun *et al.*, 1997, 1998). The heterogeneous activation of ventral visual areas can be indicative of the bizarre properties of dreams. In keeping with this hypothesis, visual imagery in dreams is absent in some patients with occipito-temporal lesions (Solms, 1997). Auditory perceptual features of dreams might be related to the activation of posterior temporal cortices. Although human functional neuroimaging evidence for this hypothesis is still lacking during REM sleep, acoustic stimulation during both NREM sleep and wakefulness is already known to be associated with similar regionally specific responses (Portas *et al.*, 2000). Motor behavior and movements

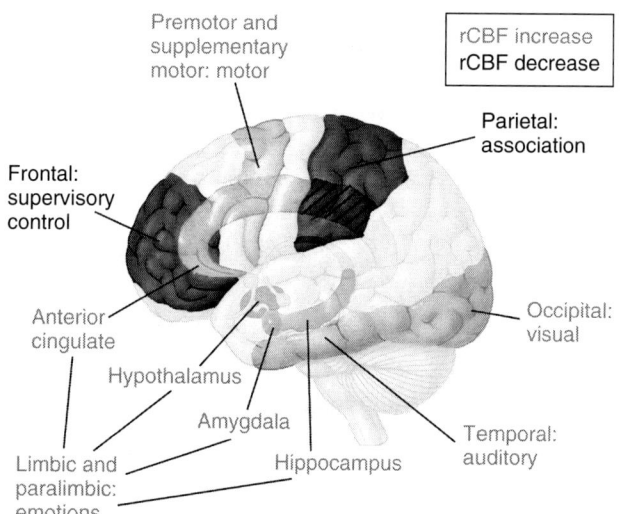

FIG. 1. Relative increases (light shade of gray) and decreases (dark shade of gray) in regional cerebral blood flow during rapid eye movement sleep [Based on references Maquet *et al.* (1996) and Schwartz and Maquet (2002)].

probably activate motor-related brain areas during REMS. Indeed, regional CBF measurements using PET have revealed a significant increase in activity of motor and supplementary motor areas during REMS after the learning of a motor task [see control group in Maquet *et al.* 2000a]. On the other hand, affect and emotional intensification (anxiety, elation, anger) which are often reported in dreams would be related to the significant activation of the limbic and paralimbic systems: amygdala, orbito-frontal cortex, and anterior cingulate cortex (Hobson *et al.*, 2000; Maquet, 2000b; Maquet and Franck, 1997; Maquet *et al.*, 1996). Given the crucial role of the limbic system in the acquisition of emotional memories, the activation pattern in the amygdala and cortical areas provides a biological basis for the processing of some types of memory during REM sleep.

The activation of mesio-temporal areas would account for the memory content commonly found in dreams. However, this mesio-temporal activity occurs in the context of a relatively low activity of ventral prefrontal cortex, which is deemed participating in memory retrieval (Rugg *et al.*, 2002). This pattern of regional brain activity would explain the peculiar aspects of episodic memory in dreams. Usually, "snips" of recent waking activity are frequently observed in dream reports. In contrast, although REM dreams can be very much story-like, complete waking life episodes, characterized by the association between specific locations, characters, objects, and actions, are seldom described as such in dream reports (Fosse *et al.*, 2003). The relative quiescence of the anterior and ventral prefrontal areas might be related to other formal character-istics of dreams, such as temporal distortions, weakening of self-reflective control, and amnesia on awakening. The hypoactivation of the frontal lobe would explain that although the dreamer has access to "day residues," probably spontaneously generated by the coordinated activity of the mesio-temporal areas and the posterior cortices, the successful retrieval of the various details of a specific past episode is hindered. Low frontal activity would also account for the deficits in working memory, and executive functions that manifest themselves in dream reports from REMS awakenings (Hobson *et al.*, 2000; Maquet, 2000b; Maquet and Franck, 1997; Maquet *et al.*, 1996).

V. Conclusions

At present, although functional neuroimaging provides a general idea about the distribution of brain activity during NREM and REM sleep, it did not directly characterize the neural correlates of dreaming. This endeavor is difficult for various reasons: (1) Dream reports are not easy to obtain reliably; (2) Func-tional neuroimaging is still difficult during steady states of NREM or REM sleep;

(3) The underlying mechanisms that activate the cortex during sleep are not necessarily comprehensively understood; and (4) The estimation of brain functional integration is still under development. Nevertheless, we are confident that, in the future, more progress will be made that will allow us to understand the neural mechanisms underlying dreaming even better.

References

Achermann, P., and Borbely, A. A. (1997). Low-frequency (<1 Hz) oscillations in the human sleep electroencephalogram. *Neuroscience* **81**(1), 213–222.

Anderer, P. *et al.* (2001). Low-resolution brain electromagnetic tomography revealed simultaneously active frontal and parietal sleep spindle sources in the human cortex. *Neuroscience* **103**(3), 581–592.

Artemidorus (1992). Oneirocritica: The Interpretation of Dreams (Tb. R. J. White, ed.), Noyes Classical Studies, Park Ridge, NJ.

Bodizs, R. *et al.* (2005). Prediction of general mental ability based on neural oscillation measures of sleep. *J. Sleep Res.* **14**(3), 285–292.

Bond, J. (1753). An Essay on the Incubus, or Nightmare. D. Wilson and T. Durham, London, p. 80.

Borbely, A. A. (2001). From slow waves to sleep homeostasis: new perspectives. *Arch. Ital. Biol.* **139** (1–2), 53–61.

Braun, A. R. *et al.* (1997). Regional cerebral blood flow throughout the sleep-wake cycle. An H2(15)O PET study. *Brain* **120**(Pt 7), 1173–1197.

Braun, A. R. *et al.* (1998). Dissociated pattern of activity in visual cortices and their projections during human rapid eye movement sleep. *Science* **279**(5347), 91–95.

Cicero De Divinatione. (2004). GF Flammarion, Paris, 382 p.

Crunelli, V., and Hughes, S. W. (2009). The slow (<1 Hz) rhythm of non-REM sleep: a dialogue between three cardinal oscillators. *Nat. Neurosci.* **13**(1), 9–17.

Dang-Vu, T. T. *et al.* (2005). Cerebral correlates of delta waves during non-REM sleep revisited. *NeuroImage* **28**(1), 14–21.

Dang-Vu, T. T. *et al.* (2008). Spontaneous neural activity during human slow wave sleep. *Proc. Natl. Acad. Sci. U.S.A* **105**(39), 15160–15165.

De Gennaro, L., and Ferrara, M. (2003). Sleep spindles: an overview. *Sleep Med. Rev.* **7**(5), 423–440.

Del Cul, A., Baillet, S., and Dehaene, S. (2007). Brain dynamics underlying the nonlinear threshold for access to consciousness. *PLoS Biol.* **5**(10), p. e260.

Doran, S. (2003). The dynamic topography of individual sleep spindles. *Sleep Res. Online* **5**(4), 133–139.

Esser, S. K., Hill, S. L., and Tononi, G. (2007). Sleep homeostasis and cortical synchronization: I. Modeling the effects of synaptic strength on sleep slow waves. *Sleep* **30**(12), 1617–1630.

Finelli, L. A., Borbely, A. A., and Achermann, P. (2001). Functional topography of the human nonREM sleep electroencephalogram. *Eur. J. Neurosci.* **13**(12), 2282–2290.

Fosse, M. J. *et al.* (2003). Dreaming and episodic memory: a functional dissociation? *J. Cogn. Neurosci.* **15**(1), 1–9.

Freud, S. (1967). L'interprétation des rêves, Presses Universitaires de France, 573 p.

Hobson, J. A., and McCarley, R. W. (1977). The brain as a dream state generator: an activation-synthesis hypothesis of the dream process. *Am. J. Psychiatr.* **134**(12), 1335–1348.

Hobson, J. A., and Pace-Schott, E. F. (2002). The cognitive neuroscience of sleep: neuronal systems, consciousness and learning. *Nat. Rev. Neurosci.* **3**(9), 679–693.

Hobson, J. A., Pace-Schott, E. F., and Stickgold, R. (2000). Dreaming and the brain: toward a cognitive neuroscience of conscious states. *Behav. Brain Sci.* **23**(6), 793–842. discussion 904–1121.

Hofle, N. *et al.* (1997). Regional cerebral blood flow changes as a function of delta and spindle activity during slow wave sleep in humans. *J. Neurosci.* **17**(12), 4800–4808.

Hong, C. C. *et al.* (2009). fMRI evidence for multisensory recruitment associated with rapid eye movements during sleep. *Hum. Brain Mapp.* **30**(5), 1705–1722.

Kajimura, N. *et al.* (1999). Activity of midbrain reticular formation and neocortex during the progression of human non-rapid eye movement sleep. *J. Neurosci.* **19**(22), 10065–10073.

Lydic, R. *et al.* (1991). Regional brain glucose metabolism is altered during rapid eye movement sleep in the cat: a preliminary study. *J. Comp. Neurol.* **304**(4), 517–529.

Madsen, P. L. *et al.* (1991a). Cerebral O2 metabolism and cerebral blood flow in humans during deep and rapid-eye-movement sleep. *J. Appl. Physiol.* **70**(6), 2597–2601.

Madsen, P. L. *et al.* (1991b). Human regional cerebral blood flow during rapid-eye-movement sleep. *J. Cereb .Blood Flow Metab.* **11**(3), 502–507.

Maquet, P. *et al.* (1990). Cerebral glucose utilization during sleep-wake cycle in man determined by positron emission tomography and [18F]2-fluoro-2-deoxy-D-glucose method. *Brain Res.* **513**(1), 136–143.

Maquet, P. *et al.* (1996). Functional neuroanatomy of human rapid-eye-movement sleep and dreaming. *Nature* **383**(6596), 163–166.

Maquet, P. *et al.* (2000a). Experience-dependent changes in cerebral activation during human REM sleep. *Nat. Neurosci.* **3**(8), 831–836.

Maquet, P. (2000b). Functional neuroimaging of normal human sleep by positron emission tomography. *J. Sleep. Res.* **9**(3), 207–231.

Maquet, P. *et al.* (2005). Human cognition during REM sleep and the activity profile within the frontal and parietal cortices: a reappraisal of functional neuroimaging data. In: Progress in Brain Research (S. Laureys, ed.), Elsevier, Amsterdam, 219–227.

Maquet, P., and Franck, G. (1997). REM sleep and amygdala. *Mol. Psychiatry* **2**(3), 195–196.

Maquet, P., and Phillips, C. (1998). Functional brain imaging of human sleep. *J. Sleep Res.* **7**(Suppl. 1), 42–47.

Maquet, P., and Ruby, P. (2004). Psychology: insight and the sleep committee. *Nature* **427**(6972), 304–305.

Massimini, M. *et al.* (2004). The sleep slow oscillation as a traveling wave. *J. Neurosci.* **24**(31), 6862–6870..

Massimini, M. *et al.* (2005). Breakdown of cortical effective connectivity during sleep. *Science* **309** (5744), 2228–2232.

McCarley, R. W., and Hoffman, E. (1981). REM sleep dreams and the activation-synthesis hypothesis. *Am. J. Psychiatr.* **138**(7), 904–912.

Miyauchi, S. *et al.* (2009). Human brain activity time-locked to rapid eye movements during REM sleep. *Exp. Brain Res.* **192**(4), 657–667.

Molle, M. *et al.* (2002). Grouping of spindle activity during slow oscillations in human non-rapid eye movement sleep. *J. Neurosci.* **22**(24), 10941–10947.

Murphy, M. *et al.* (2009). Source modeling sleep slow waves. *Proc. Natl. Acad. Sci. U.S.A* **106**(5), 1608–1613.

Nofzinger, E. A. *et al.* (1997). Forebrain activation in REM sleep: an FDG PET study. *Brain Res.* **770** (1–2), 192–201.

Peigneux, P. *et al.* (2001). Generation of rapid eye movements during paradoxical sleep in humans. *NeuroImage* **14**(3), 701–708.

Portas, C. M. *et al.* (2000). Auditory processing across the sleep-wake cycle: simultaneous EEG and fMRI monitoring in humans. *Neuron* **28**(3), 991–999.

Ramm, P., and Frost, B. J. (1983). Regional metabolic activity in the rat brain during sleep-wake activity. *Sleep* **6**(3), 196–216.

Ramm, P., and Frost, B. J. (1986). Cerebral and local cerebral metabolism in the cat during slow wave and REM sleep. *Brain Res.* **365**(1), 112–124.

Rechtschaffen, A., and Kales, A. (1968). *A manual of standardized terminology, techniques and scoring system for sleep stages of human subjects.* Brain Information Service/Brain Research Institute, University of California, Los Angeles

Rugg, M. D., Otten, L. J., and Henson, R. N. (2002). The neural basis of episodic memory: evidence from functional neuroimaging. *Philos. Trans. R. Soc. Lond. B. Biol. Sci.* **357**(1424), 1097–1110.

Sanchez-Vives, M. V., and McCormick, D. A. (2000). Cellular and network mechanisms of rhythmic recurrent activity in neocortex. *Nat. Neurosci.* **3**(10), 1027–1034.

Sastre, J. P., and Jouvet, M. (1979). Oneiric behavior in cats. *Physiol. Behav.* **22**(5), 979–989.

Schabus, M. *et al.* (2007). Hemodynamic cerebral correlates of sleep spindles during human non-rapid eye movement sleep. *Proc. Natl. Acad. Sci. U.S.A* **104**(32), 13164–13169.

Schabus, M. (2009). Still missing some significant ingredients. *Sleep* **32**(3), 291–293.

Schenck, C. H. *et al.* (1986). Chronic behavioral disorders of human REM sleep: a new category of parasomnia. *Sleep* **9**(2), 293–308.

Schwartz, S., and Maquet, P. (2002). Sleep imaging and the neuro-psychological assessment of dreams. *Trends Cogn. Sci.* **6**(1), 23–30.

Solms, M. (1997). The Neuropsychology of Dreams. A Clinico-Anatomical Study. Lawrence Erlbaum Associates, Mahwah, p. 292.

Steriade, M., and McCarley, R. W. (1990). Brainstem Control of Wakefulness and Sleep. Plenum Press, New York, p. 499.

Steriade, M., and McCarley, R. W. (2005a). Brain Control of Wakefulness and Sleep. Kluwer Academic, New York, p. 728.

Steriade, M., and McCarley, R. W. (2005b). Brain Control of Wakefulness and Sleep. Springer, New York

Steriade, M., Nunez, A., and Amzica, F. (1993a). A novel slow (<1 Hz) oscillation of neocortical neurons *in vivo*: depolarizing and hyperpolarizing components. *J. Neurosci.* **13**(8), 3252–3265.

Steriade, M., Nunez, A., and Amzica, F. (1993b). Intracellular analysis of relations between the slow (<1 Hz) neocortical oscillation and other sleep rhythms of the electroencephalogram. *J. Neurosci.* **13**(8), 3266–3283.

Steriade, M., Timofeev, I., and Grenier, F. (2001). Natural waking and sleep states: a view from inside neocortical neurons. *J. Neurophysiol.* **85**(5), 1969–1985.

Sterzer, P., and Kleinschmidt, A. (2005). A neural signature of colour and luminance correspondence in bistable apparent motion. *Eur. J. Neurosci.* **21**(11), 3097–3106.

Stickgold, R., Pace-Schott, E., and Hobson, J. A. (1994). A new paradigm for dream research: mentation reports following spontaneous arousal from REM and NREM sleep recorded in a home setting. *Conscious. Cogn.* **3**(1), 16–29.

Strauch, I., and Meier, B. (1996). In Search of Dreams. Results of Experimental Dream Research. SUNY Press, New York, p. 252.

Timofeev, I. *et al.* (2000). Origin of slow cortical oscillations in deafferented cortical slabs. *Cereb. Cortex.* **10**(12), 1185–1199.

Tong, F., Meng, M., and Blake, R. (2006). Neural bases of binocular rivalry. *Trends Cogn. Sci.* **10**(11), 502–511.

Vyazovskiy, V. V. *et al.* (2009). Cortical firing and sleep homeostasis. *Neuron* **63**(6), 865–878.

Wehrle, R. *et al.* (2005). Rapid eye movement-related brain activation in human sleep: a functional magnetic resonance imaging study. *NeuroReport* **16**(8), 853–857.

MEMORY CONSOLIDATION, THE DIURNAL RHYTHM OF CORTISOL, AND THE NATURE OF DREAMS: A NEW HYPOTHESIS

Jessica D. Payne

Department of Psychology, University of Notre Dame, Notre Dame, IN, USA

I am not sure where I am, but I'm floating with a friend in what looks like a murky ocean, or perhaps it is a gigantic lake. The water is turbid but extremely deep, and there is no place to touch bottom. Waterlogged sticks and pieces of wood float on the surface, but none look strong enough to support a human's weight. We swim for what seems a frustrated eternity, until we finally find a log that will hold us. We cling to it with our arms while our torsos and legs remain underwater. The water is cleaner "out to sea" but more cluttered with debris "toward the edge." But these terms are meaningless because there is no shore, no horizon, and no sense of space other than the clarity of the water in one direction and the unsettling presence of flotsam and jetsam in the other. I am afraid to swim in the direction of the debris because I don't know what's under it. Color is almost

101

entirely missing, except for a grayish-blue (under us, above us) and a sickly, perpetually wet brown. Sepia tones that render the dream sinister and sad. . .empty. After days/weeks (?) of paddling, we come to a strange doorway. On the other side float another two people, strangely familiar, perhaps doppelgangers on a different log. They wear no expression, but I think one of them says, "We've been here for 3 years." Panic shoots through me because I suddenly know the place is shoreless. We join forces and start looking for land more earnestly, but there is no land. Not being able to stand up or lie down is torture, and I desperately need to stretch. As this feeling grows more intense, I realize that I will soon go mad. The terror that has been building throughout the dream finally escalates to the point of waking me up.

This dream includes many characteristic features of dreaming (Hobson, 1988), including (1) sensory fragments that get woven into cognitively bizarre themes, (2) irrational content and organization, where the unities of time, place, and person are disjointed and physical laws are disobeyed, (3) the uncritical acceptance of such themes as normal within the dream, and (4) emotion so intense that it is capable of penetrating, changing, or abruptly terminating the dream state. Another feature of dreaming, that dreams are notoriously fleeting and difficult to remember, did not apply here, perhaps because of the highly emotional content in this dream (Goodenough, 1978).

While these characteristics capture the general flavor of dreams, defining precisely what dreams are, where they come from, and what, if any, purpose they serve has proven more difficult. Although there is currently no convincing explanation for why we dream, dreaming is a universal human experience that has been studied for centuries. Until the mid-1900s, however, the study of dreams relied almost exclusively on subjective reports. Not until Aserinsky and Kleitman's (1953) discovery, that dreaming is often associated with Rapid Eye Movement (REM) sleep, did the study of dreaming become more objective. Since this discovery, researchers have been searching for the neurobiological underpinnings of the bizarre, fragmented, and often highly emotional aspects of dreams. Advances in cognitive neuroscience have furthered this goal by providing a unique window on neurocognitive processes, and recent dream studies suggest that at no other time in the diurnal cycle do such large variations in thought and imagery vary so systematically with changes in the central nervous system (Antrobus, 1993; Fosse *et al.*, 2001). It thus seems promising that dreams can be explicated in terms of underlying neurobiological mechanisms (Hobson, 2009; Payne and Nadel, 2004; Wamsley *et al.*, 2010), but the search for such mechanisms is still ongoing.

The most influential contemporary theory of dreaming undoubtedly belongs to Hobson (1988, 2009; Hobson and McCarley, 1977). Hobson's theory, known as the activation-synthesis, and, more recently, AIM (activation, input, modulation) model, is a brain-based account of REM sleep dreaming, where "activation" refers to the automatic intensification and reduction of brain activity as it progresses through various sleep states. During REM sleep, the activated brainstem generates signals that randomly stimulate the cortex, a process that results in the uniquely bizarre nature of dreams (Hobson, 1988). "Synthesis," the second piece of Hobson's theory, refers to the process that weaves these bizarre and discordant dream images into a "best possible fit" narrative by the activated brain. Although Hobson has since expanded his theory (see Hobson, 2009), it has always maintained the above two pivotal ideas: activation, which is provided by the brain stem, and synthesis or integration, which is governed by the forebrain.

I. Memory Consolidation, the Diurnal Rhythm of Cortisol, and the Formal Features of Dreaming: A New Hypothesis

The purpose of this chapter is to put forward a new, brain-based dream hypothesis, one that focuses on the interrelationships among sleep, memory consolidation, and diurnal release of cortisol during late night sleep. Many have argued, as I do here, that dreams are a reflection of the memory consolidation process (e.g., Foulkes, 1985). What has not yet been considered is the influence of the neuroendocrine system and specifically stress hormone release on memories as they are consolidated during sleep (Payne and Nadel, 2004). Cortisol is a stress hormone that has widespread effects on mnemonic and emotional functioning during wakefulness (Payne and Nadel, 2004; Payne et al., 2002, 2004, 2006, 2007; see de Quervain et al., 2009 for recent review). I posit here that neural regions associated with the processing of memories and emotions (i.e., hippocampus, amygdala, regions of prefrontal cortex), via their unique modulation and activation by stress hormones, give rise to the nature of dreams. Specifically, I suggest that (1) memories will be processed differently depending on the diurnal release of cortisol, (2) memories formed under conditions of elevated cortisol share important similarities with dreams, and (3) nightly fluctuations in cortisol secretion not only help explain the formal features of dreams, but may also contribute to the general brain activation necessary for dreaming to reach conscious awareness. In this sense, HPA activation and cortisol secretion late in the sleep cycle may contribute, perhaps causally, not only to our experience of dreams, but also to dream production.

Before expanding upon these ideas, I will first review several lines of evidence. A critical claim of the current account is that sleep participates fundamentally in memory consolidation. Thus, *Section II* will examine the role of sleep in the consolidation of memories. It demonstrates that different types of memory are preferentially consolidated during different stages of sleep, and argues that this difference may stem in part from nightly fluctuations in the stress hormone cortisol. *Section III* examines the literature on stress hormones and memory during wakefulness to show that, as in sleep, cortisol has disparate effects on memory depending on emotional valence of the experience. *Section IV* argues that the features of memories formed when cortisol is elevated during wakefulness are strikingly similar to the features of late night sleep dreams—a parallel that I argue is produced by cortisol's influence on memories processed during both wakefulness and sleep. Finally, *Section V* explores how these disparate lines of evidence converge on new predictions about dreaming.

II. Sleep and Memory: The Case for Consolidation

A. STAGES OF SLEEP

There are two main types of sleep. The first, rapid eye movement (REM) sleep, occurs in roughly 90-minute cycles and alternates with four additional stages (Stages 1–4) known collectively as NREM sleep, and which comprise the second type of sleep (Aserinsky and Kleitman, 1953). Slow wave sleep (SWS) is the deepest of the NREM phases, and is characterized by high-amplitude, low frequency EEG oscillations. REM sleep, on the other hand, is lighter stage of sleep characterized by eye movements, decreased muscle tone, and low-amplitude, fast electroencephalographic (EEG) oscillations. More than 80% of SWS is concentrated in the first half of the typical 8-hour night, whereas the second half of the night contains roughly twice as much REM sleep than the first half (see Fig. 1). This domination of early sleep by SWS, and of late sleep by REM sleep likely has important functional consequences, but also makes it difficult to know which distinction is critical: NREM sleep versus REM sleep or early versus late sleep.

Neurotransmitters, particularly the monoamines (serotonin, 5-HT, and nor-epinehprine, NE) and acetylcholine, play a critical role in switching the brain from one sleep stage to another. REM sleep occurs when activity in the aminergic system has decreased enough to allow the reticular system to escape its inhibitory influence (Hobson, 1988). The release from aminergic inhibition stimulates cholinergic reticular neurons in the brainstem and switches the sleeping brain into the highly active REM state, in which acetylcholine levels are as high as in the waking state. REM sleep is also associated with higher levels of cortisol than NREM overall, especially late in the sleep cycle (Lavie, 1996; Wagner and Born, 2008; see also Fig. 3). 5-HT and NE, on the other hand, are virtually absent

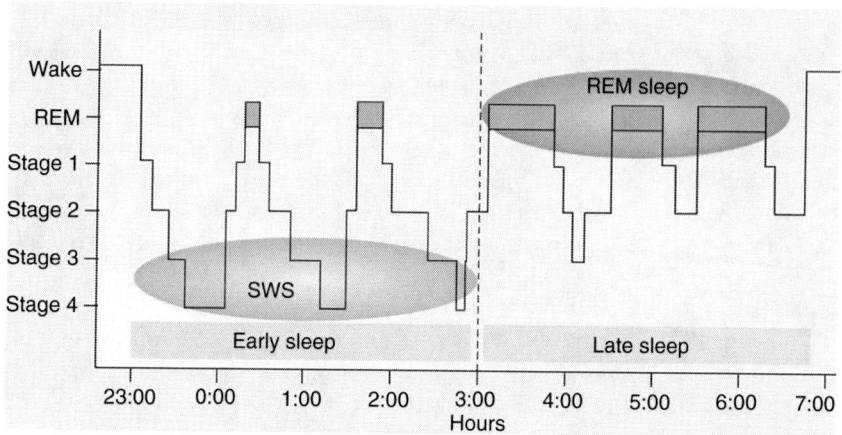

Fɪɢ. 1. Sleep histogram showing the distribution of SWS and REM sleep.

during REM. SWS, conversely, is associated with an absence of acetylcholine and low levels of cortisol but nearly normal levels of 5-HT and NE (Hobson *et al.*, 2000).

B. Stages of Memory Formation and Different Memory Types

Memory consolidation is the process by which newly acquired information, initially fragile, is integrated and stabilized into long-term memory (McGaugh, 2000). Evidence overwhelmingly suggests that sleep plays a role in the consolidation of a range of memory tasks, with the different stages of sleep selectively benefiting the consolidation of different types of memory (Diekelmann and Born, 2010; Payne *et al.*, 2008; Smith, 1995; Stickgold, 2005; Walker, 2009).

Most taxonomies break memory down into key types (Schacter and Tulving, 1994; Squire and Zola-Morgan, 1991), and several are important for our purposes here. First, there are various types of memory that we can recall explicitly, including *episodic* memories, or memories of the events in our lives, and *semantic* memories, which consist of knowledge (e.g., facts, word meanings) that has been uncoupled from place and time (Tulving, 1983). Unlike retrieving a semantic memory, retrieving an episodic memory from one's past requires access to defining contextual features of the event, such as specific details about the place of its occurrence. Because of this emphasis on space and context, *episodic* memories and *spatial* memories are closely connected. Second, there are "how to" memories for the various skills, procedures, and habits we acquire through experience. Because these memories are not so easily made explicit and are usually only evident in behavior, they are referred to as *procedural* or *implicit* memories. Third, there are *emotional memories* for the positive and negative experiences in our lives. This class of memories is mediated by a system that is particularly concerned with learning about fearful and negative stimuli, although evidence suggests it plays a role in memory for pleasant information as well (e.g., Hamann, 2001).

Each of these memory types is subserved by distinct neural systems (Schacter and Tulving, 1994). While episodic and spatial memories are governed by the hippocampus and surrounding medial temporal areas, procedural or implicit memories are thought to be independent of the hippocampus and anatomically related regions, relying instead on various neocortical and subcortical structures (Schacter and Tulving, 1994; Squire and Zola-Morgan, 1991). The emotional memory system is centered in the amygdala, a limbic structure that is richly connected to the hippocampus.

C. Sleep's Role in Episodic Memory Consolidation

Memories are processed during most, if not all, stages of sleep, and the neurochemical milieu of the brain has important consequences for memory consolidation. Although there is substantial evidence that sleep benefits the consolidation of procedural memories (see Smith, 1995, 2001; Walker and Stickgold, 2006 for excellent reviews), the evidence that sleep is critical for explicit episodic and emotional memories is most relevant to the current hypothesis on dreams.

There is a general consensus that NREM sleep, especially SWS, is essential for the consolidation of hippocampus-dependent episodic and spatial memories, whereas REM sleep is more important for procedural and emotional memory consolidation (see Ellenbogen *et al.*, 2006; Marshall and Born, 2007; Payne *et al.*, 2008 for review). In a landmark study, Plihal and Born (1997) capitalized on the unequal distribution of SWS and REM sleep to assess the recall of word pairs (an episodic memory task) and improvement in mirror-tracing (a procedural memory task) after retention intervals of early sleep (the first 3–4 hours of the sleep cycle), and late sleep (the last 3–4 hours of the sleep cycle).

Recall of word pairs improved significantly more after a 3-hour sleep period rich in SWS than after a 3-hour sleep period rich in REM or a 3-hour period of wakefulness. Mirror tracing, on the other hand, improved significantly more after a 3-hour sleep period rich in REM than after 3 hours spent either in SWS or awake. These findings dovetail nicely with older studies of sleep and episodic memory, which showed that sleep rich in SWS produced less forgetting of episodic memory materials than sleep rich in REM (Barrett and Ekstrand, 1972; Fowler *et al.*, 1973; Yaroush and Sullivan, 1971).

Similarly, using a nap paradigm, Tucker *et al.* (2006) found that naps containing only NREM sleep enhanced memory for word pairs, but did not benefit mirror tracing. Following an afternoon training session, performance on these tasks was assessed after a 6 hour delay, either with or without an intervening nap. Not only did the nap subjects recall more word pairs than the subjects who remained awake, but they also showed a weak correlation between improved recall and the amount of SWS in the nap.

These results are consistent with neurophysiological evidence derived from electrophysiological studies in rodents, which demonstrate that patterns of hippocampal place cell activity first seen during waking exploration are later re-expressed during post-learning SWS (Wilson and McNaughton, 1994; Ji and Wilson, 2007 reviewed in O'Neil *et al.*, 2010). Consistent with these findings is a PET study in humans (Peigneux *et al.* 2004), which demonstrated that learning-related hippocampal activity seen while training on a virtual maze task was again expressed during post-learning SWS, and importantly, this hippocampal reactivation during sleep strongly predicted overnight improvement on the task. Similarly, Rasch *et al.* (2007) exposed human subjects to an odor while they were learning object-location pairings in a task similar to the memory game "concentration"; subjects who were re-exposed to the odor during SWS (but not REM sleep) showed enhanced hippocampal activity and enhanced memory for the memory pairings. Likewise, in some of the strongest evidence for to date, Marshall *et al.* (2006) showed that hippocampus-dependent memories are specifically enhanced when slow oscillations (slow, $<1\,Hz$ oscillatory activity during SWS) are induced during sleep by transcranial electrical stimulation. These observations suggest that learning triggers the reactivation and reorganization of memory traces during NREM SWS, a systems-level process that in turn enhances behavioral performance.

D. Sleep's Role in Emotional Memory Consolidation

Emotional memories, which rely critically on the amygdala for their consolidation, appear to benefit most from REM sleep (see Payne and Kensinger, Current Directions in Psychological Science for review). Wagner *et al.* (2001) found that 3 hours of late night, REM-rich sleep (but not 3 hours of early night slow-wave rich sleep or 3 hours of wakefulness) facilitated memory for negative arousing narratives, an effect that could still be observed years later when the subjects were re-contacted for a follow-up memory test (Wagner *et al.*, 2006). Consistent with these findings, the amygdala and hippocampus are among the most active brain regions during REM sleep, with some evidence suggesting that they are more active during REM sleep than during wakefulness (Maquet *et al.* 1996). This suggests that emotional memory processing may be a primary function of REM sleep. Moreover, several studies have correlated features of REM sleep, including oscillatory activity in the theta frequency band range (Nishida *et al.*, 2009), with enhanced emotional memory consolidation (see Walker, 2009 for review). These findings strongly suggest a role for sleep, especially REM sleep, in the processing of memory for emotional experiences.

E. Sleep Transforms Memories in Useful Ways

The findings reviewed above provide compelling evidence that sleep plays an important role solidifying experience into long-term memory in a veridical manner, more or less true to its form at initial encoding. However, it has long been known that memories *change* with the passage of time (Bartlett, 1932), suggesting that the process of consolidation does not always yield exact representations of past experiences. On the face of it, this may seem strikingly maladaptive, yet such flexibility in memory representation allows the emergence of key cognitive abilities, such as generalization and inference (Ellenbogen *et al.*, 2007), future thought (Schacter *et al.*, 2008), and the selective preservation of useful information extracted from a barrage of incoming stimulation and experience (Payne *et al.*, 2009). Consistent with these ideas, growing evidence suggests that sleep does more than simply consolidate memories in veridical form; it also transforms them in ways rendering memories less accurate in some respects, but more useful and adaptive in the long run. Sleep leads to flexible restructuring of memory traces so that insights can be made (Wagner *et al.*, 2004), inferences can be drawn (Ellenbogen *et al.*, 2007), and integration and abstraction can occur (Payne *et al.*, 2009). In each of these cases, sleep confers a flexibility to memory that may be at times more advantageous than a literal representation of experience.

As a specific example of such qualitative changes in memory representation, recent studies demonstrate that sleep transforms the emotional memory trace. Payne *et al.* (2008) examined how the different *components* of complex negative arousing memories change across periods of sleep versus wakefulness. Emotional scenes could be stored as intact units, suffering some forgetting over time but retaining the same relative vividness for all components. Alternatively, the components of an experience could undergo differential memory processing, perhaps with a selective emphasis on what is most salient and worthy of remembering.

Participants viewed scenes depicting negative or neutral objects embedded on neutral backgrounds at 9 am or 9 pm (see Fig. 2 for e.g., stimuli). Twelve hours later, after a day spent awake or a night including at least 6 hours of sleep, they were tested on their memory for objects and backgrounds separately to examine how these individual components of emotional memories change across periods of sleep and wake. Daytime wakefulness led to forgetting of negative arousing scenes in their entirety, with both objects and backgrounds suffering forgetting at similar rates. Sleep, however, led to a selective preservation of negative objects, but not their accompanying backgrounds, suggesting that the two components undergo differential processing during sleep. This finding suggests that, rather than preserving intact representations of scenes, the sleeping brain effectively "unbinds" scenes to consolidate only their most emotionally salient, and perhaps adaptive, emotional element (see Payne *et al.*, 2009; Wagner *et al.*, 2004 for additional examples of unbinding during sleep).

Paralleling these behavioral findings, an fMRI study provided evidence that a single night of sleep is sufficient to provoke changes in the emotional memory circuitry, leading to increased activity within the amygdala and the ventromedial prefrontal cortex, and resulting in strengthened connectivity between the amygdala and both the hippocampus and the ventromedial prefrontal cortex (Payne

FIG. 2. Examples of the scenes presented to subjects. Eight versions of each scene were created by combining each of four similar objects (two neutral objects, two negative emotional objects) with each of two plausible neutral backgrounds. In this example, the two neutral central objects are cars, and the two negative central objects are cars badly damaged in an accident; the neutral backgrounds are street scenes. Two of the eight versions of the completed scene are shown.

and Kensinger, in press, Journal of Cognitive Neuroscience). These findings are consistent with a study by Sterpenich *et al.* (2009) and suggest that sleep strengthens the modulatory effect of the amygdala on other regions of the emotional memory network as memories undergo consolidation (McGaugh, 2004).

F. Neurophysiological and Neurochemical Evidence for Sleep's Role in Memory Consolidation

Each of the sleep stages is characterized by a unique collection of electrophysiological, neurotransmitter, and neuroendocrine properties that tend to overlap with the different sleep stages, but are not perfectly correlated with them. For example, SWS is associated with cortical slow oscillations (slow, <1 Hz oscillatory activity during SWS), sleep spindles (faster, 11–16 Hz, bursts of coherent brain activity, and hippocampal sharp-wave "ripple" complexes (~200 Hz)—all of which have been associated with episodic memory consolidation. Indeed, the co-occurrence of these electrophysiological events may underlie the coordinated information flow back and forth between hippocampus and neocortex as memories are integrated within neocortical long-term storage sites (e.g., Buzsaki, 1996, 1998).

There is also evidence to suggest that nocturnal changes in neurotransmitter and neurohormone levels contribute to memory consolidation. Acetylcholine, norepinephrine, serotonin, and cortisol all play important roles both in modulating sleep (Hobson *et al.*, 2000) and in memory function (Cahill and McGaugh, 1998; Hasselmo, 1999; Payne *et al.*, 2004). Cortisol, for instance, follows a marked circadian rhythm where it is at its nadir during early night, slow-wave rich sleep and reaches its zenith during late night, REM-rich sleep. Indeed, the difference between the cortisol level in the blood at sleep onset and at awakening is so great that the interpretation of cortisol blood levels is meaningless without knowing exactly when the sample was taken (Lavie, 1996; Weitzman *et al.*, 1971). Moreover, the secretion of cortisol is not continuous but comprised of gradually increasing peaks that tend to coincide with REM sleep episodes (see Fig. 3). REM sleep thus tends to co-occur with cortisol elevations (Lavie, 1996; Wagner and Born, 2008).

Interestingly, the early night reduction in acetylcholine and cortisol may be necessary for hippocampus-dependent memories to undergo effective consolidation, as experimentally elevating either substance during early sleep impairs performance on episodic memory tasks. Gais and Born (2004) trained subjects on word pair task and mirror tracing tasks before 3 hours of nocturnal sleep or wakefulness during which they received a placebo or an infusion of the cholinesterase inhibitor physostigmine (which increases cholinergic tone). When tested after 3 hours of early sleep rich in SWS, recall on the paired associates task was

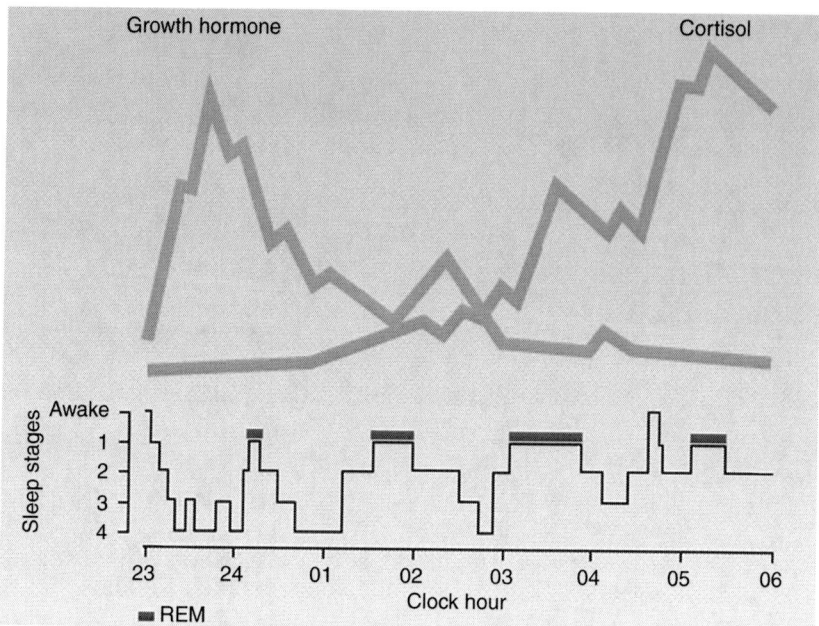

Fig. 3. The relationship between sleep-stage architecture and circulating levels of growth hormone and cortisol. Note both the linear increase in cortisol across the night and also the cortisol peaks riding on top of REM periods.

markedly impaired in the physostigmine group, while procedural memory performance was unaffected. Using a similar design, Plihal and Born (1999) showed that when cortisol was infused into the early, SWS-rich interval, retention of episodic information that is normally facilitated during this time was impaired. Thus, enhancing plasma cortisol concentrations during early sleep eradicated the benefit typically observed for episodic memory while leaving procedural memory unimpaired (see Fig. 4).

Plihal and Born (1999) concluded that as episodic, but not procedural, memory relies on hippocampal function, cortisol inhibition during early nocturnal sleep is necessary for episodic memory consolidation.[1] Thus, because cortisol release is inhibited during early night periods dense in SWS, this time window may provide the ideal physiological environment for episodic memory consolidation. REM sleep, on the other hand, is an inefficient time to consolidate episodes, due to the deleterious effect of elevated cortisol on hippocampus-dependent memory processing. Thus, the neurobiological properties of early sleep and late

---

[1] This study also specifies glucocorticoid receptors, as opposed to mineralocorticoid receptors, as being responsible for the observed effects.

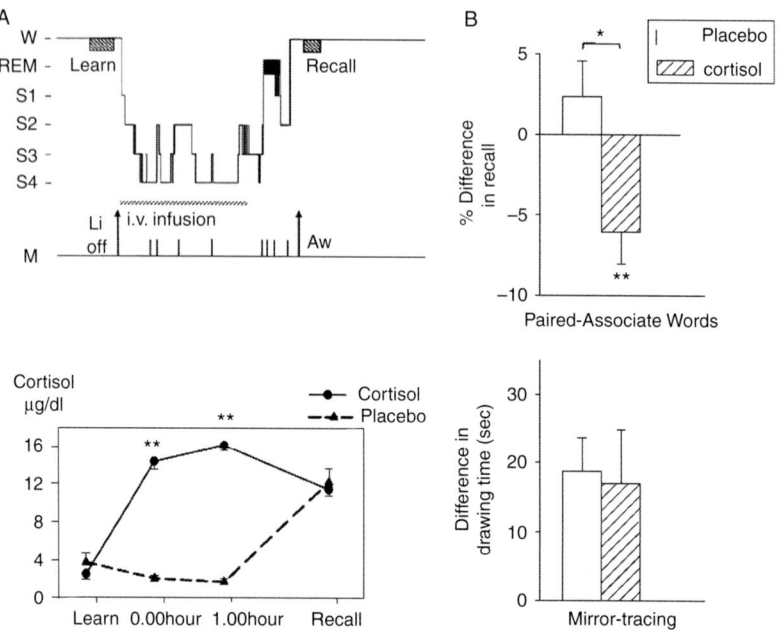

Fig. 4. (A) Top: Experimental design for examining the impact of cortisol on memory consolidation during a 3 hour period of early nocturnal sleep rich in SWS, illustrated by an individual sleep profile. Before sleep, subjects learned episodic and procedural memory tasks to a criterion. Recall was tested 15–30 minutes after awakening. Infusion of cortisol or placebo began at 11 pm and was discontinued after 2.5 hours. Bottom: Mean plasma cortisol concentrations during Placebo (dotted line) and Cortisol (solid line) conditions. (B) Cortisol infusion during sleep, compared with placebo, impaired hippocampus-dependent episodic memories for word pairs across sleep (top), but did not affect hippocampus-independent procedural memory (speed in mirror tracing, bottom). $N = 14$, $*p < .05$, $**p < .01$. Data from Plihal and Born (1999).

sleep, as opposed to SWS and REM sleep *per se*, may be essential for the consolidation of different types of memory.

In line with the above findings, cortisol elevations during wakefulness can also impair performance on episodic memory (de Quervain *et al.*, 2009). Interestingly, cortisol in the Plihal and Born (1999) study was elevated just enough to mimic the late night peak of circadian cortisol activity ($15.2 + 0.68\,\mathrm{mg/dl}$); this amount is proportionate to the cortisol typically released in response to a mild to moderate stressor (~ 10–$30\,\mathrm{mg/dl}$) and is a sufficient dose to disrupt episodic memory function during wakefulness (e.g., Kirschbaum *et al.*, 1996; Wagner and Born, 2008), particularly when administered at retrieval (de Quervain *et al.*, 2000). Thus, cortisol elevations seen during late night REM may help explain both why replay of episodic memories in REM sleep dreaming is so scarce (Baylor and

Cavallero, 2001; Fosse *et al.*, 2003) and why dreams are difficult to remember (Hobson, 1988).

III. Cortisol's Impact on Memory During Wakefulness

There is a substantial literature demonstrating that stress can produce cortisol elevations capable of altering memory function in animals and humans (de Quervain *et al.*, 2009; Kim and Diamond, 2002; Lupien *et al.*, 2009; Payne *et al.*, 2004; Roozendaal *et al.*, 2009). The hippocampus, amygdala, and memory relevant regions of the prefrontal cortex all have dense concentrations of receptors for cortisol. Elevated cortisol can impair the neuronal structure and function of the hippocampus[2] by altering hippocampal morphology, disrupting neurogenesis, and blocking the synaptic plasticity (e.g., long-term potentiation "LTP" and primed-burst potentiation "PBP") thought to underlie memory formation (McEwen, 2000). Inducing stress in the laboratory by exposing subjects to a brief, one-time stressor (e.g., a public speaking task, or the cold pressor task, which involves submerging the arm in cold water), or administering glucocorticoids directly, typically leads to impairments in episodic memory (e.g., Kirschbaum *et al.*, 1996; Payne *et al.*, 2002, 2006, 2007), particularly when cortisol is administered at retrieval (de Quervain *et al.*, 2000). As might be expected, it also disrupts spatial memory (Laurance *et al.*, unpublished data). Thus, cortisol elevations, including those consistent with increases seen during late night sleep, are capable of disrupting episodic memory function.

However, recent studies suggest that cortisol facilitates consolidation of emotional relative to neutral episodic memories (Payne *et al.*, 2007), and even emotional relative to neutral features within a complex episode (Payne *et al.*, 2006). This is similar to the previously discussed sleep and episodic memory finding by Wagner *et al.* (2001), showing that emotionally laden episodic memories were *facilitated* relative to neutral memories during late night, REM-rich sleep. Indeed, substantial evidence in animals and humans demonstrates that cortisol can selectively impair hippocampus-dependent neutral memories, while leaving emotional memories intact (Payne *et al.*, 2006) or even enhancing them (Buchanan and Lovallo, 2001; de Quervain *et al.*, 2009; Jelicic *et al.*, 2004; Payne *et al.*, 2007).

Paralleling these functional effects, animal research demonstrates that hippocampal structural plasticity suffers under high levels of cortisol, while amygdala

---

[2] It is important to note that extremely low levels of cortisol can disrupt hippocampal function as well. It thus appears that a moderate level of circulating glucocorticoids enhances memory whereas maximal or minimal levels disrupt it, perhaps explaining why some studies of stress and memory fail to find memory disruption.

plasticity is enhanced (Vyas *et al.*, 2002). Human neuroimaging studies likewise show that cortisol elevations impair hippocampal activity, while facilitating activity in the amygdala. For example, Pruessner *et al.* (2008) showed that acute stress induced prior to an encoding task caused significant deactivation in the hippocampus, and degree of hippocampal deactivation was significantly correlated with the cortisol stress response. van Stegeren *et al.* (2007), on the other hand, showed that elevated cortisol levels correlated with intensified amygdala activation at encoding and better memory for emotional information later on. Together, these findings provide a basis for the claim that stress hormone modulation of hippocampal activity underlies the damaging effects of stress on neutral episodic memory, while their modulation of the amygdala underlies the enhancing effects of stress on emotional events (see Payne *et al.*, 2004).

Cortisol's opposing effects on neutral and emotional memories are mainly observed in studies examining the memory consolidation phase specifically. For example, direct cortisol administration before or after encoding selectively enhances memory for emotionally arousing, but not neutral material (Buchanan and Lovallo, 2001; Cahill *et al.*, 2003; Okuda *et al.* 2004; Payne *et al.*, 2007). Moreover, cold pressor stress enhances memory for emotional slides but does not affect memory for neutral slides. Consistent with these findings, endogenous cortisol levels at encoding correlate with enhanced memory consolidation only in individuals who were emotionally aroused. Furthermore, Payne *et al.* (2007) showed that psychosocial stress (elicited by a public speaking task called the Trier Social Stress Test, or TSST), which elevates cortisol (Fig. 5, right), enhances memory for an emotionally arousing story while impairing memory for a closely matched neutral story (see Fig. 5, left). Interactions between cortisol and amygdala activity are likely key to determining this selectivity (de Quervain *et al.*, 2009).

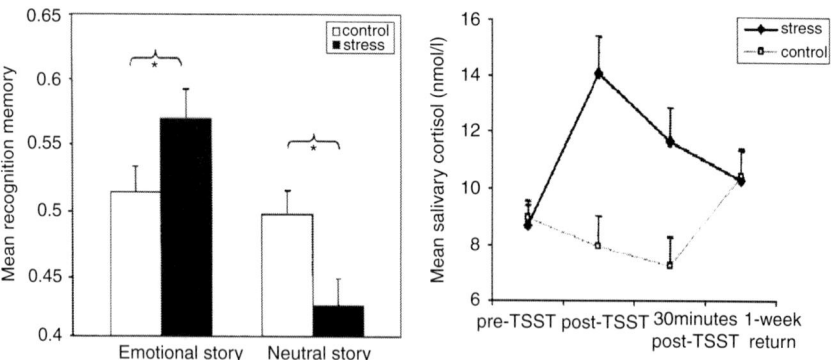

FIG. 5. Stress enhances the consolidation of negative emotional information, but disrupts the consolidation of neutral information (left). The Trier Social Stress Test (TSST), a social stressor involving public speaking, increases cortisol levels compared to a control condition (right). From Payne *et al.* (2007).

IV. A Clinical View of Memory Under Stress

Before connecting these studies of cortisol and waking memory with the realm of sleep and dreams, we will next examine the clinical characteristics of memories formed under stressful or traumatic conditions. A discussion of stress and memory from a clinical viewpoint serves a twofold purpose: It deepens our understanding of the cortisol-based memory deficits reviewed above, and serves as a bridge between the waking relationship between cortisol and memory and the relationship cortisol might have to sleep and dreaming. A close look at these clinical characteristics reveals memories that share much in common with late-sleep dreams; they are often bizarre, fragmented, lacking in spatial and temporal context, highly emotional, and difficult to remember.

A. CORTISOL, THE HIPPOCAMPUS, AND FRAGMENTED MEMORIES OF TRAUMATIC EXPERIENCE

Retrieving an episodic memory requires one to access and integrate multiple fragments of experience stored in disparate memory systems—what happened, who engaged in which actions, what it all looked like, sounded like, smelled like, etc. In addition to this episodic "content," there is also spatial-contextual information. Because all events by definition occur *someplace*, it can be difficult if not impossible to divorce memories of episodes from memories of context, a notion that led us to propose that context serves as an organizing frame to which the various elements of an episodic memory trace are attached (Nadel and Payne, 2002).

Consider a hypothetical traumatic war experience: a memory of this experience might include smells (the jungle, unwashed bodies), sounds (gunfire and explosions), sights (the flash of a sniper's weapon, the sight of a seriously wounded friend), tactile feelings (the humidity, pain from a wound), actions (diving for cover, returning fire), and emotions (fear, anger, guilt). Each of these independent features is stored in the relevant part of the brain, typically, but not exclusively, in the neocortex. The hippocampus and adjacent medial temporal regions are critical for binding these disparate fragments of an episode from multiple brain regions into a unified memory trace (e.g., Cohen and Eichenbaum, 1994; Schacter and Tulving, 1994). Regions of the prefrontal cortex may also share, with the hippocampus, responsibilities for binding in episodic memory (e.g., Mitchell *et al.*, 2000).

The idea that memories are disaggregated during storage and then re-aggregated during retrieval has several implications for understanding interactions between stress and memory (Jacobs and Nadel, 1998). By disrupting the hippocampus and prefrontal cortical-based contextual representation system, stress may impair memory for time, space, and other contextual information. By disrupting the hippocampus and prefrontal cortical-based binding function, stress may leave the various aspects of a memory trace disconnected. Thus, memories formed under high levels of stress and cortisol may be disjointed, fragmented, and lacking in detail and spatial or temporal context, thus creating specific deficits in episodic memory.

Indeed, clinical reports suggest that memory for stressful experiences lack coherence, context, and episodic detail and can be highly fragmented (Bremner, 1999; Murray *et al.*, 2002; van der Kolk, 1991, 1997, 1998; van der Kolk and Fisler, 1995; Verfaellie and Vasterling, 2009). In a clinical sense, the term "fragmented" means that the various bits and pieces of experience are not related to one another as a whole, and that the various features of memory fail to be bound together to produce a "good gestalt," or a coherent episode (Jacobs and Nadel, 1998; Payne *et al.*, 2004; van der Kolk, 1997, 1998). For example, memory fragmentation is an important feature of post-traumatic stress disorder (PTSD), in which patients describe gaps in their memories, not only of the trauma, but of other personal experiences as well (Bremner, 1999).

Rather than simply retrieving fragments and reporting them as such, some individuals make educated guesses about memory in a process we have called narrative smoothing (Jacobs and Nadel, 1998; Payne *et al.*, 2004). Constructing narratives is, to an extent, a normal function of human memory. The hippocampal system, rather than creating inflexible representations that serve as permanent records of events, stores representations from which episodic memories are recreated from the various attributes or features of the event (Cohen and Eichenbaum, 2004). In this way, memory is fundamentally reconstructive. But in the presence of high levels of stress hormones, this normal process can be taken to an extreme. In both clinical settings and the laboratory, recombining these fragments in an attempt to make sense of one's experience can lead to inconsistencies in one's recall or even blatantly "false" memories (e.g., Murray *et al.*, 2002; Payne *et al.*, 2002, 2004, 2006).

B. Cortisol, the Amygdala, and Enhanced Memory for Emotional Aspects of Traumatic Events

The amygdala is highly active during emotionally charged, stressful experiences and the memories dependent on this region are highly resistant to forgetting (LeDoux, 2002). This "stamping in" of emotional memories is at least partly mediated by stress hormones such as cortisol (and norepinephrine—see McGaugh, 2000, 2004 for review) and is another core feature of PTSD. Hormones released during times of stress thus have opposing effects on different neural structures, and consequently opposing effects on the types of memories subserved by these structures. This is highlighted in the following quote from a Vietnam veteran suffering from PTSD:

> Parts of my memory are very vivid, especially the emotional parts. At other times, there are these weird holes. I can't make the memories make sense together like in a story. I don't know how to say it. It's like my memories [for the event] are fragments...puzzle pieces, but I can't figure out how to put the puzzle together.

In the absence of an intact hippocampus-based memory system, the amygdala-based system is left to store emotional information that is not bound to the spatio-temporal context within which the relevant events occurred. This results in a pool of isolated memory fragments, many of which are emotional, that have been encoded without a coherent spatio-temporal frame to organize them (Jacobs and Nadel, 1998). Thus, the stress encountered during traumatic events does not lead to a complete eradication of memory. Rather, it leads to the storage of fragments that lack a spatial/contextual framework to bind and define them as belonging to an individual episode. This binding failure leads to one of two outcomes. As in the case of the veteran above, the memory fragments surface as disconnected images, feelings, or sensations (referred to as "body memories" by Van der Kolk, 1991), or they elicit narrative smoothing, where retrieved memory fragments are cobbled together by a narrative based upon gist, inference and educated guesswork, often guided by preserved emotional information (Jacobs and Nadel, 1998; Nadel and Jacobs, 1998; Payne et al., 2004). Although these narratives can be logical and similar to the real experience, more often they are bizarre and distorted. Indeed, we have previously argued that schematization of disconnected features underlies many cases of false memory reported in the media and the scientific literature (e.g., Jacobs and Nadel, 1998; Nadel and Payne, 2002b; Payne et al., 2004).

The above account offers a neurobiological explanation for (1) why memories laid down under high levels of stress can be fragmented or forgotten, (2) why preserved fragments are often emotional and remembered exceptionally well, and (3) how memories can emerge as coherent, if distorted and bizarre, emotionally guided reconstructions of personal experience. If high levels of cortisol can influence memory processing in this manner during wakefulness, it follows that similar influences may be at work during sleep. I suggest that as memories undergo sleep-based consolidation, the presence of elevated cortisol during late-night sleep may contribute to the subjective characteristics of dreaming. As readers will no doubt have noted, dream features are highly similar to the features of waking memories that were formed in the presence of high levels of stress.

V. Tying It All Together: Toward a New Hypothesis of Dreaming

It is generally assumed that long-term memory consolidation involves inter-actions among multiple brain systems, modulated by various neurotransmitters and neurohormones. I propose that the characteristics of dreams are best under-stood in the context of this neuromodulatory impact on brain systems. Although a number of neurotransmitters and neurohormones are likely involved, I argue that the relationship between late night elevations in cortisol and explicit memory consolidation have important consequences for dreams: it produces fragmented dreams, gives dreams their uniquely bizarre flavor, accounts for their emotional nature, and explains not only why veridical replay of episodic memories during dreaming is rare, but also why dreams are so fleeting and difficult to remember. While many researchers have argued, as I argue here, that sleep is important for the consolidation of memories, I suggest that memory consolidation proceeds differently depending on changes in the neurochemical milieu of the sleeping brain. This, in turn, means that our experience of dreams should differ depending on when they occur.

A. CORTISOL AND THE FORMAL FEATURES OF DREAMING

1. *Dreams are Fragmented*

Neurochemical properties, rather than sleep stages *per se*, determine how memories are processed, and the degree to which they are experienced as coherent "units" as opposed to fragments. As in wakefulness, cortisol elevations during late night sleep disrupt the ability of the hippocampus and PFC to contribute contextual information to a memory and bind its elements into a coherent whole. This leads to activation of memory fragments in the absence of contribution from the structures that normally contextualize and connect them. It is these memory fragments, which are stored in dispersed neocortical regions, that compose the disconnected sounds and images and bizarre plot lines that constitute dreams. When we become conscious of this altered memory proces-sing, we experience a typical REM sleep dream.

Although dreams can be extremely fragmented, they are not experienced as completely random sequences of associated images. Rather, they exhibit varying degrees of thematic coherence (Cipolli, 1995; Foulkes, 1985). The cortisol notion of dreaming accounts for narratization in dreams in the same way it accounts for narratization in wakefulness. When the sleeping brain is confronted with frag-mented information, it automatically attempts to synthesize them into narrative themes.

The neurohormonal milieu of during late night sleep can explain why we experience memories in our dreams as far removed from the waking episodes upon which they are based, rather than re-experiencing actual episodes (Baylor and Cavellero, 2001). This idea also helps resolve a paradox in dream research—that memory function is profoundly impaired in dreams and yet memories must be processed during sleep because dreams consist of memory fragments (Hobson *et al.*, 2000). Indeed, Hobson and colleagues have argued that a deficiency of memory in dreaming goes "a long way" toward explaining the characteristics of dreaming (Hobson *et al.*, 2000). The cortisol account explains how memory processing under specific conditions gives rise to dream character- istics. Memories are processed during sleep and give rise to dreams, and are also profoundly altered in sleep periods most associated with dreaming. By the cortisol account, these things are not mutually exclusive.

Not all dreams with these features have to be REM dreams. While such dreams may be most prevalent during REM sleep, NREM dreams occur more frequently than was once believed (Nielsen, 2000). Foulkes (1985) and Solms (2000) have both argued against a simple REM sleep = dreaming perspective, and are proponents of NREM dreaming. By simply changing the question asked of awakened subjects from "Did you dream?" to "Did you experience any mental content?," Foulkes was able to show a far higher percentage of dream reports from NREM stages than original studies had suggested (e.g., Aserinsky and Kleitman, 1953). These dream reports after NREM awakenings led Foulkes and others to conclude that the stream of consciousness never ceases during sleep and that the brain engages in cognitive activity during all sleep stages (also see Antrobus, 1991, 1993).

Some NREM dreams are similar in content to REM dreams, and the majority of these come from NREM periods occurring early in the morning (Cicogna *et al.*, 1998; Kondo *et al.*, 1989), during the peak phase of the diurnal rhythm when cortisol levels are at their zenith. Interestingly, Antrobus has demonstrated that dreaming is associated with two independent sources of neural activation. The first, as Hobson has consistently shown, is associated with the REM–NREM cycling that is governed by the brainstem. The second is asso- ciated with the rising morning phase of the diurnal rhythm. This is the period during which cortisol levels are at their highest, and one tantalizing but spec- ulative possibility is that cortisol contributes directly to neural activation as the sleep cycle nears its end. Kondo *et al.* (1989) have shown that the two sources of activation are additive, so that late REM episodes are associated with more vividness of visual imagery, more bizarreness, and more discontinuities of person, place and time than are earlier REM episodes (Antrobus, 1991). There is also evidence that NREM dreaming not only increases but that NREM dream reports become more vivid and disjointed during the early morning hours (Antrobus, 1991; Cicogna *et al.*, 1998). In fact, most of the characteristic features of dreaming intensify across the night in both REM and NREM sleep (Wamsley *et al.*, 2007).

Note that these are precisely the qualities of dreaming we would expect to intensify as cortisol levels become elevated in the morning. I argue that a simple increase in cortical arousal alone would be unlikely to produce the bizarre and fragmented features of dreaming—if anything, an activated brain (i.e., during wakefulness) is thought to process information much more smoothly and effectively. Thus, it is much more likely that escalating cortisol, as it takes its toll on ongoing memory processes, helps shape the features of our dreams and perhaps contributes to general cortical arousal as well. Antrobus (1991) notes that although dream features (such as fragmentation and bizarreness) are regarded as the most salient characteristics of dreaming, they are not particularly prominent in laboratory REM reports *unless they are obtained in the late morning hours* (Kondo *et al.*, 1989). Given this, it seems likely that the rising morning phase of the diurnal rhythm, which is associated with elevated cortisol, contributes seminally to many of the hallmark features of dreams.

2. *Sensory Dream Fragments Produce Bizarre Cognitive Themes*

Though fragmentation and bizarreness are sometimes treated as separate features of dreaming, the cortisol hypothesis sees them as intimately intertwined. The recombination of activated memory fragments into unusual narratives is what produces the experience of bizarreness. Why, however, are some dreams particularly fragmented and bizarre whereas others are less so? By the current view, memory processing that occurs when cortisol is elevated should lead to dreams that are more fragmented and/or bizarre, while memory consolidation occurring when cortisol levels are low should be less so. For this reason, waking up earlier in the night (before cortisol levels are at their peak) should result in dreams that are fairly straightforward, whereas waking up later should result in dreams that are prototypically bizarre.

In line with this, some evidence suggests that dreams during these earlier periods look more like actual episodic memories. NREM and SWS dreams contain more mentation and replay of actual episodic experience than REM dreams (Foulkes, 1962), and NREM dreams are typically less bizarre than REM dreams. Thus, early night NREM dreams are typically associated with coherent episode memories, while REM dreams are associated with a mixture of episodic, personal semantic, and semantic fragments that combine to produce bizarre plot lines (e.g., Cavallero *et al.*, 1992; Cicogna *et al.*, 1991). Comparisons between REM and NREM dreams later in the night, however, reveal that many of these differences disappear and NREM and REM dreams begin to look similar. This finding fits with the notion that NREM dreams begin to resemble REM dreams as cortisol levels rise late in the night, becoming increasingly more fragmented and bizarre.

Antrobus (1991, 1993), Foulkes (1985), and Hobson (1988, 2009) have proposed related explanations of bizarreness. For example, much like the current proposal, Foulkes (1985) views dreaming as a process of semantic and episodic memory activation. He believes this activation proceeds in a largely diffuse and arbitrary fashion, but that dreams nonetheless consist of predictable features derived from memory. Dream content thus loosely reflects the recent or remote past of the dreamer, not as a simple replay of past events but rather as a plausible variation on such events. Similarly, Hobson ascribes narratization to forebrain structures including the hippocampus, amygdala, and medial prefrontal cortex, and, like the current account, attributes bizarreness to the unnatural fitting together of dream fragments.

Schwartz and Maquet (2002) also have an interesting account of dream bizarreness, although they focus solely on two aspects of bizarre dreams. They point to similarities between dream reports and neuropsychological syndromes in which faces and places are misidentified. Fregoli syndrome, for instance, involves the delusional misidentification of faces, or hyperidentification of faces, where an unknown face is recognized as familiar, and typically develops after damage to frontal and temporal regions. Schwartz and Maquet (2002) suggest that Fregoli-like delusions in dreams imply neuronal processes during sleep that simulta-neously engage unimodal visual regions that produce facial percepts (e.g., the fusiform gyrus), and distinct multimodal associative areas in the temporal lobe that allow retrieval of facial identity—areas that are clearly active during REM sleep. Corresponding deactivations in dorsal frontal regions would dampen supervisory control functions that would normally signal a mismatch between facial appearance and facial identity. This, they argue, produces the delusive quality of Fregoli-like representations in dreams (Schwartz and Maquet, 2002).

Misidentifications during dreams also extend to places. Familiar places in dreams often share little if anything in common with real places in waking life. For example, a dream that takes place in one's old house may look nothing like the place where one used to live. These authors attribute misidentified places to the parahippocampal cortex (which is adjacent to the fusiform face area). They note that passive viewing of spatial scenes and layouts activates this region during wake, and thus suggest that it is likely involved in the "reduplicative paramnesia" for places seen in dreams and in patients with combined damage to temporal and prefrontal areas—a disorder that likely results from faulty integration of a particular place with semantic information about place identity (Schwartz and Maquet, 2002).

Both of these phenomena can be considered binding deficits, and can be understood as the consequence of disrupted hippocampal, parahippocampal, and prefrontal (and perhaps fusiform given its location in the temporal lobe and proximity to the parahippocampal region) processing under conditions of ele-vated cortisol. As mentioned, these regions are needed to integrate dispersed features into a coherent memory or percept, and this binding occurs at various levels. It involves the ability to bind a particular face with a particular identity, and also the ability to bind the retrieved place or person representation with the appropriate time, situation, and so on. Both processes are clearly altered in dreaming, as they are in amnesic patients with damage to the hippocampus and surrounding areas of the medial temporal lobe (and sometimes PFC regions as well).

As one example, amnesics make a substantial number of "memory conjunction errors," which are produced in paradigms used to test mnemonic binding (Kroll et al., 1996; Reinitz et al., 1992, 1996). Memory conjunction errors occur when components of previously presented information are inappropriately recombined into episodes that never occurred, or into stimuli that were never presented. For example, subjects are shown facial stimuli, which are later recombined to make new faces, or words, which are later recombined into new words. For instance, subjects might be shown "SPANIEL" and "VARNISH" and then asked if they recognize the unpresented conjunction word "SPANISH," or shown two faces and then asked if they recognize a new face combining features of both.

While healthy subjects make only a moderate number of conjunction errors, they jump dramatically when such stimuli are shown to amnesic patients. Amnesics have particular difficulty conjoining features into an internal representation of the correct configuration that defines a memory trace. Thus, they falsely recognize recombinations of previously presented features, even though they can correctly discriminate them from new items that were not previously studied. This deficit is not surprising, considering that amnesics have damage to areas so critical for efficient binding (e.g., Eichenbaum and Bunsey, 1995; Reinitz et al., 1996). To the extent that hippocampal processing is altered by elevations in cortisol, as in late night sleep, misidentifications and recombinations would be expected. Indeed, the fact that memory fragments are activated in the absence of hippocampus and PFC-provided context virtually insures that such misidentifications and recombinations will occur in dreams.

By the current view then, dream fragmentation and bizarreness result from intensified HPA activity, particularly late in the sleep cycle. Episodic memories become activated in an increasingly unbound or disconnected manner as cortisol levels rise. High levels of cortisol render the patterns of reactivation incomplete, which may produce the fragmented quality of dreams. As in waking studies of stress hormones and memory, the disconnected or "unbound" fragments are woven into the bizarre storylines that constitute dreams. Narrative smoothing may begin during sleep, as the brain attempts to impose meaning on fragmented memory traces activated in circuits that include the hippocampus and amygdala, but it likely intensifies upon awakening as we impose additional structure and logic upon remembered fragments and themes. This is where we find ourselves reasoning through our dreams, thinking things like, "I'm not sure where I was, but it must have been Tucson because it looked like the desert," or "I assume I was with John although it didn't look like him." Our attempts to synthesize and impose narrative structure on the fragments produce the bizarreness of dreams.

Of course, on some occasions narratization fails or simply isn't complete, and in these cases we are left with severely fragmented dreams and with gaps in our morning recall. If we recall none of these fragments or rudimentary themes, we say that we did not dream.

VI. The Emotional Nature of Dreams

We are biased to consolidate emotional information during both wakefulness and sleep (Buchanan and Lovallo, 2001; Payne *et al.*, 2006, 2007, 2008; Wagner *et al.*, 2001), ensuring that we remember information and events that have adaptive value. The stress response system has evolved to ensure that we remember emotionally salient information, in some cases indefinitely (e.g., LeDoux, 2002). Stress hormones, including cortisol and norepinephrine, are thought to modulate the amygdala/hippocampus interactions that allow such memories to be stored in long-lasting format (Buchanan and Lovallo, 2001; McGaugh, 2000). Because high levels of cortisol facilitate amygdala function but impair hippocampal and PFC function, emotional memories are enhanced at the expense of other (neutral) aspects of memory (e.g., Buchanan and Lovallo, 2001; Jacobs and Nadel, 1998; LeDoux, 2002; Payne *et al.*, 2006).

Similar to the study by Payne *et al.*, 2007, (see Fig. 5), Buchanan and Lovallo (2001) have demonstrated that a low dose of cortisol (20 mg) administered during wakefulness enhances highly negative (high emotion) relative to neutral (low emotion) pictures compared to placebo (see Fig. 6).

Interestingly, these data are similar to the sleep data of Wagner *et al.* (2001), which demonstrate that emotional memory formation is selectively enhanced

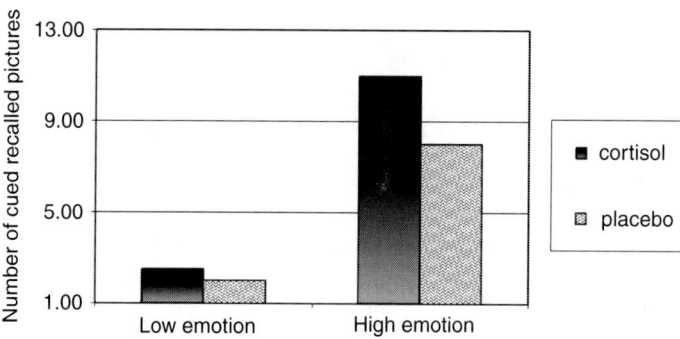

FIG. 6. Cortisol administration enhances the consolidation of highly emotional pictures, but not neutral (low-emotion) pictures. From Buchanan and Lovallo (2001).

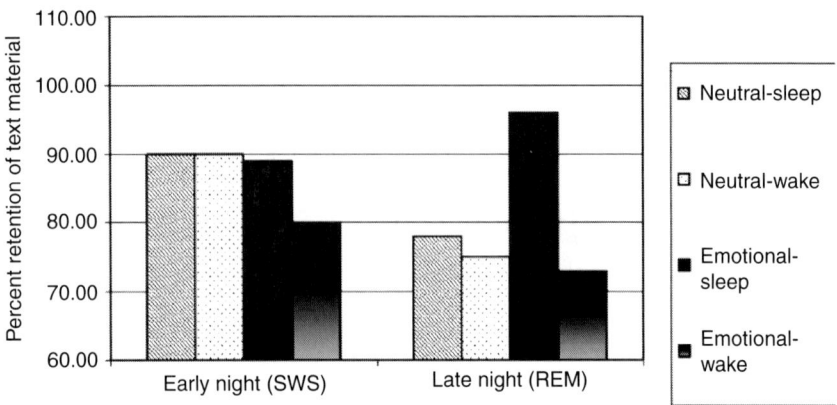

FIG. 7. Emotional information is selectively enhanced by late night, REM-rich sleep (black bar, right), but not by early night sleep rich in SWS. From Wagner *et al.* (2001).

after REM-rich sleep periods late in the night when cortisol activity is at a maximum (see Fig. 7).

It is illustrative to note the similarities between the above two figures. During both wakefulness and sleep, emotional memory is enhanced when cortisol levels are elevated. Thus, cortisol activity may directly contribute to both the emotional memory enhancement seen in late night, REM-rich sleep, and to the emotionality of REM sleep dreams (Casagrande *et al.*, 1996; Foulkes, 1962).

Many theorists have pointed out that dreams are often emotional. They are frequently biased toward negative emotions, although positive emotions, such as sexual pleasure and elation are also featured (Hartmann and Basile, 2003; Merritt *et al.*, 1994). The amygdala plays a role in all of these emotions, although research suggests that it may be preferentially involved in negative emotions like fear and anxiety (LeDoux, 2002), which may be the most common dream emotions. Dreams also incorporate instinctual programs (Revonsuo, 2000), such as the stress response system's "fight or flight" programs (McEwen, 2007), which, like emotion, can act as powerful synthesizers of dream fragments. Thus, via its effects on amygdala activation and associated selective preservation of emotional memory, cortisol might inspire an emotionally guided narratization process. Newell and Cartwright (2000) argue that emotion is a primary shaper of dream narratives. Fragments may be synthesized according to one's emotional state, and within an emotional framework provided by preserved emotional memory themes. Emotion might continue to inspire a search for meaning and personal salience upon awakening, as emotional themes connect with waking agendas to influence dream reconstruction. In this sense, dreams may well have some "meaning."

A. ACCEPTANCE OF BIZARRE IMAGES AS COMMONPLACE WITHIN THE DREAM

When dreaming, we often accept bizarre and even outlandish themes as completely normal, never questioning them until we awaken (except in the case of lucid dreaming, see LaBerge et al., 1981; LaBerge, 1990). Hobson and others have speculated that selective inactivation of the frontal cortices during REM sleep may be responsible for this phenomenon (Maquet, 2001). Together with evidence that the frontal lobes contribute to reality monitoring (Johnson, 1991; Schnider, 2001) and executive functions (Smith and Jonides, 1999) during wake, these massive frontal deactivations likely contribute to the intellectual complacency exhibited in dreams. Reality monitoring refers to our ability to compare dream or other fictitious events with waking reality and deem them "not real" (Johnson et al., 1984). Interestingly, damaged frontal lobes are associated with failures in reality monitoring and also bizarre confabulations that often resemble dreams (Kopelman, 1987; Schnider, 2001; Schnider and Ptak, 1999; Shallice, 1999).

The frontal lobes have received increased attention in the stress literature (e.g., Lupien and LePage, 2001). Both Type I (mineralocorticoid, "MR") and Type II (glucocorticoid, GR) cortisol receptors are present in cortical regions, with a preferential and dense distribution in the prefrontal cortex. Hence, stress may exert effects not only on hippocampal neurons, but also on neurons critical for the normal function of the PFC. Given evidence that both hippocampus and PFC are components of a memory circuit underlying memory retrieval (e.g., Fletcher et al., 1998) and the binding of memory elements (e.g., Mitchell et al., 2000), this would mean that both parts of the binding system are affected by cortisol. Thus, given the high concentrations of cortisol receptors in PFC regions, late night cortisol impairment of both hippocampal and prefrontal cortical function may contribute to the binding deficits and impaired reality monitoring seen in dreams.

B. DREAMS ARE DIFFICULT TO RECALL AND EPISODIC MEMORY "REPLAY" IS RARE IN DREAMS

Dreams are notoriously difficult to recall (e.g., Goodenough, 1978). Hobson asserts that the difficulty of dream recall is a thorny issue, but one that any dream theory must explain in the face of the robust activation of memory regions in REM sleep (Hobson et al., 2000). According to Hobson, dream amnesia is largely attributable to the aminergic demodulation of REM sleep (Hobson, 1988; Hobson et al., 2000).

Because elevated cortisol disrupts memory retrieval (de Quervain et al., 2000), it seems reasonable to assume that it plays a role in dream amnesia upon awakening, impairing recall of dreams in the same way it impairs recall during wake (see *Section III*). This is not the only memory impairment concerning dreams, however, as there is evidence for memory deficits within dreams themselves. Although a widely held belief is that dreams incorporate events from our waking lives, this has not been borne out in the laboratory. Complete events are rarely re-experienced in dreams, with recent estimates suggesting that only 1–2% of dream reports accurately reflect waking-life experience (Fosse *et al.*, 2003). Yet in spite of this, individual features (or fragments) of events are incorporated into dreams, although these often get woven into scenarios bearing little resemblance to real waking events (e.g., Baylor and Cavallero, 2001).

Interestingly, while the phenomenological experience of episodic memory processing during dreams is only loosely related to the original experience, activation of isolated features of a memory may be enough to boost memory consolidation. Recently we examined the influence of post-learning dream content on improved performance on a spatial navigation task (Wamsley *et al.*, 2010). We found that dreaming of the maze task did indeed benefit post-sleep performance, even though the dream reports never consisted of exact, veridical replays of the original learning experience. Instead, while subjects' dream reports were unquestionably related to the maze, they primarily consisted of remote memories and themes connected to the task, or of isolated fragments and thoughts of the maze-navigation experience.

This lack of exact replay mirrors that observed in animal studies of neuronal-level reactivation (e.g., Ji and Wilson, 2007; Wilson and McNaughton, 1994) and in human brain imaging studies (e.g., Piegneux *et al.*, 2004) where the reactivation observed is not precisely identical to that seen during encoding. For example, although patterns of neural reactivation seen in rodent sleep statistically resemble the activity of these networks during prior waking task performance, the activity patterns are never identical to those observed during prior wake, more typically containing only fragments of waking experience. These findings fit with the fact that that complete episodes are not typically replayed in dreams. Elevated cortisol is likely to produce precisely these unbound fragments of episodic memories in the phenomenological experience of dreams.

C. Predictions

The ideas presented above are quite testable and lead to several interesting predictions. Cortisol administration during early sleep should modify the nature of dreams via its impact on memory processing; specifically, it should (1) intensify dream fragmentation and bizarreness relative to the administration of a placebo and (2) reduce the differences between early and late night REM sleep dreams (such that they would be equally fragmented and bizarre). The assumption here is that cortisol levels account for the differences between dreams from early versus late REM sleep episodes. As such, dreams reported from late REM sleep periods (when cortisol is at its peak) should be more bizarre and more fragmented than early REM periods (when cortisol is low). There is already evidence that REM dreams are generally more bizarre than NREM dreams (Antrobus et al., 1995; Casagrande et al., 1996; Foulkes, 1962), a finding that might be informed by cortisol elevations during late-night sleep that is largely composed of REM sleep. However, early and late REM episodes have not, to my knowledge, been directly compared on measures of fragmentation and bizarreness. Cortisol administration during early sleep should also (3) increase the emotionality of dreams, both during REM and NREM sleep stages and (4) if administered during early NREM sleep specifically, reduce the likelihood of thinking about or re-experiencing recent episodes.

D. CAVEATS

Several of the neurotransmitters that fluctuate across the sleep cycle are also know to affect memory function during wakefulness (e.g., acetylcholine, see Hasselmo, 1999; norepinephrine, see Cahill and McGaugh, 1998), and there has been much speculation about their influence on memory processing during sleep (Hobson et al., 2000; Solmes, 2000). Moreover, these neurotransmitters likely interact with cortisol during sleep. For example, acetylcholine and cortisol may interact to modulate dreams. Also, because the PFC is important for memory function generally, its deactivation during REM sleep (Braun et al., 1997; Maquet et al., 1996) likely contributes directly to impaired memory processing during late night sleep. Thus, although I have suggested that diurnal elevation in cortisol may help explain the nature of dreams, I do not mean to suggest that cortisol is the only factor affecting their structure and content.

VII. Concluding Remarks

In sum, converging evidence suggests that cortisol's impact on the hippocampus, amygdala, and prefrontal cortex can account for many formal features of dreaming, including dream fragmentation and bizarreness, the emotional nature of dreams, the lack of episodic memory replay in dreams, and the tendency for dreams to be forgotten upon awakening. Memory consolidation, a process that includes activation and reprocessing of memory traces, as well as their integration with pre-existing knowledge and experience, appears to play a critical role in dreaming: We dream when we become aware of these activated memory traces, and we experience them as fantastical as they are woven into confabulatory, bizarre, but also enormously creative, story lines during late night REM sleep.

An interesting question is whether the process of weaving unbound fragments into creative narratives is functional. Obviously, when high levels of stress impact memory during wakefulness, fragmentation and the resulting reconstruction may be truly disruptive, as when reconstruction produces memories that are flagrantly "false" (Schacter *et al.*, 2003). Many people think of such memory inaccuracies as strictly negative events that lead to false recollections at best and devastating confabulations at worst (Loftus, 1996). While it is true that accurate recall is adaptive in many cases and imperative to our survival in a few (e.g., accurately remembering that a certain mushroom is poisonous), there may be a positive side to the process that disrupts memory binding and produces fragmentation, especially during the protected state of sleep.

When the bonds attaching the components of a memory are broken, this information can be flexibly recombined. In other words, memories are "unbound," which in turn allows novel configurations of knowledge and events. This has already been suggested in studies of sleep and memory showing that the sleeping brain goes beyond simple memory solidification to actually transform memories in ways that are useful and adaptive for future behavior. Such transformation, whether it be selectively consolidating emotional foreground information at the expense of neutral information in the background (Payne *et al.*, 2008), or suppressing a well-learned problem-solving strategy to allow insight into a novel shortcut (Wagner *et al.*, 2004), requires unbinding, as rigid representations do not allow for the flexible recombinations necessary for these effects. In wakefulness these recombinations may lead us to misremember episodes, but during sleep, they may take us down different paths to creative insights and novel ideas, and allow us to role-play and test possible future scenarios before we ever encounter them. Although each of these ideas awaits experimental confirmation, cortisol-induced fragmentation during dreams might be ultimately beneficial for human cognition.

In conclusion, I argue that dreams are a reflection of the memory consolidation process, which serves not only to strengthen the neural traces of recent events and integrate them with older memories and previously stored knowledge, but also to recombine and restructure features of experience. All new ideas are based upon previously stored information, and nearly every definition of creativity includes combining this information in novel and useful ways. Unbinding the individual features of a memory trace may be essential for these processes to occur, and cortisol's impact on the dreaming brain may be the ideal time to safely make new connections and to test out new ideas for the future.

Dreams are but interludes which fancy makes
When monarch reason sleeps, this mimic wakes.
Compounds a medley of disjointed things
A mob of cobblers and a court of Kings.

—John Dryden (1700)

References

Antrobus, J. S. (1991). Dreaming: cognitive processes during cortical activation and high afferent thresholds. *Psychol. Rev.* **98**, 96–121.

Antrobus, J. S. (1993). Dreaming: Could we do without it? In: The Functions of Dreaming (A. Moffitt, M. Kramer, and R. Hoffman, eds.), New York, SUNY Press.

Antrobus, J. S., Kondo, T., and Reinsel, R. (1995). Dreaming in the late morning: summation of REM and dirural cortical activation. *Conscious Cogn.* **4**, 275–299.

Azerinsky, E. and Kleitman, N. (1953). Regularly occurring periods of eye motility and concurrent phenomena during sleep. *Science* **118**, 273–274.

Barrett, T. R. and Ekstrand, B. R. (1972). Effect of sleep on memory. 3. Controlling for time-of-day effects. *J. Exp. Psychol.* **96**(2), 321–327.

Bartlett, F. C. (1932). Remembering: A Study in Experimental and Social Psychology. Cambridge, Cambridge University Press.

Baylor, G. W. and Cavallero, C. (2001). Memory sources associated with REM and NREM dream reports throughout the night: a new look at the data. *Sleep* **24**, 165–170.

Behrens, C. J., van den Boom, L. P. *et al.* (2005). Induction of sharp wave-ripple complexes in vitro and reorganization of hippocampal networks. *Nat. Neurosci.* **8**, 1560–1567.

Braun, A. R., Balkin, T. J., Wesenten, N. J., Carson, R. E., Varga, M., Baldwin, P., Selbie, S., Belenky, G., and Herscovitch, P. (1997). Regional cerebral blood flow throughout the sleep-wake cycle. An H2(15)O PET study. *Brain* **120**, 1173–1197.

Bremner, J. D. (1999). Does stress damage the brain? *Biol. Psychiatry* **45**, 797–805.

Buchanan, T. W. and Lovallo, W. R. (2001). Enhanced memory for emotional material following stress-level cortisol treatment in humans. *Psychoneuroendocrinology* **26**, 307–317.

Buzsaki, G. (1996). The hippocampo-neocortical dialogue. *Cerebr. Cortex* **6**, 81–92.

Buzsaki, G. (1998). Memory consolidation during sleep: a neurophysiological perspective. *J. Sleep Res.* **7(Suppl.)**, 17–23.

Cahill, L. and McGaugh, J. L. (1998). Mechanisms of emotional arousal and lasting declarative memory. *Trends Neurosci.* **21**, 294–299.

Cahill, L., Gorski, L., and Le, K. (2003). Enhanced human memory consolidation with post-learning stress: Interaction with the degree of arousal at encoding. Learning and Memory, **10**, 270–274.

Casagrande, M., Violani, C., Lucidi, F., Buttinelli, B., and Bertini, M. (1996). Variations in sleep mentation as a function of time of night. *Int. J. Neurosci.* **85**, 19–30.

Cavallero, C., Cicogna, P., Natale, V., Occhionero, M., and Zito, A. (1992). Slow wave sleep dreaming. *Sleep* **15**, 562–566.

Cicogna, P., Cavallero, C., and Bosinelli, M. (1991). Cognitive aspects of mental activity during sleep. *Am. J. Psychol.* **104**, 413–425.

Cicogna, P., Natale, V., Occhionero, M., and Bosinelli, M. (1998). A comparison of mental activity during sleep onset and morning awakening. *Sleep* **21**(5), 462–470.

Cipolli, C. (1995). Sleep, dreams, and memory: An overview. *J. Sleep Res.* **4**, 2–9.

Cohen, N. J. and Eichenbaum, H. (1994). Memory, amnesia, and the hippocampal system. Cambridge MA, MIT Press.

Cohen, N. and Eichenbaum, H. (2004). From Conditioning to Conscious Recollection: Memory Systems of the Brain. Oxford, UK, Oxford University Press.

de Quervain, D. J. F., Roozendaal, B., Nitsch, R. M., McGaugh, J. L., and Hock, C. (2000). Acute cortisone administration impairs retrieval of long-term declarative memory in humans. *Nat. Neurosci.* **3**, 313–314.

Diekelmann, S. and Born, J.(2010). The memory function of sleep. *Nat. Rev. Neurosci.* **11**, 114–126.

Eichenbaum, H. and Bunsey, M. (1995). On the binding of associations in memory: clues from studies on the role of the hippocampal regions in paired-associate learning. *Curr. Dir. Psychol. Sci.* **4**, 19023.

Ellenbogen, J. M., Hu, P., Payne, J. D., Titone, D., and Walker, M. P. (2007). Human relational memory requires time and sleep. *Proc. Nat. Acad. Sci.* **104**, 7723–7728.

Ellenbogen, J. M., Payne, J. D., and Stickgold, R. (2006). Sleep's role in declarative memory consolidation: Passive, Permissive, Active or None? *Curr. Opin. Neurobiol.* **16**, 716–722.

Fletcher, P. C., Shallice, T., Frith, C. D., Frackowiak, R. S., and Dolan, R. J. (1998). The functional roles of prefrontal cortex in episodic memory. II. Retrieval. *Brain* **121**, 1249–1256.

Fosse, M. J., Fosse, R., Hobson, J. A., and Stickgold, R. (2003). Dreaming and episodic memory: A functional dissociation? *J. Cognit. Neurosci.* **15**, 1–9.

Fosse, R., Stickgold, R., and Hobson, J. A. (2001). Brain-Mind States: Reciprocal variation in thoughts and hallucinations. *Psychol. Sci.* **12**, 30–36.

Foulkes, W. D. (1962). Dream reports from different stages of sleep. *J. Abnorm. Soc. Psychol.* **65**, 14–25.

Foulkes, D. (1985). Dreaming: A Cognitive-Psychological Analysis. Erlbaum, NJ, Hillsdale.

Fowler, M. J., Sullivan, M. J., and Ekstrand, B. R. (1973). Sleep and memory. *Science* **179**, 302–304.

Gais, S. and Born, J. (2004). Low acetylcholine during slow-wave sleep is critical for declarative memory consolidation. *Proc. Nat. Acad. Sci. U.S.A.* **101**(7), 2140–2144.

Goodenough, D. R. (1978). Dream recall: history and current status of the field.In:Mind in Sleep: Psychology and Psychophysiology (A. M. Arkin, J. S. Antrobus, and S. J. Ellman, eds.), New Jersey, Lawrence Erlbaum.

Hamann, S. B. (2001). Cognitive and neural mechanisms of emotional memory. *Trends Cogn. Sci.* **5**, 394–400.

Hartmann, E. and Basile, R. (2003). Dream imagery becomes more intense after 9/11/01. *Dreaming* **13**, 61–66.

Hasselmo, M. E. (1999). Neuromodulation: acetylcholine and memory consolidation. *Trends Cogn. Sci.* **3**, 351–359.

Hobson, J. A. (1988). The Dreaming Brain. New York, Basic Books.

Hobson, J. A. (2009). REM sleep and dreaming: toward a theory of protoconsciousness. *Nat. Rev. Neurosci.*

Hobson, J. A. and McCarley, R. (1977). The brain as a dream state generator: An activation-synthesis hypothesis of the dream process. *Am. J. Psychiat.* **134**, 1335–1348.

Hobson, J. A., Pace-Schott, E., and Stickgold, R. (2000). Dreaming and the brain: Towards a cognitive neuroscience of conscious states. *Behav. Brain Sci.* **23**, 793–842.

Hobson, J. A., Stickgold, R., and Pace-Schott, E. (1998). The neuropsychology of REM sleep dreaming. *NeuroReport* **9**, R1–R14.

Jacobs, W. J. and Nadel, L. (1998). Neurobiology of reconstructed memory. *Psychol. Public Policy Law* **4**, 1110.

Jelicic, M., Geraerts, E., Merckelbach, H., and Guerrieri, R. (2004). Acute stress enhances memory for emotional words, but impairs memory for neutral words. *Int. J. Neurosci.* **114**, 1343–1351.

Ji, D. and Wilson, M. A. (2007). Coordinated memory replay in the visual cortex and hippocampus during sleep. *Nat. Neurosci.* **10**, 100–107.

Johnson, M. K. (1991).Reality monitoring: Evidence from confabulation in organic brain disease patients.In:Awareness of Deficit After Brain Injury (G. P.Prigatano and D. L.Schacter, eds.), New York, Oxford, pp. 176–197.

Johnson, M. K., Kahan, T. L., and Raye, C. L. (1984). Dreams and reality monitoring. *J. Exp. Psychol. Gen.* **113**, 329–344.

Kim, J. J. and Diamond, D. M. (2002). The stressed hippocampus, synaptic plasticity and lost memories. *Nat. Rev. Neurosci.* **3**(6), 453–462.

Kirschbaum, C., Wolf, O. T., May, M., Wippich, W., and Hellhammer, D. H. (1996). Stress and treatment induced elevations of cortisol levels associated with impaired declarative memory in healthy adults. *Life Sci.* **58**, 1475–1483.

Kondo, T., Antrobus, J., and Fein, G. (1989). Later REM activation and sleep mentation. *Sleep Res.* **18**, 147.

Kopelman, M. D. (1987). Two types of confabulation. *J. Neurol. Neurosurg. Psychiat.* **50**, 1482–1487.

Kroll, N. E. A., Knight, R. T., Metcalfe, J., Wolf, E. S., and Tulving, E. (1996). Cohesion failure as a source of memory illusions. *J. Mem. Lang.* **35**, 176–196.

LaBerge, S. (1990). Lucid dreaming: psychophysiological studies of consciousness during REM Sleep. In: Sleep and Cognition (R. R.Bootzin, J. F. Kihlstrom, and D. L. Schacter, eds.), Washington, DC, American Psychological Association, pp. 109–126.

LaBerge, S., Nagel, L., Dement, W., and Zarcone, V. (1981). Lucid dreaming verified by volitional communication during REM sleep. *Perceptual Mot. Skills* **52**, 727–732.

Lavie, P. (1996). The Enchanted World of Sleep. New Haven, CT, Yale University Press.

LeDoux, J. E. (2002). Emotion circuits in the brain. *Annu. Rev. Neurosci.* **23**, 155–184.

Loftus, E. (1996). Memory distortion and false memory creation. *Bull. Am. Acad. Psychiatr. Law* **24**, 281–295.

Lupien, S. J. and LePage, M. (2001). Stress, memory and the hippocampus: Can't live with it, can't live without it. *Behav. Brain Res.* **127**, 137–158.

Lupien, S. J., McEwen, B. S., Gunnar, M. R., and Heim, C. (2009). Effects of stress throughout the lifespan on the brain, behavior, and cognition. *Nat. Rev. Neurosci.* **10**, 434–445.

Maquet, P. *et al.*, (1996). Functional neuroanatomy of human rapid-eye-movement sleep and dreaming. *Nature* **383**, 163–166.

Maquet, P. (2001). The role of sleep in learning and memory. *Science* **294**, 1048–1052.

Marshall, L. and Born, J. (2007). The contribution of sleep to hippocampus-dependent memory consolidation. *Trends Cogn. Sci.* **11**, 442–450.

Marshall, L., Helgadottir, H., Molle, M., and Born, J. (2006). Boosting slow oscillations during sleep potentiates memory. *Nature* **444**, 610–613.

McEwen, B. S. (2000). The neurobiology of stress: From serendipity to clinical relevance. *Brain Res.* **866**, 172–189.

McEwen, B. (2007). Physiology and neurobiology of stress and adaptation: Central role of the brain. *Physiol. Rev.* **87**, 873–904.

McGaugh, J. L. (2000). Memory—A century of consolidation. *Science* **287**, 248–251.

McGaugh, J. L. (2004). The amygdala modulates the consolidation of memories of emotionally arousing experiences. *Annu. Rev. Neurosci.* **27**, 1–28.

Merritt, J. M., Stickgold, R., Pace-Schott, E., Williams, J., and Hobson, J. A. (1994). Emotion profiles in the dreams of men and women. *Conscious. Cogn.* **3**, 46–60.

Mitchell, K. J., Johnson, M. K., Raye, C. L., and D'Esposito, M. (2000). fMRI evidence of age-related hippocampal dysfunction in feature binding in working memory. *Brain Res.* **10**, 197–206.

Murray, J., Ehlers, A., and Mayou, R. A. (2002). Dissociation and post-traumatic stress disorder: two prospective studies of road traffic accident survivors. *Br. J. Psychiatry* **180**, 363–368.

Nadel, L. and Jacobs, W. J. (1998). Traumatic memory is special. *Curr. Dir. Psychol. Sci.* **7**, 154–157.

Nadel, L. and Payne, J. D. (2002a). The relationship between episodic memory and context: Clues from memory errors made while under stress. *Physiol. Res.* **9**, 74–89.

Nadel, L. and Payne, J. D. (2002b). The hippocampus, wayfinding, and episodic memory. In The Neural Basis of Navigation: Evidence From Single Cell Recording (P. Sharp, ed.). (pp. 235–247). Kluwer Academic Publishers, Boston.

Newell, P. T. and Cartwright, R. D. (2000). Affect and cognition in dreams: a critique of the cognitive role in adaptive dream functioning and support for associative models. *Psychiatry* **63**, 34–44.

Nielsen, T. A. (2000). A review of mentation in REM and NREM sleep: "Covert" REM sleep as a possible reconciliation of two opposing models. *Behav. Brain Sci.* **23**, 851–866, .

Nishida, M., Pearsall, J., Buckner, R. L., and Walker, M. P. (2009). REM sleep, prefrontal theta and the consolidation of human emotional memory. *Cerebr. Cortex* **19**, 1158–1166.

O'Neil, J. Pleydell-Bouverie, B., Dupret, D., and Csicsvari, J. (2010). Play it again: reactivation of waking experience and memory. *Trends Neurosci.* in press.

Payne, J. D., Jackson, E. D., Hoscheidt, S., Ryan, L., Jacobs, W. J., and Nadel, L. (2007). Stress administered prior to encoding impairs neutral but enhances emotional long-term episodic memories. *Learn. Mem.* **14**, 861–868.

Payne, J. D., Jackson, E. D., Ryan, L., Hoscheidt, S., Jacobs, W. J., and Nadel, L. (2006). The impact of stress on memory for neutral vs. emotional aspects of episodic memory. *Memory* **14**(1), 1–16.

Payne, J. D. and Kensinger, E. A. Sleep leads to qualitative changes in the emotional memory trace: evidence from fMRI. J. Cogn. Neurosci.

Payne, J. D. and Kensinger, E.A. Sleep's role in the consolidation of emotional episodic memories. Curr. Dir. Psychol. Sci.

Payne, J. D. and Nadel, L. (2004). Sleep, dreams, and memory consolidation: The role of the stress hormone cortisol. *Learn. Mem.* **11**, 671–678.

Payne, J. D., Nadel, L., Allen, J. J. B., Thomas, K. G. F., and Jacobs, W. J. (2002). The effects of experimentally induced stress on false recognition. *Memory* **10**, 1–6.

Payne, J. D., Nadel, L., Britton, W. B., and Jacobs, W. J. (2004). The biopsychology of trauma and memory. In Emotion and Memory (D.Reisberg and P. Hertel, eds.). Oxford University Press, Oxford, UK.

Payne, J. D., Schacter, D. L., Tucker, M. A., Wamsley, E., Huang, L., Walker, M. P., and Stickgold, R. (2009). The role of sleep in false memory formation. *Neurobiol. Learn. Mem.* **92**, 327–334.

Payne, J. D., Stickgold, R., Swanberg, K., and Kensinger, E. K. (2008). Sleep preferentially enhances memory for emotional components of scenes. *Psychol. Sci.* **19**(8), 781–788.

Peigneux, P., Laureys, S. *et al.* (2004). Are spatial memories strengthened in the human hippocampus during slow wave sleep? *Neuron* **44**, 535–545.

Peterson, L. R. and Peterson, M. J. (1959). Short-term retention of individual verbal items. *J. Exp. Psychol.* **58**, 193–198.

Plihal, W. and Born, J. (1997). Effects of early and late nocturnal sleep on declarative and procedural memory. *J. Cognit. Neurosci.* **9**(4), 534–547.

Plihal, W. and Born, J. (1999). Memory consolidation in human sleep depends on inhibition of glucocorticoid release. *NeuroReport* **10**, 2741–2747.

Pruessner, J. C., Dedovic, K., Khalli-Mahani, N., Engert, V., Pruessner, M., Buss, C., Renwick, R., Dagher, A., Meaney, M. J., and Lupien, S. (2008). Deactivation of the limbic system during acute psychosocial stress: Evidence from Positron Emission Tomography and Functional Magnetic Resonance Imaging studies. *Biol. Psychiatry*, **63**, 234–240.

Rasch, B., Buchel, C., Gias, S., and Born, J. (2007). Odor cues during slow-wave sleep prompt declarative memory consolidation. *Science* **315**, 1426–1429.

Reinitz, M. T., Lammers, W. J., and Cochran, B. P. (1992). Memory conjunction errors: Miscombination of stored stimulus features can produce illusions of memory. *Mem. Cogn.* **20**, 1–11.

Reinitz, M. T., Verfaellie, M., and Milberg, W. P. (1996). Memory conjunction erros in normal and amnesic subjects. *J. Mem. Lang.* **35**, 286–299.

Revonsuo, A. (2000). The reinterpretation of dreams: An evolutionary hypothesis of the function of dreaming. *Behav. Brain Sci.* **23**, 877–901.

Roozendaal, B., McEwen, B. S., and Chattarji, S. (2009). Stress, memory and the amygdala. *Nat. Rev. Neurosci.* **10**, 423–433.

Schacter, D. L., Addis, D. R., and Buckner, R. L. (2008). Episodic simulation of future events. *Ann. N. Y. Acad. Sci.* **1124**, 39–60.

Schacter, D. L., Chiao, J. Y., and Mitchell, J. P. (2003). The seven sins of memory: Implications for self. *Ann. N. Y. Acad. Sci.* **1001**, 226–239.

Schacter, D. L. and Tulving, E. (1994). Memory Systems. Cambridge, MA, MIT Press.

Schnider, A. (2001). Spontaneous confabulation, reality monitoring, and the limbic system—a review. *Brain Res. Rev.* **36**, 150–160.

Schnider, A. and Ptak, R. (1999). Spontaneous confabulators fail to suppress currently irrelevant memory traces. *Nat. Neurosci.* **2**, 677–681.

Schwartz, S. and Maquet, P. (2002). Sleep imaging and the neuropsychological assessment of dreams. *Trends Cogn. Sci.* **6**, 23–30.

Shallice, T. (1999). The origin of confabulations. *Nat. Neurosci.* **2**, 588–590.

Smith, C. (1995). Sleep states, memory processes and synaptic plasticity. *Behav. Brain Res.* **78**, 49–56.

Smith, C. (2001). Sleep states and memory processes in humans: procedural versus declarative memory systems. *Sleep Med. Rev.* **5**, 491–506.

Smith, E. E. and Jonides, J. (1999). Storage and executive processes in the frontal lobes. *Science* **283**, 1657–1661.

Solms, M. (2000). Dreaming and REM sleep are controlled by different brain mechanisms. *Behav. Brain Sci.* **23**, 843–850.

Squire, L. R. and Zola-Morgan, S. (1991). The medial temporal lobe memory system. *Science* **253**, 1380–1386.

Sterpenich, V., Albouy, G., Darsaud, A., Schmidt, C., Vandewalle, G. *et al.* (2009). Sleep promotes the neural reorganization of remote emotional memory. *J. Neurosci.* **29**, 5143–5152.

Stickgold, R. (2005). Sleep-dependent memory consolidation. *Nature* **437**, 1272–1278.

Tucker, M. A. *et al.* (2006). A daytime nap containing solely non-REM sleep enhances declarative but not procedural memory. *Neurobiol. Learn. Mem.* **86**, 241–247.

Tucker, M. A., Hirota, Y. *et al.* (2006). A daytime nap containing solely non-REM sleep enhances declarative but not procedural memory. *Neurobiol. Learn. Mem.* **86**(2), 241–247.

Tulving, E. (1983). Elements of Episodic Memory. Oxford, UK, Oxford University Press.

Van der Kolk, B. A. (1991). The biological response to psychic trauma: Mechanisms and treatment of intrusion and numbing. *Anxiety Res.* **4**, 199–212.

Van der Kolk, B. A. (1997). The psychobiology of posttraumatic stress disorder. *Clin. Psychiatry* **58**, 16–24.

Van der Kolk, B. A. (1998). Trauma and memory. *Psychiatry. Clin. Neurosci.* **52**, S97–S5109.

Van der Kolk, B. A. and Fisler, R. (1995). Dissociation and the fragmentary nature of traumatic memories. Overview and exploratory study. *J. Trauma. Stress* **8**, 505–525.

Van Stegeren, A. H., Wolf, O. T., Everaerd, W., Rombouts, S. A. R. B. (2007). Interaction of endogenous cortisol and noradrenaline in the human amygdala. *Prog. in Brain Res.* **167**, 263–268.

Verfaellie, M. and Vasterling, J. (2009). Memory in PTSD: A Neurocognitive Approach. In Post-Traumatic Stress Disorder: Basic Science and Clinical Practice. New York, Humana Press.

Vyas, A., Mitra, R., Rao, S. B. S., and Chatterji, S. (2002). Chronic stress induces contrasting patterns of dendritic remodeling in hippocampal and amygdaloid neurons. *J. Neurosci.* **22**, 6810–6818.

Wagner, U. and Born, J. (2008). Memory consolidation during sleep: Interactive effects of sleep stages and HPA regulation. *Stress* **11**, 28–41.

Wagner, U., Gais, S., and Born, J. (2001). Emotional memory formation is enhanced across sleep intervals with high amounts of rapid eye movement sleep. *Learn. Mem.* **8**, 112–119.

Wagner, U., Gais, S., Haider, H., Verleger, R., and Born, J. (2004). Sleep inspires insight. *Nature* **427**, 352–355.

Wagner, U., Hallschmid, M., Rasch, B., and Born, J. (2006). Brief sleep after learning keeps emotional memories alive for years. *Biol. Psychiatry* **60**, 788–790.

Walker, M. P. (2009). The role of sleep in cognition and emotion. *Ann. N. Y. Acad. Sci.* **1156**, 168–197.

Walker, M. P. and Stickgold, R. (2006). Sleep, memory and plasticity. *Ann. Rev. Psychol.* **57**, 139–166.

Wamsley, E. J., Hirota, Y., Tucker, M. A., Smith, M. R., and Antrobus, J. S. (2007). Circadian and ultradian influences on dreaming: A dual rhythm model. *Brain Res. Bull.* **71**(4), 347–354.

Wamsley, E. J., Tucker, M., Payne, J. D., Benavides, J. A., and Stickgold, R. (2010). Dreaming of a learning task is associated with enhance sleep-dependent memory consolidation. *Curr. Biol.* **20**, 850–855.

Weitzman, E. D., Fukushima, D., Nogeire, C., Roffwarg, H., Gallagher, T. F., and Hellman, L. (1971). Twenty-four hour pattern of the episodic secretion of cortisol in normal subjects. *J. Clin. Endocrinol. Metabol.* **33**, 14–22.

Wilson, M. A. and McNaughton, B. L. (1994). Reactivation of hippocampal ensemble memories during sleep. *Science* **265**, 676–679.

Yaroush, R., Sullivan, M. J. *et al.* (1971). Effect of sleep on memory. II: Differential effect of first and second half of the night. *J. Exp. Psychol.* **88**, s361–366.

CHARACTERISTICS AND CONTENTS OF DREAMS

Michael Schredl

Sleep laboratory, Central Institute of Mental Health, Mannheim, Germany

Dreams have been studied from different perspectives: psychoanalysis, academic psychology, and neurosciences. After presenting the definition of dreaming and the methodological tools of dream research, the major findings regarding the phenomenology of dreaming and the factors influencing dream content are briefly reviewed. The so-called continuity hypothesis stating that dreams reflect waking-life experiences is supported by studies investigating the dreams of psychiatric patients and patients with sleep disorders, i.e., their daytime symptoms and problems are reflected in their dreams. Dreams also have an effect on subsequent waking life, e.g., on daytime mood and creativity. The question about the functions of dreaming is still unanswered and open to future research.

I. Introduction

Mankind has been fascinated with dreaming since the dawn of history (Van de Castle, 1994). Three major scientific approaches to the study of dreams can be differentiated: psychoanalytic, neurophysiological, and psychological. Although this chapter is based on the findings of the third approach, the other two approaches were mentioned briefly.

INTERNATIONAL REVIEW OF
NEUROBIOLOGY, VOL. 92
DOI: 10.1016/S0074-7742(10)92007-2

135

From the psychoanalytic perspective, described initially by Sigmund Freud in his fundamental book "Die Traumdeutung" (*The Interpretation of Dreams*) published in 1899, the dream is conceptualized as expression of the person's inner life, i.e., "the interpretation of dreams is the royal road to a knowledge of the unconscious activities of the mind" (Freud, 1987). Although psychoanalysis has brought together a large variety of interesting clinical material and may be helpful to patients on a case-by-case basis, psychoanalytic dream theories remain speculative and they are at least partly not in concordance with recent research findings (Schredl, 2008c).

The discovery of rapid eye movement (REM) sleep was the starting point of the neurophysiological approach for studying the dream state. In 1953, Eugene Aserinsky and Nathaniel Kleitman published their finding that in the course of the night the sleeper goes through different sleep stages, among others periods with REMs (Aserinsky and Kleitman, 1953). After awakenings from these sleep stages, a vivid and pictorial dream was reported very often. The initial euphoria hoping to find a direct access to the dream world subsided very quickly, because REM sleep measured by electroencephalogram, electrooculogram, and electromyogram recordings and dreams (elicited by interviews) are two distinct realms (physiological level vs. psychological level) which correspond to a specific amount but might not share similar functions.

The third approach is based on methods of academic psychology. The basic methodology was provided by Calvin S. Hall and his coworker Robert Van de Castle who published the book "The content analysis of dreams" in 1966. Their extensive research carried out since the 1940s led to the development of a comprehensive coding system for dream content analysis and to "norms" which were derived from 1000 dreams of college students. The major advantage of this approach is that specific dream characteristics can be quantified in a reliable fashion so that hypotheses can be tested by using common statistical methods. For example, Hall and Van de Castle (1966) have shown that men dreamed significantly more often of physical aggression and sex than women. Subsequent studies with different samples (Hall *et al.*, 1982; Schredl *et al.*, 1998a) were able to replicate this finding. Psychological approach concentrates specifically on the following questions: How are the phenomenological characteristics of dreams? How is dream content related to waking-life experiences? And how do dreams affect subsequent waking life? The ultimate goal of this kind of paradigm is to answer the question whether dreaming serves a function or several functions independently from the functions of sleep, especially those formulated for REM sleep.

II. Definitions and Methodological Issues

Since each approach to dreaming focuses on different aspects, a clear definition of dreaming within the area of psychological dream research seems necessary. The following definition attempts to cover the consensus of the researchers in the field:

A dream or a dream report is the recollection of mental activity which has occurred during sleep. (Schredl, 1999, p 12)

It is important to notice that dreaming as a mental activity during sleep is not directly measurable, two boundaries have to be crossed (the sleep–wake transition and time) before the person can report the subjective experiences that occurred during sleep. These issues point to the problem of validity, i.e., the question whether the dream report is an appropriate account of the actual dream experience (see Section III). The second question which has been raised by Maury (1861) is whether the dream report reflects mental activity during sleep or is merely produced during the awakening process. Modern research combining physiological approaches with dream content analysis, however, has been able to demonstrate that dream reports are accounts of mental activity during sleep since physiological parameters (e.g., eye movements, heart rate) during REM sleep at least partially match with dream contents elicited upon awakening (cf. Erlacher and Schredl, 2008). In addition, the incorporation of stimuli applied during sleep into dreams (Schredl, 1999; Strauch and Meier, 2004) corroborates the assumption that dreaming occurs during sleep.

During the initial phase of psycho-physiological dream research, REM sleep was considered to be the physiological concomitant of dreaming (Dement and Kleitman, 1957). However, Foulkes (1962) showed that dream reports can be elicited after awakening out of all stages of sleep. A recent review (Nielsen, 2000) showed that recall rates are somewhat lower for non-rapid eye movement (NREM) awakenings (43.0%) than for REM awakenings (81.9%). Although Nielsen (2000) tried to connect NREM dreaming with REM sleep by postulating covert REM processes to be responsible for dream recall out of NREM sleep, Wittmann and Schredl (2004) pointed out the logical errors of this assumption and argued that mental activity is presumably present continuously during sleep. Dreaming therefore is the subjective experienced correlate of the continuous brain activity. Differences regarding formal characteristics of dreaming and waking cognition as well as between dream reports of different sleep stages can be explained by factors such as cortical activation, blockade of external sensory input, and neuromodulation as described by the AIM model (Hobson et al., 2000).

Nightmares are a subgroup of REM dreams with strong negative emotions but whether these actually cause awakening is not yet known (cf. Nielsen and Levin, 2007). Night terrors, on the other hand, occur out of NREM sleep and the person often does not remember the incident in the morning (Stuck et al., 2009). The third dream phenomenon associated with fear is called posttraumatic re-enactment or posttraumatic nightmares which are special since they seem to occur in REM sleep as well as in NREM sleep (cf. Wittmann et al., 2007). Lucid dreams in which the dreamer is aware that she/he is dreaming offer fascinating opportunities to study the body–mind relationship during sleep because the dreamer can carry out prearranged tasks during the dream and mark their beginning and end by distinct eye movements which can be measured electrically (cf. Hobson, 2009).

Naturally, dream recall is a prerequisite for dream content studies. The large amount of research in this area (overview: Schredl, 2007) showed that personality factors such as "thin boundaries," absorption, openness to experiences, creativity, visual memory, and sleep behavior (frequent nocturnal awakenings) are associated with heightened dream recall. But a recent large-scale study (Schredl et al., 2003b) including these factors indicated that the variance explained by these factors is rather small (below 10%). At present, the reasons of the large interindividual differences or intraindividual fluctuations in dream recall are poorly understood. More in-depth research, for example applying event-related potential paradigms investigating the awakening process, is necessary.

Several approaches have been used to collect dream reports (see Table I). The most convenient way to collect large samples of dream reports is the so-called most recent dream approach (Domhoff, 1996). The participants are asked to write down (as completely as possible) the last dream they remember. The advantage of this retrospective collecting method is that dreaming is not affected by the method but, on the other hand—depending how long ago the dream was recalled in the first place—the participant might have problems remembering the dream fully. For example, research has demonstrated that intense, bizarre dreams are more often reported in such settings than mundane dreams (Schredl, 1999). Similar effects have been found for dreams reported during an interview or patients' dreams recorded by the therapist after the therapy session (Hopf, 1989). To minimize recall bias, dream diaries are an appropriate tool (Schredl, 2002). In order to optimize controllability of the experimental situation and increase the amount of dream material, dream reports collected during laboratory awakenings are the "gold standard." The major drawback of this paradigm, however, is the strong effect of the setting on dream content; i.e., up to 50% of the dreams include laboratory references (Schredl, 2008a). Other studies (e.g., Weisz and Foulkes, 1970) found that aggressive and sexual elements occur in laboratory dreams less often than in home dreams, a finding which is interpreted as an "inhibitory" effect of the lab setting (including video taping of the sleeper, technical staff presence). Some

TABLE I
METHODS OF DREAM COLLECTION

Method	Example study
Questionnaire	Domhoff (1996)
Interview	Parekh (1988)
Dream diary	Schredl (2002)
Laboratory awakenings	Strauch and Meier (2004)

researchers (e.g., Hobson and Stickgold, 1994) have tried to combine the advantages of the different methods by using ambulatory measurement units to collect dreams and physiological parameters in the home setting.

III. Dream Content Analysis

The main goal of dream content analysis is the quantification of specific aspects of the dream (e.g., number of dream persons, types of interactions, settings) in order to perform statistical analyses (Hall and Van de Castle, 1966; Schredl, 2010). The following fictive example illustrates the procedure. A clinical psychologist formulates the hypothesis that depressed patients dream more often about rejection than healthy persons. The researcher develops a scale measuring rejection (occurrence vs. not present in the dream content). Dream reports are collected from the two groups and ordered randomly so that external judges applying the content scale do not know whether the dream is a patient's dream or a control dream. After the rating procedure, the dream reports are reassigned to the two groups and the difference in percentage of dreams including at least one rejection of the dream can be tested statistically.

Within their book "Dimensions of dreams," Winget and Kramer (1979) have compiled 132 scales and rating systems. The most elaborated coding system was published by Hall and Van de Castle (1966). Domhoff (1996) presents an overview over the dream content analytic studies and resumes that this coding system is the most widely used research tool. However, it seems more appropriate to use global rating scales, e.g., for measuring emotional intensity because of the validity problem (see below). In some cases, it is also useful to use self-rating scale applied by the dreamer herself/himself, e.g., for asking about temporal references of specific dream elements.

The most important methodological issues in dream content analysis are reliability and validity. The interrater reliability designates the agreement between two or more external judges rating the same dream material using the same scales. High indices indicate that the scale can be applied easily and the findings are not biased by the raters' subjective point of view. Explicit cutoff points for sufficient interrater reliability have not yet been published. The exact agreements for the Hall and Van de Castle system vary between 61 and 98% (Hall and Van de Castle, 1966); for ordinal rating scales, the coefficients are typically between 0.70 and 0.95 (Schredl, 1999). Only one study, however, has systematically studied the effect of training on interrater reliability (Schredl et al., 2004). The findings indicate that training (coding 100–200 dreams and discussing differences) is valuable in improving reliability. In addition to the issue of interrater reliability indicating the quality of the scales, the common reliability problem has also to be considered in dream

TABLE II
DREAM EMOTIONS (N=133 DREAMS; SCHREDL AND DOLL, 1998)

Category	Self-rating (%)	Rating by judge (%)	Hall and Van de Castle (%)
No emotions	0.8	13.5	57.9
Balanced emotions	12.0	9.0	6.8
Predominantly negative emotions	50.4	56.4	26.3
Predominantly positive emotions	36.8	21.1	9.0

content research. Schredl (1998b) reported that up to 20 dream reports per participant are necessary to measure reliably interindividual differences in dream content which is necessary, for example, if dream characteristics are related to differences in personality measures.

At first glance, the validity issue seems very easy to solve; a scale designed for measuring aggression reflects the amount of aggression within the dream report. But one has to keep in mind that the researcher is genuinely interested in the dream experience itself. So, the question arises whether the dream report represents the dream experience sufficiently and, thus, can be measured by the dream content analytic scale applied to the dream report. The following example regarding the measurement of dream emotions will illustrate this line of thinking. Schredl and Doll (1998)applied three different methods of measuring dream emotions: the emotion scales of Hall and Van de Castle (1966) measuring only explicitly mentioned emotions, the four-point global rating scales designed by Schredl (1999) allowing coding emotions when it is obvious from the dream action, and similar four-point scales rated by the dreamer. The dreams were sorted into four groups (see Table II).

Two findings are striking: first, the emotions are markedly underestimated by the external judges and second, the ratio of positive and negative emotions differs, depending on the measurement technique: a balanced ratio for the self-ratings and predominantly negative for the external ratings (Schredl and Doll, 1998). A similar underestimation was found for the number of bizarre elements within the dream (Schredl and Erlacher, 2003). These two studies clearly indicate that some aspects of the dream experiences might not be measured validly if dream content analytic scales were applied to dream reports only. More research is needed to estimate the effect of this validity problem on dream content analytic findings.

IV. Phenomenology of Dreams

In the analysis of large samples of dream reports, different dream aspects like bizarreness, emotions, and perception have been characterized. Over 90% of the dreams included the dream ego and dreaming is experienced in a similar way as

TABLE III
REALISM/BIZARRENESS OF DIARY DREAMS (SCHREDL *ET AL.*, 1999b)

Category	Frequency ($N = 365$) (%)
Possible in waking life, everyday experiences	29.0
Possible in waking life, uncommon elements	49.5
One or two bizarre (impossible) elements	27.4
Several bizarre elements	4.1

waking life, the only exception is lucid dreams which occur very rarely (Schredl, 1999). About 20% of the laboratory dream reports collected by Strauch and Meier (2004) included bizarre elements, whereas about 30% were realistic (could have happened in the exact same way in waking life) and 50% were fictional, e.g., possible in real life but unlikely to happen in the dreamer's everyday life. For diary dream reports, the percentages were comparable (see Table III). Using a broader definition of bizarreness, e.g., including incongruencies obtained by comparing the dream element with waking life (e.g., a street of the home town with a new building), the number of bizarre elements increases drastically (Hobson *et al.*, 1987). Strauch and Meier (2004) and Schredl and Doll (1998) found a balance between positive and negative emotions in larger samples of laboratory and home dreams in healthy persons. Studies reporting predominantly negative dreams (e.g., Hall and Van de Castle, 1966) have to be considered with caution due to methodological problems (see above).

In Table IV, three studies investigating sensory perceptions are depicted. Visual perceptions are present in every dream. Auditory perceptions are very common, whereas tactile, gustatory, and olfactory perceptions and pain are quite rare. Colors are not very often reported spontaneously (25% of 1612 dream

TABLE IV
SENSORY PERCEPTIONS IN DREAMS

Modality	Laboratory dreams[a] ($N = 635$) (%)	Laboratory dreams[b] ($N = 107$) (%)	Diary dreams[c] ($N = 3372$) (%)
Visual	100	100	100
Auditory	76	65	53
Vestibular	–	8	–
Tactile	1	1	–
Gustatory	1	1	<1
Olfactory	<1	1	1
Pain	–	–	1

[a] Snyder (1970)
[b] McCarley and Hoffman (1981)
[c] Zadra *et al.* (1997)

reports; Schredl, 2008b) but Rechtschaffen and Buchignani (1983, 1992) who instructed their participants to compare their dream images with 129 colored pictures with different intensities and contrasts found that colors of elements which are prominent in the dream are comparable to the colors experienced in waking life, solely the background in dreams is less intense than one would expect in waking life. As the percentage of black-and-white dreams correlated negatively with the performance in color memory tasks (Schredl *et al.*, 2008), this might be merely a problem of recall and not reflect uncolored dreaming.

V. Factors Influencing Dream Content

This section will review very briefly the studies investigating the effect of waking-life experiences on subsequent dream content. In Table V, different methodological approaches for studying the relationship between waking life and dreaming are depicted.

The most detailed study regarding temporal references of dream elements were carried out by Strauch and Meier (2004). Fifty dreams stemming from REM awakenings of five subjects included 80 key role characters, 39 extras (person playing a minor role in the dream), 74 settings, and 298 objects. Strauch and Meier (2004) did not only ask about the last occurrence in waking life but also when the dream element had appeared in waking thought. Over 50% of the references were from the previous day and less than 10% were older than 1 year. The major drawback of this approach is the limited memory capacity of the subjects; it is difficult to remember completely all waking-life experiences let alone all thoughts occurring during the preceding days. Another problem is that of multiple correspondences, e.g., if the mother was present in the dream, it could not easily be determined whether this refers to a childhood experience or a recent telephone conversation.

TABLE V

PARADIGMS TO STUDY THE EFFECT OF WAKING-LIFE EXPERIENCES ON DREAM CONTENT

Research paradigms
Assessing temporal references of dream elements retrospectively
Experimental manipulation of the pre-sleep situation
Field studies
Fluctuations over time within persons
Differences between persons

The experimental approach manipulates the pre-sleep situation, most often by showing an exciting film (e.g., Foulkes and Rechtschaffen, 1964). Dreams in the night after such a film will be compared to a control condition (e.g., neutral film). Interestingly, the effect of films (even if they are strongly negatively toned) on subsequent dreams is quite small (overview: Schredl, 1999). The effect of "real" stress like intense psychotherapy or awaiting a major surgery is much stronger (Breger *et al.*, 1971). The strong effects of traumata such as war experiences (Schredl and Piel, 2006), kidnapping (Terr, 1981) or sexual abuse (Krakow *et al.*, 1995) on dreams even years later also indicate that the emotional intensity affects the incorporation rate of waking-life events into dreams. Intraindividual fluctuations over time can affect dream content considerably.

Another approach for investigating the effects of waking life on dreams is to look at differences in dreams of specific groups of persons who differ in particular aspects of their waking life, e.g., males and females. Gender differences in dream content, e.g., heightened physical aggression in male dreams, are paralleled by similar differences in waking-life behavior found in meta-analyses (Schredl *et al.*, 1998a).

The last paradigm presented in this section uses the method of correlating waking-life parameters with dream content variables. For example, the amount of time spent with a particular waking-life activity (e.g., driving a car, reading, spending time with the partner) was positively correlated with number of occurrences of these elements within dreams (Schredl and Hofmann, 2003).

In Table VI, the factors which might affect the continuity between waking life and dreaming and, thus, are of importance for a mathematical model (see next section) have been compiled from the literature by Schredl (2003a).

Many studies (e.g., Botman and Crovitz, 1989; Strauch and Meier, 2004) have shown an exponential decrease of the incorporation rate of waking-life experiences into dreams with elapsed time between experience and the subsequent dream. Also, the differences in effects of experimental stress and "real" stress and trauma research (see above) indicate that emotional involvement (EI) affects the incorporation rate. Three studies (Hartmann, 2000; Schredl, 2000a; Schredl and Hofmann, 2003) have shown that focused thinking activity (reading,

TABLE VI
FACTORS WHICH AFFECT THE CONTINUITY BETWEEN WAKING LIFE
AND DREAMING

Factors
Exponential decrease with time
Emotional involvement
Type of waking-life experience
Personality traits
Time of the night (time interval between sleep onset and dream onset

working with a computer) during dreams occurs less frequently than unfocused activities such as talking with friends. These results also indicate that the type of activity is of importance for the continuity between waking life and dreaming. The time of the night or the time interval between sleep onset and dream onset has affected the incorporation rate of waking-life experiences in two studies (Roussy *et al.*, 1998; Verdone, 1965); dreams of the second part of the night comprise more elements of the distant past, while dreams of the first part of the night incorporate mostly recent daytime experiences (cf. Lauer *et al.*, 1987; Roffwarg *et al.*, 1978). The last factor which has been studied rarely is the interaction between personality traits and incorporation of waking-life experiences into dreams. It seems plausible that personality dimensions such as field dependence (This concept describes how people are influenced by inner (field-independent) or environmental (field-dependent) cues in orienting themselves in space and the extent to which they select information within the environment) or thin boundaries moderate the magnitude of continuity between waking and dreaming (Baekeland *et al.*, 1968; Schredl *et al.*, 1996).

VI. The "Continuity Hypothesis" of Dreaming

Many researchers (e.g., Hall, 1966; Schredl, 1999; Strauch and Meier, 2004) are advocating the so-called continuity hypothesis of dreaming which simply states that dreams reflect waking-life experiences. However, for deriving specific hypotheses that can be tested empirically, the continuity hypothesis in its general formulation is too imprecise. In order to advance the research in this field, Schredl (2003a) postulated a mathematical model that is based on the published findings and seems to be promising for further empirical testing (see Table VII). The multiplying factor includes the effects of EI, type of the waking-life experience, and the interaction between personality traits and incorporation rates. The

TABLE VII

MATHEMATICAL MODEL FOR THE CONTINUITY BETWEEN WAKING LIFE AND DREAMING (SCHREDL, 2003a)

Incorporation rate $= a$ (EI, TYPE, PERS) $\times\ e^{-b(TN) \times t} +$ Constant

a (EI, TYPE, PERS)	Multiplying factor which is a function of emotional involvement (EI), type of the waking-life experience (TYPE), and the interaction between experience and personality traits (PERS)
b (TN)	Slope of the exponential function which is itself a function of the time interval between sleep onset and dream onset (TN)
t	Time interval between waking-life experience and occurrence of the dream incorporation

relationships between these factors should be determined by future studies. The slope of the exponential function may be moderated by the time interval between sleep onset and dream onset (time of the night, TN).

VII. Dreams and Psychopathology

Two motives have stimulated investigating the relationship between dreaming and mental disorders. First, the dream state itself was conceptualized by several theorists (e.g., Hobson, 1997) as a mental disorder and, in reverse, hallucinations of schizophrenic patients have been thought of as breakthroughs of dreams into the waking state (e.g., Noble, 1950). Second, many clinicians since Freud have attempted to use dreams in the diagnosis and treatment of their patients (e.g., Whitman et al., 1970). This research is based on the continuity hypothesis (see above) which predicts that dreams of patients with mental disorders reflect their specific waking symptomatology.

The literature reviews showed that the majority of empirical studies support the continuity hypothesis (Kramer and Roth, 1978; Kramer et al., 2000; Mellen et al., 1993). On the one hand, it was found that hallucinations of schizophrenic patients are not dreams experienced during the waking state (e.g., Fischman, 1983) and that the concept of dreaming as a mental disorder is not very helpful (cf. States, 1998). On the other hand, dreams of schizophrenic patients are typical for this disorder, i.e., the dreams are more bizarre (Schredl and Engelhardt, 2001) and are characterized by aggression and negative emotions (Lusignan et al., 2009). For depressive patients, Beck and Hurvich (1959) and Beck and Ward (1961) have found an increased amount of "masochistic" themes in their dreams. Subsequent studies (overview: Schredl and Schnitzler, 1999) confirmed that dreams of depressive patients are more negatively toned and include unpleasant experiences more often (these were the dream characteristics underlying the definition of "masochistic" dream content according to Beck and Hurvich (1959)) than healthy controls. Schredl and Engelhardt (2001) were able to demonstrate that severity of depressive symptomatology was directly correlated with the intensity of negative dream emotions, irrespective of the patients' diagnoses, supporting a dimensional and not categorical relationship between waking life symptoms and dream content. In addition, severe depressed patients dreamed more often about aggression and death (Schredl and Engelhardt, 2001).

Regarding dream recall, many studies failed to show marked differences among various diagnostic groups, such as schizophrenia, eating disorders, and healthy controls (Kramer and Roth, 1978). An exception is depression where patients have a reduced dream recall frequency (Armitage et al., 1995; Riemann *et al.*, 1990); the reduction was again related to symptom severity (Schredl, 1995). The explanations for the reduced dream recall in depressed patients, however, remain unclear. The question whether the typical sleep architecture of depressive patients (phase advanced REM sleep), cognitive impairment often found in severely disturbed patients or intrinsic alterations related to depression, is responsible for the reduced dream recall is yet to be answered.

VIII. Dreams and Sleep Disorders

This section will focus on several sleep disorders which have been studied in relation to dreaming: insomnia, sleep apnea syndrome, narcolepsy, and restless legs syndrome. For other diagnoses, like idiopathic hypersomnia, or NREM parasomnias, such as sleep walking or night terrors, systematic dream content analytic studies are lacking. However, extensive reviews are available for nightmares (Spoormaker *et al.*, 2006), REM sleep behavior disorder (Schenck *et al.*, 1993), and dreaming in posttraumatic stress disorder (Wittmann *et al.*, 2007).

Schredl *et al.* (1998b) found an elevated dream recall frequency in insomnia patients in contrast to healthy controls; a finding which was no longer significant if number of nocturnal awakenings (self-report measure) was statistically controlled. This parallels the correlation between nocturnal awakenings and increased dream recall in healthy persons (Schredl et al., 2003b). Percentage of dream recall after REM awakenings carried out in the laboratory did not differ between insomnia patients and controls (Ermann *et al.*, 1993). Ermann (1995) and Schredl *et al.* (1998b) found more negatively toned dreams in patients with insomnia. In addition, occurrence of problems within the dream was directly correlated with the number of waking-life problems the patients reported in the questionnaire. Nightmare frequency was also found to be higher in insomnia patients compared to healthy controls (Schredl, 2009b). Therefore, dreams might reflect the topics which are at least partially responsible for the development and maintenance of the primary insomnia.

In sleep apnea, the findings regarding the dream recall frequency are inconsistent (Schredl, 2009a). In the 19th century, nightmares were thought due to decreased flow of oxygen (e.g., due to pillow blocking of the mouth and nose; see Boerner, 1855). However, parameters like minimal oxygen saturation nadir or respiratory disturbance index do not correlate with dream recall frequency (Schredl and Schmitt, 2009; Schredl *et al.*, 2006). Furthermore, a heightened nightmare frequency in sleep apnea patients has not been found (Schredl, 2009a). In this patient group, heightened nightmare frequency is only associated with psychiatric comorbidities like mood disorder (Schredl and Schmitt, 2009; Schredl *et al.*, 2006) and, thus, are inline with the research regarding psychopathology and dream content (see above). Only very few dream reports include references to the dramatic physiological changes due to the occurrence of breathing pauses:

> During the dream I felt tied up or chained. I saw thick ropes around my arms and was not able to move. I experienced the fear of suffocation without being able to cope with the situation. Powerlessness and also resignation came up. (Patient with sleep apnea, male, 39 years, respiratory disturbance index (RDI): 68.1 apneas per hour, maximal drop of blood oxygen saturation: 43%). (Schredl, 1998c, p 295)

Overall, the low incidence of breathing-related dream topics might be explained by adaptation, i.e., the increase of number and severity of sleep apneas over months and years might explain why this stimuli is rarely incorporated into dreams, whereas external stimuli are at least sometimes incorporated into dreams (Schredl, 2009a). It would be intriguing to conduct a systematic study where the sleeper wears a mask which would allow transient occlusion of airflow and thus would allow testing whether a "novel" apnea is more often incorporated into dreams.

Narcolepsy is a sleep disorder characterized by a disinhibition of the REM sleep regulations systems, and, therefore, the findings of increased dream recall frequency (Schredl, 1998a) and higher occurrence of nightmares (Mayer et al., 2002) are not astonishing. In addition, dream content is more bizarre (Schredl, 1998a) and more negatively toned (Fosse et al., 2002), whereas dream reports of the first REM period were longer in patients compared to healthy controls (Cipolli et al., 2008). These findings clearly reflect the hyperactive REM sleep regulation system on the psychological level and, thus, help in the diagnostic process of these patients.

Lastly, Schredl (2001) found a negative correlation between number of periodic limb movements associated with arousals and dream recall frequency in a sample of 131 restless legs patients. Taken together that high respiratory disturbance indices are related to less bizarre dreams in sleep apnea patients (Schredl et al., 1999a), one might hypothesize that frequent micro-arousal might interfere with the dreaming process itself. Systematic studies in this area, however, are lacking.

To summarize, the findings in patients with sleep disorders indicate that physiological sleep processes as well as the waking-life experiences of the patients affect their dreaming.

IX. Effect of Dreams on Waking Life

Whereas the amount of studies investigating the effect of waking life on dreams is considerable, research looking into the effect of dreams on waking life is encountered quite rarely. Three major topics have been studied: (1) effect of nightmares on daytime mood, (2) creative inspiration by dreams, and (3) dreams and psychotherapy.

Schredl (2000b) has found that "dreams affect the mood of the following day" is the effect most often reported of dreams on waking life. Carrying out a carefully designed diary study, Köthe and Piotrowsky (2001) reported that days after experiencing a nightmare are rated much lower on scales of anxiety, concentration, and self-esteem than days after non-nightmare nights. The hypothesis of Belicki (1992) that the effects of nightmares on waking-life are over-estimated by persons with high neuroticism scores had not been supported by the findings of Schredl et al. (2003a). The major factor contributing to nightmare distress is nightmare frequency which is best explained by current stress and personality factors (neuroticism, thin boundaries, etc.; see Schredl, 2003b). In addition, the fact that anxiety phenomena can be perpetuated by avoidance behaviors (Schredl, 2009c) is very important regarding the therapy of nightmares.

The most effective treatment strategy for reducing nightmare frequency and their effects of waking life is imagery rehearsal therapy (IRT), developed by Barry Krakow and coworkers (Krakow and Zadra, 2006). Patients write down a recent (less intense) nightmare and they are asked to change the dream in any way they wish and to write down the altered version. Lastly, they are instructed to rehearse the new "dream" once a day over a 2-week period. Five randomized controlled trials of IRT performed with chronic nightmare sufferers (e.g., Krakow et al., 2001) showed the efficiency of confronting the nightmare anxiety (by writing down the dream) and coping with this anxiety by creating a new action pattern. Long-term follow-ups have shown stable treatment effects over time (Krakow et al., 1993).

Many examples of creative inspiration from dreams have been reported over the years (overviews: Barrett, 2001; Van de Castle, 1994): "Wild strawberries" (a film by Ingmar Bermann), the story of Dr. Jekyll and Mr. Hide by Robert Louis Stevenson, the pop song "Yesterday" by Paul McCartney, and the paintings of Salvador Dali all provide excellent examples. Kuiken and Sikora (1993) and Schredl (2000b) found that 20 and 28% of the participants (student samples), respectively, reported creative inspirations from dreams at least twice a year. In a large-scaled study with over 1000 participants (Schredl and Erlacher, 2007), about 7.8% of the recalled dreams included a creative aspect. Reported were dreams stimulating art, giving an impulse to try something new (approaching a person, traveling, etc.) or helping to solve a problem (e.g., mathematical problems, etc.). The factors that are associated with the frequency of creative dreams in this study were dream recall frequency itself, the "thin boundaries" personality dimension, a positive attitude toward creative activities and visual imagination (Schredl and Erlacher, 2007).

Although dream work is quite common in modern psychotherapy (Crook and Hill, 2003; Schredl et al., 2000) and despite the extensive literature on case reports since Freud's "The interpretation of dreams," systematic research on the efficiency of dream work is limited to the research efforts of one group. For over 10 years, Clara Hill and her coworkers carried out studies to measure the effectiveness of single dream interpretation sessions (Hill *et al.*, 1993), dream groups over 6 weeks (Falk and Hill, 1995), or dream interpretation within short-term psychotherapy (Diemer *et al.*, 1996). The basis for their work is a cognitive-experiential model of dream interpretation that includes the three stages: exploration, insight, and action (Hill, 2004). Reviewing the research, Hill and Goates (Hill and Goates, 2004) cited three sources of evidence for the beneficial effects of dream work: (1) Clients have reported on post-session measures that they gained insight, (2) judges rated clients' levels of insight into written dream interpretations as higher after dream sessions than before, and (3) clients identified gaining awareness or insight as the most helpful component of dream sessions. Similar studies for other therapeutic approaches to dream work are overdue.

The findings cited within this section clearly indicate that there is not only a continuity from waking life to dreaming but also from dreaming to waking life. A recent study (Schredl and Reinhard, 2009–2010) even demonstrated second-order effects regarding the continuity between waking life and dreaming, i.e., dreams that have been strongly affected by waking-life effects had a stronger effect on daytime mood than equal intense dreams that were not rated to be related to current concerns.

X. Conclusion and Future Directions

The reported findings clearly indicate that dream content is affected by the waking-life experiences of the dreamer, the so-called continuity hypothesis of dreaming. Future research has to clarify what kinds of factors are important for this continuity between waking and dreaming. In addition, more research is needed to study the direct effects of dreaming on waking life and more indirect effects like telling dreams or working with dreams. Duffey *et al.* (2004), for example, reported a positive effect of regular dream sharing on marital satisfaction. In order to increase the scientific credibility of dream research, several methodological issues regarding dream reporting (e.g., mood congruency effect) and dream content analysis (e.g., validity) have to be studied very carefully.

One of the major goals of psychological dream research is of course to determine the possible function(s) of dreaming. But investigating this venue is limited by a methodological issue, namely, how to differentiate between a direct beneficial effect of dreams during the night on subsequent waking life from the effect of thinking about the dream during waking life. The following example will illustrate this problem. Cartwright *et al.* (1984) found that divorcing women dreaming about their ex-husband are more psychologically adapted than women who dreamed about other topics. The authors concluded that working through the divorce issue within in the dream serves an adaptive function. But one might argue that the women who reported the ex-husband dreams began to think about the dream and, therefore, were able to cope better with the stressful divorce. One cannot differentiate between the effect of the dreamed dream and the effect of the recalled, reported dream (necessarily processed by the waking mind), i.e., we do not know and might never know whether unremembered dreams serve any function.

References

Armitage, R. *et al.* (1995). Dream recall and major depression: a preliminary report. *Dreaming* **5**, 189–198.

Aserinsky, E., and Kleitman, N. (1953). Regularly occurring periods of eye motility and concomitant phenomena during sleep. *Science* **118**, 273–274.

Baekeland, F. *et al.* (1968). Presleep metation and dream reports: I. Cognitive style, contiguity to sleep and time of the night. *Arch. Gen. Psychiatry* **19**, 300–311.

Barrett, D. (2001). The committee of sleep: how artists. Scientists, and Athletes use Dreams for Creative Problem-Solving—and How You can too. Crown, New York.

Beck, A. T., and Hurvich, M. S. (1959). Psychological correlates of depression: I. frequency of "masochistic" dream content in a private practice sample. *Psychosom. Med.* **1**, 50–55.

Beck, A. T., and Ward, C. H. (1961). Dreams of depressed patients. *Arch. Gen. Psychiatry* **5**, 462–467.

Belicki, K. (1992). Nightmare frequency versus nightmare distress: relation to psychopathology and cognitive style. *J. Abnorm. Psychol.* **101**, 592–597.

Boerner, J. (1855). Das Alpdrücken: Seine Begründung und Verhütung. Carl Joseph Becker, Würzburg.

Botman, H. I., and Crovitz, H. F. (1989). Dream reports and autobiographical memory. *Imagin. Cogn. Pers.* **9**, 213–224.

Breger, L. *et al.* (1971). The Effect of Stress on Dreams. International Universities Press, New York.

Cartwright, R. D. *et al.* (1984). Broken dreams: a study of the effects of divorce and depression on dream content. *Psychiatry.* **47**, 251–259.

Cipolli, C. *et al.* (2008). Story-like organization of REM-dreams in patients with narcolepsy-cataplexy. *Brain Res. Bull.* **77**, 206–213.

Crook, R. E., and Hill, C. E. (2003). Working with dreams in psychotherapy: the therapist's perspective. *Dreaming* **13**, 83–93.

Dement, W. C., and Kleitman, N. (1957). The relation of eye movements during sleep to dream activity: an objective method for the study of dreaming. *J. Exp. Psychol.* **53**, 339–346.

Diemer, R. A. *et al.* (1996). Comparison of dream interpretation, event interpretation, and unstructured sessions in brief therapy. *J. Consult. Psychol.* **43**, 99–112.

Domhoff, G. W. (1996). Finding Meaning in Dreams: A Quantitative Approach. Plenum Press, New York.

Duffey, T. H. *et al.* (2004). The effects of dream sharing on marital intimacy and satisfaction. *J. Couple Relatsh Ther.* **3**, 53–68.

Erlacher, D., and Schredl, M. (2008). Do REM (lucid) dreamed and executed actions share the same neural substrate? *Int. J. Dream Res.* **1**, 7–14.

Ermann, M. *et al.* (1993). Spontanerwachen und Träume bei Patienten mit psychovegetativen Schlafstörungen. *Psychother. Psychosom. Medizinische Psychol.* **43**, 333–340.

Ermann, M. (1995). Die Traumerinnerung bei Patienten mit psychogenen Schlafstörungen: Empirische Befunde und einige Folgerungen für das Verständnis des Träumens. In: Traum und Gedächtnis: Neue Ergebnisse aus psychologischer, psychoanalytischer und neurophysiologischer Forschung. (W. Leuschner and S. Hau, eds.), LIT Verlag, Münster, pp. 165–186.

Falk, D. R., and Hill, C. E. (1995). The effectiveness of dream interpretation groups for women undergoing a divorce transition. *Dreaming* **5**, 29–42.

Fischman, L. G. (1983). Dreams, hallucinogenic drug states and schizophrenia: a psychological and biological comparison. *Schizophr Bull.* **9**, 73–94.

Fosse, R. *et al.* (2002). Emotional experience during rapid-eye-movement sleep in narcolepsy. *Sleep* **25**, 724–732.

Foulkes, D. (1962). Dream reports from different stages of sleep. *J. Abnorm. Soc. Psychol.* **65**, 14–25.

Foulkes, D., and Rechtschaffen, A. (1964). Presleep determinants of dream content. *Percept. Mot. Skills* **19**, 983–1005.

Freud, S. (1987). Die Traumdeutung (1900). Fischer Taschenbuch, Frankfurt.

Hall, C. S. (1966). The Meaning of Dreams. McGraw-Hill, New York.

Hall, C. S. *et al.* (1982.). The dreams of college men and women in 1959 and 1980: a comparison of dream contents and sex differences. *Sleep* **5**, 188–194.

Hall, C. S., and Van de Castle, R. L. (1966). The Content Analysis of Dreams. Appleton-Century-Crofts, New York.

Hartmann, E. (2000). We do not dream of the 3 R's: implications for the nature of dream mentation. *Dreaming* **10**, 103–110.

Hill, C. E. *et al.* (1993). Are the effects of dream interpretation on session quality, insight and emotion due to the dream itself, to projection or to the interpretation process? *Dreaming* **3**, 269–280.

Hill, C. E. (2004). Introduction to "Dream work in therapy." In: Dream Work in Therapy: Facilitating Exploration, Insight, and Action (C. E. Hill, ed.), American Psychological Association, Washington, pp. 3–15.

Hill, C. E., and Goates, M. K. (2004). Research on the Hill cognitive-experiential dream model. In: Dream Work in Therapy: Facilitating Exploration, Insight, and Action (C. E. Hill, ed.), American Psychological Association, Washington, pp. 245–288.

Hobson, J. A. *et al.* (1987). Dream bizarreness and the activation-synthesis hypothesis. *Hum. Neurobiol.* **6**, 157–164.

Hobson, J. A. (1997). Dreaming as delirium: a mental status examination of our nightly madness. *Semin. Neurol.* **17**, 121–128.

Hobson, J. A. *et al.* (2000). Dreaming and the brain: toward a cognitive neuroscience of conscious states. *Behav. Brain Sci.* **23**, 793–842.

Hobson, A. (2009). The neurobiology of consciousness: lucid dreaming wakes up. *Int. J. Dream Res.* **2**, 41–44.

Hobson, J. A., and Stickgold, R. (1994). Dreaming: a neurocognitive approach. *Conscious Cogn.* **3**, 1–15.

Hopf, H. H. (1989). Wie "objektiv" sind unsere Notizen? Ein Vergleich von Patiententraumprotokollen mit von Therapeuten niedergeschriebenen Protokollen. *Kind und Umwelt.* **63**, 42–52.

Köthe, M., and Pietrowsky, R. (2001). Behavioral effects of nightmares and their correlations to personality patterns. *Dreaming* **11**, 43–52.

Krakow, B. *et al.* (1993). Imagery rehearsal treatment of chronic nightmares: with a thirty month follow-up. *J. Behav. Ther. Exp. Psychiatry* **24**, 325–330.

Krakow, B. *et al.* (1995). Nightmares and sleep disturbance in sexually assaulted women. *Dreaming* **5**, 199–206.

Krakow, B. *et al.* (2001). Imagery rehearsal therapy for chronic nightmares in sexual assault survivors with posttraumatic stress disorder: a randomized controlled trial. *J. Am. Med. Assoc.* **286**, 537–545.

Krakow, B., and Zadra, A. (2006). Clinical management of chronic nightmares: imagery rehearsal therapy. *Behav. Sleep Med.* **4**, 45–70.

Kramer, M. (2000). Dreams and psychopathology. In: Principles and Practice of Sleep Medicine (M. H. Kryger. *et al.*, eds.), W. B. Saunders, Philadelphia, pp. 511–519.

Kramer, M., and Roth, T. (1978). Dreams in psychopathologic patient groups. In: Sleep Disorders: Diagnosis and Treatment (R. L. Williams and I. Karacan, eds.), John Wiley & Sons, New York, pp. 323–349.

Kuiken, D., and Sikora, S. (1993). The impact of dreams on waking thoughts and feelings. In: The Functions of Dreaming (A. Moffitt *et al.*., eds.), State University of New York Press, Albany, pp. 419–476.

Lauer, C. *et al.* (1987). Shortened REM latency: a consequence of psychological strain? *Psychophysiology* **24**, 263–271.

Lusignan, F. -A. *et al.* (2009). Dream content in chronically-treated persons with schizophrenia. *Schizophr. Res.* **112**, 164–173.

Maury, A. (1861). Le sommeil et les reves. Didier, Paris.

Mayer, G. *et al.* (2002). Untersuchung zur Komorbidität bei Narkolepsiepatienten. *Dtsch. Med. Wochenschr.* **127**, 1942–1946.

McCarley, R. W., and Hoffman, E. (1981). REM sleep dreams and the activation-synthesis hypothesis. *Am. J. Psychiatry.* **138**, 904–912.

Mellen, R. R. *et al.* (1993). Manifest content in the dreams of clinical populations. *J. Ment. Health Couns.* **15**, 170–183.

Nielsen, T. A. (2000). A review of mentation in REM and NREM sleep: "covert" REM sleep as a possible reconciliation of two opposing models. *Behav. Brain Sci.* **23**, 851–866.

Nielsen, T. A., and Levin, R. (2007). Nightmares: a new neurocognitive model. *Sleep Med. Rev.* **11**, 295–310.

Noble, D. (1950). A study of dreams in schizophrenia and allied states. Am. J. Psychiatry **107**, 612–616.

Parekh, H. (1988). Träume der "Gesunden"—Inhaltsanalyse von manifesten Traumtexten aus einer Zufallsstichprobe einer Großstadtpopulation. Dissertation für Klinische Medizin Mannheim, Universität Heidelberg.

Rechtschaffen, A., and Buchignani, C. (1983). Visual dimensions and correlates of dream images. *Sleep Res.* **12**, 189.

Rechtschaffen, A., and Buchignani, C. (1992). The visual appearance of dreams. In: The Neuropsychology of Sleep and Dreaming (J. S. Antrobus and M. Bertini, eds.), Lawrence Erlbaum, Hillsdale, pp. 143–155.

Riemann, D. *et al.* (1990). Investigations of morning and laboratory dream recall and content in depressive patients during baseline conditions and under antidepressive treatment with trimipramine. *Psychiat. J. Univ. Ott.* **15**, 93–99.

Roffwarg, H. P. *et al.* (1978). The effects of sustained alterations of waking visual input on dream content. In: The Mind in Sleep: Psychology and Psychophysiology (A. M. Arkin, et al., eds.), Lawrence Erlbaum, Hillsdale, New Jersey, pp. 295–349.

Roussy, F. *et al.* (1998). Temporal references in manifest dream content: confirmation of increased remoteness as the night progresses. *Sleep Suppl.* **21**, 285.

Schenck, C. H. *et al.* (1993). REM sleep behavior disorder: an update on a series of 96 patients and a review of the world literature. *J Sleep Res.* **2**, 224–231.

Schredl, M. (1995). Traumerinnerung bei depressiven Patienten. *Psychother. Psychosom. Medizinische Psychol.* **45**, 414–417.

Schredl, M. *et al.* (1996). Dreaming and personality: thick vs. thin boundaries. *Dreaming* **6**, 219–223.

Schredl, M. (1998a). Dream content in patients with narcolepsy: preliminary findings. *Dreaming* **8**, 103–107.

Schredl, M. (1998b). The stability and variability of dream content. *Percept. Mot. Skills* **86**, 733–734.

Schredl, M. (1998c). Träume und Schlafstörungen: Empirische Studie zur Traumerinner-ungshäufigkeit und zum Trauminhalt schlafgestörter PatientInnen. Tectum, Marburg.

Schredl, M. *et al.* (1998a). Gender differences in dreams: do they reflect gender differences in waking life? *Pers Individ Dif.* **25**, 433–442.

Schredl, M. *et al.* (1998b). Dreaming and insomnia: dream recall and dream content of patients with insomnia. *J. Sleep Res.* **7**, 191–198.

Schredl, M. (1999). Die nächtliche Traumwelt: Eine Einführung in die psychologische Traum-forschung. Kohlhammer, Stuttgart.

Schredl, M. *et al.* (1999a). Dream content of patients with sleep apnea. *Somnologie* **3**, 319–323.

Schredl, M. *et al.* (1999b). Dream content and personality: thick vs. thin boundaries. *Dreaming* **9**, 257–263.

Schredl, M. (2000a). Continuity between waking life and dreaming: are all waking activities reflected equally often in dreams. *Percept. Mot. Skills* **90**, 844–846.

Schredl, M. (2000b). The effect of dreams on waking life. *Sleep Hypn.* **2**, 120–124.

Schredl, M. *et al.* (2000). The use of dreams in psychotherapy: a survey of psychotherapists in private practice. *J. Psychother. Pract. Res.* **9**, 81–87.

Schredl, M. (2001). Dream recall frequency and sleep quality of patients with restless legs syndrome. *Eur. J. Neurol.* **8**, 185–189.

Schredl, M. (2002). Questionnaire and diaries as research instruments in dream research: methodo-logical issues. *Dreaming* **12**, 17–15.

Schredl, M. (2003a). Continuity between waking and dreaming: a proposal for a mathematical model. *Sleep Hypn.* **5**, 38–52.

Schredl, M. (2003). Effects of state and trait factors on nightmare frequency. *Eur. Arch. Psychiatr. Clin. Neurosci.* **253**, 241–247.

Schredl, M. *et al.* (2003a). Nightmare frequency, nightmare distress and neuroticism. *North Am. J. Psychol.* **5**, 345–350.

Schredl, M. *et al.* (2003b). Factors of home dream recall: a structural equation model. *J Sleep Res.* **12**, 133–141.

Schredl, M. *et al.* (2004). The effect of training on interrater reliability in dream content analysis. *Sleep Hypn.* **6**, 139–144.

Schredl, M. *et al.* (2006). Nightmares and oxygen desaturations: is sleep apnea related to heightened nightmare frequency? *Sleep Breath* **10**, 203–209.

Schredl, M. (2007). Dream recall: models and empirical data. In: The New Science of Drea-ming—Volume 2: Content, Recall, and Personality Correlates (D. Barrett and P. McNamara, eds.) Praeger, Westport, pp. 79–114.

Schredl, M. (2008a). Laboratory references in dreams: methodological problem and/or evidence for the continuity hypothesis of dreaming? *Int. J. Dream Res.* **1**, 3–6.

Schredl, M. (2008b). Spontaneously reported colors in dreams: correlations with attitude towards creativity, personality and memory. *Sleep Hypn.* **10**, 54–60.

Schredl, M. (2008c). Traum. Reinhardt/UTB, München.

Schredl, M. *et al.* (2008). Do we think dreams are in black and white due to memory problems? *Dreaming* **18**, 175–180.

Schredl, M. (2009a). Dreams in patients with sleep disorders. *Sleep Med. Rev.* **13**, 215–221.

Schredl, M. (2009b). Nightmare frequency in patients with primary insomnia. *Int. J. Dream Res.* **2**, 85–88.

Schredl, M. (2009c). Nightmares. In: Encyclopedia of Neuroscience Vol. 6. (L. R. Squire, ed.) Academic Press, Oxford, pp. 1145–1150.

Schredl, M. (2010). Dream content analysis: basic principles. *Int. J. Dream Res.* **3**, 65–73.

Schredl, M., and Doll, E. (1998). Emotions in diary dreams. *Conscious Cogn.* **7**, 634–646.

Schredl, M., and Engelhardt, H. (2001). Dreaming and psychopathology: dream recall and dream content of psychiatric inpatients. *Sleep Hypn.* **3**, 44–54.

Schredl, M., and Erlacher, D. (2003). The problem of dream content analysis validity as shown by a bizarreness scale. *Sleep Hypn.* **5**, 129–135.

Schredl, M., and Erlacher, D. (2007). Self-reported effects of dreams on waking-life creativity: an empirical study. *J. Psychol.* **141**, 35–46.

Schredl, M., and Hofmann, F. (2003). Continuity between waking activities and dream activities. *Conscious Cogn.* **12**, 298–308.

Schredl, M., and Piel, E. (2006.). War-related dream themes in Germany from 1956 to 2000. *Polit. Psychol.* **27**, 299–307.

Schredl, M., and Reinhard, I. (2009–2010). The continuity between waking mood and dream emotions: direct and second-order effects. *Imagin. Cogn. Pers.* **29**, 271–282.

Schredl, M., and Schmitt, J. (2009). Dream recall frequency and nightmare frequency in patients with sleep disordered breathing. *Somnologie* **13**, 12–17.

Schredl, M., and Schnitzler, M. (1999). Träume und Depression—Eine Literaturübersicht. *Psycho* **25**, 693–696.

Snyder, F. (1970). The phenomenology of dreaming. In: The Psychodynamic Implications of the Physiological Studies on Dreams (L. Madow and L. H. Snow, eds.), Charles C. Thomas, Springfield, pp. 124–151.

Spoormaker, V. I. *et al.* (2006). Nightmares: from anxiety symptom to sleep disorder. *Sleep Med. Rev.* **10**, 19–31.

States, B. O. (1998). Dreaming as psychosis: re-reading Allan Hobson. *Dreaming* **8**, 137–148.

Strauch, I., and Meier, B. (2004). Dem Traum auf der Spur: Zugang zur modernen Traumforschung (2.Auflage). Hans Huber, Bern.

Stuck, B. A. *et al.* (2009). Praxis der Schlafmedizin. Springer, Heidelberg.

Terr, L. C. (1981). Psychic trauma in children: observations following the Chowchilla school-bus kidnapping. *Am. J. Psychiatry* **138**, 14–19.

Van de Castle, R. L. (1994). Our Dreaming Mind. Ballantine, New York.

Verdone, P. (1965). Temporal reference of manifest dream content. *Percept. Mot. Skills* **20**, 1253–1268.

Weisz, R., and Foulkes, D. (1970). Home and laboratory dreams collected under uniform sampling conditions. *Psychophysiology* **6**, 588–596.

Whitman, R. M. *et al.* (1970). The varying uses of the dream in clinical psychiatry. In: The Psychodynamic Implications of the Physiological Studies on Dreams (L. Madow and L. H. Snow, eds.), Charles C. Thomas, Springfield, pp. 24–46.

Winget, C., and Kramer, M. (1979). Dimensions of Dreams. University of Florida Press, Gainesville.

Wittmann, L. *et al.* (2007). The role of dreaming in posttraumatic stress disorder. *Psychother. Psychosom.* **76**, 25–39.

Wittmann, L., and Schredl, M. (2004). Does the mind sleep? An answer to "What is a dream generator?" *Sleep Hypn.* **6**, 177–178.

Zadra, A. L. *et al.* (1997). The prevalence of auditory, olfactory, gustatory and pain experiences in 3372 home dreams. *Sleep Res.* **26**, 281.

TRAIT AND NEUROBIOLOGICAL CORRELATES OF INDIVIDUAL DIFFERENCES IN DREAM RECALL AND DREAM CONTENT

Mark Blagrove* and Edward F. Pace-Schott†

*Department of Psychology, School of Human and Health Sciences, Swansea University, Wales, UK
†Department of Psychology, University of Massachusetts Amherst, Amherst, MA, USA

Individuals differ greatly in their dream recall frequency, in their incidence of recalling types of dreams, such as nightmares, and in the content of their dreams. This chapter reviews work on the waking life correlates of these differences between people in their experience of dreaming and reviews some of the neurobiological correlates of these individual differences. The chapter concludes that despite there being trait-like aspects of general dream recall and of dream content, very few psychometrically assessed correlates for dream recall frequency and dream content have been found. More successful has been the investigation of correlates of frequency of particular types of dreams, such as nightmares and lucid dreams, and also of how waking-life experience is associated with dream content. There is also potential in establishing neurobiological correlates of individual differences in dream recall and dream content, and recent work on this is reviewed.

I. Individual Differences in DRF

A. MEMORY AND DRF

A simple explanation of why individuals differ in their ability to remember dreams is because they differ in some more general memory ability. There is some support for an association of DRF with visual memory (Butler and Watson, 1985; Cory et al., 1975; Schredl and Montasser, 1996–1997a) but also evidence against it for verbal and visual material (Cohen, 1971) and short- or long-term story narrative recall (Blagrove and Akehurst, 2000).

The salience hypothesis holds that some individuals have more vivid or striking dreams and that it is this characteristic of the dream that leads to it being better recalled (Cohen, 1974a). In addressing differences between people in dream recall, the aim for testing this hypothesis would be to correlate dream recall with vividness-related characteristics of waking mental imagery or with vividness-related characteristics of dreams. For example, Hiscock and Cohen (1973) found high dream recallers have a generalized capacity for visualization in waking life (including visualizing clear, vivid, and controllable images), which they proposed may contribute to the quality of the dreaming experience and, consequently, to its recallability. Notably, persons with posterior cortical lesions who suffer from inability to visualize in waking may also have non-visual dreams (Solms, 1997). Cohen and MacNeilage (1974) found that verbal reports of the dreams of frequent recallers were more salient than dream reports of low-frequency recallers, but with this difference between the groups being greater for non-rapid eye movement (NREM) than rapid eye movement (REM) dreams. Here, salience was calculated as the sum of scores of vividness, bizarreness, emotionality, and activity for each dream report. This hypothesis, however, has the problem of how to assess dream salience independently of dream reports.

B. PERSONALITY AND DRF

Several personality variables have been found to be related to DRF, but the literature is inconsistent. Absorption refers to how involved one can be with experiences, such as watching a film or having a daydream (Tellegen and Atkinson, 1974). Some studies have found its significant association with DRF (Levin and Young, 2001; Schredl et al., 1997; Spanos et al., 1980–1902), but others have not (Levin et al., 2003; Schredl et al., 2003).

Support for an association of DRF with creativity is provided by Bone and Corlett (1968) using the remote associates test and Fitch and Armitage (1989) using the Torrance Tests of Creative Thinking. However, Domino (1976) studied the dream content of creative and non-creative participants and found only a negligible difference in dream recall between the two groups. Importantly, assignment to groups here was performed by the students' teachers on the basis of students' "originality, adaptiveness to reality, and elaboration of original insight," which may have less of a verbal intelligence confound than the remote associates and original uses tests mentioned above. It is thus not clear if creativity is associated with higher dream recall.

Schechter et al. (1965) found that the proportion of dream recallers was greater among art students than science students, who had greater recall than engineering students. The art students were also higher on the creativity measure used, but it is not clear if the difference in dream recall is due to differences in the disciplines studied rather than differences in creativity between the disciplines.

Giesbrecht and Merckelbach (2006) and Watson (2003) found that individuals high in absorption, imagination, and fantasy proneness are particularly likely to remember their dreams and to report other vivid nocturnal experiences, these results being consistent with a salience model of dream recall and a continuity model of human consciousness. Support is also given by Tonay (1993) and Levin and Young (2001–2002), although Levin et al. (2003) found a small correlation that was only significant for women.

One of the most researched correlates of DRF is thin and thick boundariness. Hartmann (1989) defined boundariness as the degree of separation of intrapsychic components of the mind and found that boundary thinness was correlated positively with frequency of dream recall, a result that was replicated by Hartmann (1991), who reported a correlation of $r = .40$ between thin boundariness and dream recall, and by Hartmann et al. (1991), Schredl and Engelhardt (2001), and Schredl et al. (1999). Although Funkhouser et al. (2001) found no significant correlation between boundary score and dream recall, telling one's dreams resulted in a small increase in boundary score, i.e., in the direction of thinner boundaries, which raises the possibility of the direction of causality in the boundariness–DRF correlations being that dream recall results in thinner boundaries.

Watson (2003) found that dream recall was associated with openness to experience (OE) but not with the other Big Five traits. However, only small correlations with OE were found by Schredl (2002) and Schredl *et al.* (2003), the latter finding a much larger correlation of openness and boundariness with attitude toward dreams (ATD). This finding indicates that Schonbar's (1965) "lifestyle" hypothesis could be revised as positive ATD, and not high DRF, being part of a broader "lifestyle" of acceptance of inner processes.

It may be that people recall more dreams because they have a favorable attitude toward them, in that an unfavorable attitude results in dreams not being attended to on waking, with the dreams hence being forgotten. Cernovsky (1984) devised a 16-item measure of ATD. The measure has three subscales: the person's own ATD, the person's perception of the attitudes of their significant others toward dreams, and the person's perception of the attitudes of people in general toward dreams. DRF correlated significantly with the first subscale ($r = .32$) and with the whole 16-item scale ($r = .31$).

The relationship of ATD and DRF has been confirmed by Tonay (1993). However, Schredl *et al.* (2002) found that the relationship between ATD and DRF appears to be not as strong as previously reported when items with direct reference to dream recall are not included in the scale. Similarly, Domhoff (1968) found that ATD, as measured by a semantic differential scale, did not differ between recallers and nonrecallers. In general the literature supports a small correlation between DRF and ATD; however, the direction of causation is unclear. It may be that a positive attitude increases the remembering of dreams or it may be that frequent dream recall causes greater appreciation of dreams.

Tart (1962), Farley *et al.* (1971) and Blagrove and Akehurst (2000) found no relationship between dream recall and neuroticism. There are, however, some small correlations in Bone (1968), Lang and O'Connor (1984) and Watson (2001). Whereas Bone (1968) found a correlation between extraversion and dream recall in females, Farley *et al.* (1971) and Blagrove and Akehurst (2000) found no relationship between these variables.

C. Psychopathology and DRF

1. *Depression*

In Barrett and Loeffler (1992), depressed participants recalled fewer dreams, had significantly shorter dream length, displayed less anger in their dreams, and had fewer characters, especially strangers, in their dreams. A trend for lower dream recall in depressed psychiatric inpatients, and significantly shorter dreams, was also found by Schredl and Engelhardt (2001) and Armitage *et al.* (1995). In the latter study most dreams were short and bland. The relationship between depression and dream recall, however, seems to be in the other direction for non-clinical depression. Robbins and Tanck (1988–1989) had undergraduates keep a 10 day diary and found depression tended to be higher preceding nights for which dreams were recalled and described than for nights in which participants reported no dreaming.

2. *Anxiety and Dissociation*

Connor and Boblitt (1970) found a significant correlation between anxiety and dream recall ($r = .44$), but the review of correlational studies in Blagrove and Haywood (2006) finds a median correlation of only 0.26 from such studies. Watson (2001) used the Iowa Sleep Experiences Scale that combined DRF with various dream intensity and content items, such as frequency of flying dreams, of vivid dreams, and of dreams on waking and on going to sleep. Dissociation subscales had moderate correlations with Sleep Experiences score ($r = .42$–.57), and schizotypy subscales had small to moderate correlations with Sleep Experiences ($r = .31$–.47). Watson interprets the results as showing that Sleep Experiences are related to absorption, imagination, daydreaming, and fantasy, with high dream recall and sleep experience scoring individuals being able to pass more easily between different states of consciousness, as in Hartmann's concept of thin boundaries.

D. Trait and Neurobiological Variables Associated with Frequency of Types of Dreams

1. *Lucid Dreaming*

Lucid dreaming is the ability to recognize while dreaming that one is dreaming. Frequency of lucid dreaming has been associated with internal Locus of Control and Need for Cognition (Blagrove and Hartnell, 2000; Patrick and Durndell, 2004), creativity (Blagrove and Hartnell, 2000; Brodsky *et al.*, 1990–1991), and Stroop task performance (Blagrove *et al.*, in press). There is some support for lucid dreaming frequency being related to field independence (Gackenbach *et al.*, 1985; Gruber *et al.*, 1995; Patrick and Durndell, 2004), but this was not found by Blagrove and Tucker (1994). Blagrove and Wilkinson (2010) Lucid dreaming frequency is not related to the ability to quickly identify the changes that occur in change blindness tasks; (Blagrove and Wilkinson, 2010), nor to OE or any of the other Big Five dimensions (Schredl and Erlacher, 2004).

Assessments of how many people report having had a lucid dream at least once in their lifetime include 82% of participants in a student sample in Schredl and Erlacher (2004), 58% in a conservative summary estimate in Snyder and Gackenbach's (1988) review, and 47–92% in Erlacher *et al.*'s (2008) review. The latter review opens the possibility of the incidence of lucid dreaming having increased over recent decades. This may be due to increased public knowledge of the existence of lucid dreams or, for some young people, due to frequent video game use (Gackenbach, 2009).

Although most work on explanations for individual differences in lucid dreaming have addressed personality correlates, there may also be electroencephalographic (EEG) correlates. Lucid dreaming has EEG differences from waking as well as from REM sleep, particularly in frontal areas (Voss *et al.*, 2009) and in lucid dreaming, a tendency toward the greatest increase in EEG beta power was observed in the left parietal lobe (P3), an area of the brain considered to be related to semantic understanding and self-awareness (Holzinger *et al.*, 2006).

2. *Nightmares*

Nightmare frequency is associated with fantasy proneness and psychological absorption (Levin and Fireman, 2001–2002) as well as thin boundariness (Cowen and Levin, 1995; Hartmann, 1989; Levin *et al.*, 1998; Schredl *et al.*, 1996–1999). An association between nightmare frequency and neuroticism was found by Berquier and Ashton (1992) and Schredl (2003) but not by Roberts and Lennings (2006) or Chivers and Blagrove (1999). Although an association with psychosis proneness was found by Hartmann *et al.* (1981, 1987), Levin (1998), Levin and Raulin (1991), Claridge *et al.* (1997), and Roberts and Lennings (2006), this association was not found by Chivers and Blagrove (1999) or Levin and Fireman (2002) or Berquier and Ashton (1992). Additionally, Kales *et al.* (1980) found no evidence for a specific relationship between nightmares and psychosis proneness, as opposed to psychopathology in general.

Potential for artistic achievement has also been linked to frequent nightmares (Hartmann *et al.*, 1981, 1987), although Chivers and Blagrove (1999) did not find an association between self-assessed creativity and nightmare frequency and Levin *et al.* (1991) found mixed support for this relationship. Nightmare frequency has also been associated with Dissociative Disorder (Agargun *et al.*, 2003), early-life maltreatment (Schäfer and Bader, 2009), borderline personality disorder (Semiz *et al.*, 2008), and heart rate variability (Nielsen *et al.*, 2010). Simard *et al.* (2008) found the prevalence of frequent bad dreams in preschool children to be 1.3–3.9%, with its prevalence being trait-like and associated with an anxious temperament measured as early as at 5 months.

In adolescents, Nielsen *et al.* (2000) found that anxiety at ages 13 and 16 years was related to the presence of disturbing dreams at those ages, and anxiety at 13 years was predictive of disturbing dreams at 16 years. Using a more general measure of waking-life acute stress, Chivers and Blagrove (1999) found nightmare frequency correlated significantly with General Health Questionnaire score. Anxiety, depression, neuroticism, and stress were all found to be associated with frequency of nightmares as well as with the confounding variable of distress caused by nightmares (Blagrove et al., 2004). Furthermore, Blagrove and Fisher (2009) found that thin boundariness, Symptom Checklist—Global Severity, adverse life events, and childhood adversity were associated with the incidence of nightmares on a night-by-night basis as a function of state anxiety or depression.

E. DISCUSSION OF PART I

The above review shows that most traits that have been proposed as correlates of DRF have been found to have, at best, only weak relationships with DRF. Schredl et al. (2003) came to a similar conclusion, with the four factors they found to be significantly related to DRF—personality (OE, thin boundaries, absorption), creativity, nocturnal awakenings, and ATD—explaining only 8.4% of the total variance in DRF. Levin et al. (2003) similarly concluded that DRF is largely independent of stable personality traits and is better understood in terms of expectancy and attitudinal factors.

It may even be that these small relationships are inflated and spurious. Beaulieu-Prévost and Zadra (2007) show that two of the strongest personality relationships with retrospective DRF, boundariness and absorption, become negligible when dream recall is assessed prospectively, and even ATD had only a marginal relationship with prospective dream recall. Inflation of correlations may also occur due to expectancy effects when a study only assesses a small number of personality variables (as noted by Schredl et al., 2003) or where a retrospective dream recall questionnaire is completed in the same session as the personality measure (as noted by Levin and Young, 2001–2002).

Although there remains the possibility that some rarely utilized variable such as length or detail of dream reports may be more strongly related to personality traits than is DRF, as suggested and reviewed by Blagrove (2007), the general conclusion from the literature is that the correlations between trait measures and dream recall are at best small and may even then be spuriously inflated. It may thus be that state factors are the main factors affecting dream recall. Such state factors may be, for example, nocturnal awakening, focusing on dreams in the morning (Schredl and Montasser, 1996–1997b), pre-sleep mood (Cohen, 1974b), physiological arousal during dreaming (Farley et al., 1971), individual differences in likelihood of waking from REM sleep (Webb and Kersey, 1967), increased high-frequency beta incidence (Rochlen et al., 1998), or EEG spectral power values (Takeuchi *et al.*, 2001). There may even be individual differences in ponto-geniculo-occipital (PGO) stimulation, a characteristic of mammalian REM sleep. Stuart and Conduit (2009) describe the debate in dream research as to whether phasic waves or cortical arousal during sleep underlies the biological mechanisms of dreaming. They found that auditory stimulation during REM sleep was related to an increase in EEG arousal, a decrease in the amplitude and frequency of eye movements, and a decrease in the frequency of visual imagery reports on awakening. These results provide phenomenological support for PGO-based theories of dream production, and it may be speculated that individuals differ in the activity of this PGO system.

A simpler possibility is put forward by Schredl and Reinhard (2008), who describe how frequent dream recall should in theory be related to longer sleep duration, because of associated longer REM sleep. Indeed they found that changes in an individual's sleep length, and also differences between people in usual sleep duration, both affect dream recall. Along the same lines Pagel and Shocknesse (2007) found a decline in polysomnographic sleep quality was associated with a decline in reported dream and nightmare recall frequency, which, they hypothesize, is due to disrupted sleep not being able to produce dreams.

F. Do the Individual Difference Variables Act at Dream Production or Recall?

It may be difficult to ascertain whether an individual difference variable that correlates with dream recall acts at the level of dream production, or of recall, or at both levels. For example, on the dream salience explanation for differences in DRF, a variable that is associated with the ability to form vivid images may also be associated with the ability to recall images. The problem with explanations involving individual difference variables is that in almost all cases an argument can be made for how they could act at dream recall as well as at dream production. For example, creative people might dream more or they might dream just as much as non-creative people but be better at recalling those dreams. One series of studies, though, that may have successfully excluded the recall confound, is described in Foulkes (1999). Here, in children, level of dream recall was related to spatial reasoning as assessed by the block design task. Foulkes claims that the abilities assessed by the spatial reasoning task are clearly concerned with dream formation rather than with the interpersonal matters of dream recall and telling. Similarly, an association between score on block design and recall of REM dreams of adults in the sleep laboratory was found by Butler and Watson (1985), to which the same argument can be applied. Notably, damage to the inferior parietal lobe, especially on the right, disrupts spatial cognition and may result in global cessation of dreaming (Solms, 1997).

Also in favor of the view that some people may produce more dreams than others is the finding by Dement and Kleitman (1957) that people differ in their ability to recall dreams when woken in the sleep laboratory. However, it could still be argued that, although these differences between people occur in a situation where one is suddenly woken and asked to report a dream, there may still be a memory effect operating. This argument follows from the results of Conduit *et al.* (2004) where, compared to a REM sleep condition, in NREM stage 2, participants were less able to judge correctly whether they had responded to a tone presented during sleep. Stage 2 sleep thus has a memory recall deficit in comparison to REM sleep. Conduit *et al.* concluded that the low level of dream recall from stage 2 sleep may be due to deficiencies in recall, rather than because dreams are produced at a lower level, or more rarely, in stage 2 sleep than in REM sleep. It is arguable that if stage 2 sleep has a dream recall deficiency in comparison to REM sleep, then there may be individual differences in this dream recall deficiency, both for REM and for NREM dreams.

II. Individual Differences in Dream Content

A. DREAM CONTENT AND WAKING-LIFE EXPERIENCES

There is considerable evidence that daytime experiences are incorporated into dreams. For example, in Wamsley *et al.* (2010) 30% of NREM dream reports were related to a highly interactive video arcade game that participants had been playing, with the incorporations becoming more abstracted from the original experience as time into sleep increased. Also, in Najam *et al.* (2006), an earthquake survivor group had more vivid, unpleasant, horrifying, and hostile dreams than a control group.

From dream content thus being specific to the individual or group of individuals, the question therefore arises of whether this dream content can be stable over time. Domhoff (1996) reviews evidence of trait-like aspects of dream content, with data from dreams collected in the laboratory, or at home over a period of weeks, or in personal dream diaries that an individual has kept, often over many years. He summarizes this work as showing "amazing consistency" in dreaming of "types of characters, social interactions, objects, and activities," and with the dream content being "relatively free of major changes in life circumstances." He also provides considerable evidence for correspondences between dreams and waking life, such that there are differences between individuals (such as famous writers) in these dream characteristics, and differences between groups, such as cross-cultural or sex differences. This follows on from the work of Hall (1947), where judges were able to identify postulated waking-life conflicts of dream diarists, and Hall (1953), which showed from a sample of 10,000 dreams how they are the embodiment of the person's whole personality and daily problems. Dream content can even be predictive of future circumstances: Cartwright and Wood (1993) showed that masochistic dream content was predictive, in women undergoing divorce, of less improvement at follow-up and more need for emotional support. There is therefore evidence for connections between waking life and dream content. However, the evidence is less compelling on the narrower issue of whether there are relationships between standard psychometric personality tests and dream content.

B. Dream Content and Psychometrically Assessed Personality Traits

There have been findings of relationships between dream content and psychometrically assessed waking-life traits such as nurturance and dominance (Evans and Singer, 1994–1995) and coping style (Rim, 1986, 1988). Also, Bruni et al. (1999) found, for children aged 9–13 years, correlations between neuroticism and negative dream emotions $(r = .34)$ and between extraversion and sexual interactions in the dream $(r = .24)$. Furthermore, there is a major series of relationships between dream content and boundariness. Hartmann et al. (1991) found that the dreams of thin-boundaried individuals were more vivid, more emotional, and had more interaction between characters, compared with dreams of thick-boundaried individuals. These results were supported by Schredl et al. (1996) as well as Hartmann et al. (1998). Schredl et al. (1999) also found that thin boundaries correlate with level of negative emotions in dreams $(r = .31)$, with emotional intensity of dreams $(r = .27)$, with how favorably, i.e., meaningfully, dreams are regarded $(r = .29)$, and with creative content $(r = .21)$, and level of dreamt verbal interaction with others in the dream $(r = .20)$.

However, many studies do not find such relationships. For example, no relationships were found between dream content and each of extraversion or neuroticism (Howarth, 1962), emotional style (Woods et al., 1977), or attributional style for causal inferences (Volpe and Levin, 1998). Furthermore, note should be taken that Hartmann et al. (1998) found that many correlations between boundariness and dream content became insignificant when the number of words in the dream report was partialled out. Hartmann et al. did this procedure because number of words in the dream correlated highly with boundariness $(r = .49)$. However, this procedure of partialling out the number of words in dream reports is opposed by some researchers (e.g., Hobson et al., 2000; Hunt et al., 1993).

Where significant relationships are claimed, the dream content is often not a simple depiction or reflection of the waking-life trait, but rather is symbolic of waking-life variables, such as waking-life anxiety being correlated with symbolic sexual dream content (Robbins *et al.*, 1985), anxiety and depression correlated with recurring teeth-loss dreams (Coolidge and Bracken, 1984), and introversion correlated with occurrence of everyday, less archetypal dreams (Cann and Donderi, 1986). These researchers do provide justifications for these hypotheses, but as many combinations of relationships may be possible there is a potential for type 1 errors.

It should be remarked, however, in justification for investigating such indirect relationships, that correspondences between waking-life variables and dream content can be quite tangential, such as greater vividness of dreams, but not greater violent content, occurring after seeing a violent film (Foulkes and Rechtschaffen, 1964), and greater intensity of the central image of a dream, but no significant change toward more negative emotions, occurring after abuse (Hartmann *et al.*, 2001) and after 9/11 (Hartmann and Basile, 2003). There can even be paradoxical effects, such that it is repressed rather than consciously acknowledged thoughts that are preferentially incorporated into dreams (Wegner *et al.*, 2004). Nevertheless, the potential for spurious correlations remains.

A further problem is that when relationships are found between dream content and waking traits, the significant findings may be due to response biases, especially when a retrospective account of one's dreams is made at the same time as the personality tests are completed. For example, Bernstein and Belicki (1995–1996) found that prospective dream diary content was inconsistent over time and was unrelated to any of 10 personality traits assessed, whereas a retrospective Dream Content Questionnaire did show relationships with personality traits. Openness and Absorption were found to be correlated with dream bizarreness ($r = .38$ and $.34$, respectively), and extraversion was correlated with number of dream characters ($r = .27$). These authors suggest that self-concept may influence responses to retrospective dream questionnaires, which are thus an inaccurate measure of actual dream content, resulting in spuriously enhanced correlations with trait variables. A similar point is made by Bernstein and Roberts (1995).

A direct test of this hypothesized problem with retrospective assessment of dream content has been made by Beaulieu-Prévost and Zadra (2005). They showed that when memories of past dreams are readily available (i.e., for individuals whose DRF is high), people's beliefs about their general dream content are closely related to their actual dream experiences as measured by a dream diary. However, when such memories are not easily available (i.e., for individuals whose DRF is low), people's beliefs about their dream content are influenced by their current emotional state.

A further problem arises in that some of the significant associations between dream content and waking-life variables involve a comparison of psychiatric patient groups with controls, rather than groups differing on a trait measure but within the normal population. Examples of findings of such studies are that people with schizophrenia have dreams with more strangers, more aggression directed at themselves, and more bizarreness, than do people without schizophrenia (Cohen, 1979), the association of serious reactive depression with masochism in dreams and hostility in the dream environment (Hauri, 1976), and fewer emotional elements in the dream content of individuals with Autistic Spectrum Disorders (Daoust *et al.*, 2008). Similarly, in comparing dreams of patients with a personality disorder to normative data on dream content, the former group has more estrangement in their dreams, fewer interactions, and more emotionality (Guralnik *et al.*, 1999). Schredl and Engelhardt (2001) caution that the environmental setting for patients is different from that of healthy controls and is thus a confounding variable in many of these studies.

A further problem with waking-life trait–dream content relationships is that it is not clear if the trait itself has an effect at dream production or at dream recall. For example, dream bizarreness has been found to be related to waking-life creativity by Sylvia *et al.* (1978) and Domino (1976), with partial support for this by Schechter *et al.* (1965). However, these results do not show conclusively that creative people really have more bizarre dreams because, as stated by Domino (1976), the creative participants may instead just be better able to report or to tolerate and remember less logical dream content. This possibility is supported by the results of Wood *et al.* (1989–1990), who noted that many creativity tests may just be assessing language fluency. They found that partialling out verbal ability or partialling out length in words of dream reports resulted in the correlation between creativity and dream bizarreness becoming insignificant.

It is acknowledged that Hunt *et al.* (1993) do caution that controlling for dream report length may be a methodological error that falsely dilutes a defining dimension of dreaming, namely its bizarreness, and thus conclude that creative or imaginative people really do have more bizarre dreams. However, the point here is that it is necessary to distinguish the content of the dream from the recall of that dream content. A similar consideration occurs with the results of Murzyn (2008), where people who had access to black and white media before color media were found to experience more gray scale dreams than do people with no such exposure. This section can be summarized as showing either the influence of waking life on dream content or that waking-life experience affects how we describe our dreams.

C. DISCUSSION OF PART II

The main conclusion from the above review is that there is limited evidence for waking life psychometrically assessed personality traits being associated with dream content variables. Domhoff (1996,Chapter 3) cites some studies that do show such correspondences, but points out that the studies generally have low sample sizes, no mention of effect sizes, lack of replication studies, and, above all, have so many waking and dream variables that multiple comparisons are occurring. He concludes that, because of the "meager findings with personality tests," "dream content may not be about 'personality' in the usual sense of the term. Instead, dream content may provide us with different information about people than most personality tests do." He concludes that "dream content reveals conceptions and concerns" of the dreamer.

There is much evidence that dreams incorporate the waking-life concerns of the individual, such as is seen in the work of Hall (e.g., Hall and Van de Castle, 1966) and Domhoff (e.g., Domhoff, 1996, 2003), and also in that, for example, waking-life concerns can be matched by independent judges with dream content (Kramer, 2007; Nikles et al., 1998; Propper et al., 2007). However, this review has shown that there is little evidence for psychometrically assessed personality traits being associated with dream content. This could be because a trait may only be predictive of dream content if that trait is important to that person's overall personality (Cohen, 1979). For such individuals there may then be correspondences between that psychometric personality trait and dream content, but the inclusion in a correlational study of individuals for whom the trait is not important may result in the diluting and diminishing of any trait–dream content relationship.

III. Neurobiology of Individual Difference Variables Relevant to Dreaming

A. NEUROBIOLOGY OF INDIVIDUAL DIFFERENCES

Individual differences in dream frequency, form, content, and, especially, themes and emotional tone may arise from the same neurobiological substrate as individual differences in waking personality. Following early speculations on the brain bases of personality (e.g., Cloninger et al., 1993), modern neuroimaging techniques have allowed much to be learned about brain changes associated with personality disorders, especially borderline and antisocial disorders (e.g., Koenigsberg et al., 2009; Yang and Raine, 2009), as well as personality change associated with brain injury (Velikonja et al., 2009; Warriner and Velikonja, 2006). However, of greater relevance to the study of personality in dreams are recent functional neuroimaging studies that have begun to reveal the neuronal bases of normal inter-individual variation in personality traits. The "Big Five" personality traits (Costa and McCrae, 1992; Deyoung et al., 2010) and, in particular, the dimensions of extraversion and neuroticism have been most widely studied (Canli, 2004; Deckersbach *et al.*, 2006; Kim *et al.*, 2008). These two dimensions are closely related to individual differences in the processing of reward and threat, respectively, that, in turn, relate to basic tendencies toward approach and avoidance (Deyoung *et al.*, 2010). Interestingly, these two dimensions are closely paralleled by contemporary neurocognitive theories of dream function such as the appetitive (approach) model of Solms (2000) and the threat-rehearsal (withdrawal) theory of Revonsuo (2000), as well as theories that explain the ubiquity of both tendencies in dreams based upon widespread limbic activation in REM (Hobson *et al.*, 2000; Pace-Schott, 2010).

From a dimensional point of view, the discovery of brain–personality relationships that predispose individuals to psychiatric disorders (Canli, 2004; Canli *et al.*, 2009) leads logically to the notion that there ought to be similar neural correlates of personality underlying individual differences at more normative points in specific trait spectra. Similarly, the predictive power of infant temperament in regard to individual differences in personality traits arising later in development (Kagan *et al.*, 1995; Schwartz *et al.*, 1999) as well as functional and structural neuroimaging of such trait differences (Schwartz *et al.*, 2003, 2010) lends further support to the concept of neurally based correlates of personality that, in turn, may contribute to stable inter-individual differences.

In comparison to measures derived from personality inventories such as the NEO Personality Inventory-Revised (NEO-PI) (Costa and McCrae, 1992) or the Temperament and Character Inventory (TCI; Cloninger *et al.*, 1994), neural and behavioral endophenotypes are believed to have closer correspondence to the neurochemical and genetic bases of inter-individual differences. Notably, such endophenotypes may reveal longitudinally stable intra-individual measures that contrast with the high variability and poor replicability seen in cross-sectional, inter-individual measures (Canli, 2004). Endophenotypes may be revealed by behavioral probes such as affective valence-modulated attention (e.g., emotional stroop, dot probe tasks) as well as by characteristic responses of neural circuits when exposed to emotional stimuli (Canli, 2004, 2008; Canli *et al.*, 2009; Congdon and Canli, 2008). At the most basic biological level, stable inter-individual differences in personality traits as well as behavioral and neural endophenotypes may be associated with specific gene polymorphisms (genetic variation occurring in greater than 1% of the population). For example, the short variant of a polymorphism in a specific region of the gene coding for the serotonin transporter (5-HT transporter-linked polymorphic region or 5-HTTLPR) has been widely linked to personality traits of neuroticism and Harm Avoidance as well as to affective disorders (Canli, 2008; Canli *et al.*, 2009). Potential neural endophenotypes of waking personality traits have been extensively reported in recent functional neuroimaging studies. Although only a brief overview is provided here, the potential relevance to inter-individual variability in dream content, and especially its emotional components, is readily apparent.

B. Amygdala

A subcortical structure associated with affective labeling of stimuli, the amygdala has been most often associated with experience of threat and fear, although amygdala responses are also associated with encoding appetitive responses (Sergerie et al., 2008). For example, amygdala activation in response to emotionally positive pictures has been correlated with NEO-PI extraversion, whereas amygdala activity in response to negative pictures correlated with NEO-PI neuroticism (Canli, 2004; Canli et al., 2001). Similarly, amygdala responses to happy faces correlated with extraversion, but not with any of the other 4 Big Five personality traits (Canli et al., 2002). In contrast, NEO-PI neuroticism was correlated with activation of the amygdala in a Stroop-like task when the affective tone of presented words conflicted highly (versus little) with that of simultaneously presented facial expressions (Haas et al., 2007).

C. Medial Prefrontal Cortex

Cortical regions crucially linked with trait measures of emotion regulation and emotional memory of both positive and negative valences are medial regions of the prefrontal cortex (mPFC), especially its ventromedial aspect (vmPFC), that includes ventral portions of the anterior cingulate cortex (ACC). For example, along with the amygdala, activation in the ACC to positive pictures was shown to correlate with NEO-PI extraversion (Canli et al., 2001). In contrast, neuroticism was associated with increased duration of mPFC activation following exposure to pictures of faces expressing negative emotion (Haas et al., 2008) as well as with activation of the subgenual anterior cingulate cortex (sgACC; Brodmann area 25) during an emotional conflict task (Haas et al., 2007). This posterior-most ventral portion of the ACC, the sgACC, is a key bridge between cortical limbic areas and subcortical autonomic structures (Ongur and Price, 2000), and structural and functional abnormalities in this region have been widely linked to depression, including familial, i.e., heritable and hence trait-like, forms of depression (Drevets et al., 2008; Greicius et al., 2007). In contrast, both neuroticism (Kim et al., 2008) and TCI Harm Avoidance, a personality dimension closely related to neuroticism (Hakamata et al., 2009), have been shown to negatively correlate with resting metabolism in more *anterior* portions of mPFC. Similarly, structural neuroanatomical studies have linked extraversion with mPFC areas well anterior to the sgACC such as orbitofrontal (Brodmann Area 11) cortex (Deyoung et al., 2010).

D. Amygdala–Prefrontal Circuits

The neural correlates of regulation of negative emotional states such as fear include mPFC–amygdala circuitry in which amygdala activation is associated with fear expression that is then inhibited by mPFC input (Milad *et al.*, 2006; Phelps and Ledoux, 2005). Both the ability to recall extinction of a conditioned fear and the thickness of vmPFC have been linked with NEO-PI measures of extraversion (Milad *et al.*, 2005; Rauch *et al.*, 2005). Uniquely human cognitive strategies for emotion regulation such as reappraisal, that involve more dorsal and lateral PFC (Ochsner and Gross, 2005), may recruit the vmPFC–amygdala circuitry involved in more primitive mammalian mechanisms such as extinction (Delgado *et al.*, 2008). Notably, diffusion tensor imaging has shown that trait anxiety as measured by the Spielberger State-Trait Anxiety Index (Spielberger *et al.*, 1990) is negatively correlated with the structural integrity of amygdala–PFC fiber pathways (Kim and Whalen, 2009).

E. Reward Systems

Extraversion has been associated with brain reward systems (Depue and Collins, 1999; Deyoung *et al.*, 2010). The hypothetical personality dimension Behavioral Activation (BAS; Carver and White, 1994) is associated with reward circuitry in the ventral striatum, amygdala, and ACC, regions that are also associated with trait aggression (Beaver *et al.*, 2008). Similarly, a component of this construct, Reward Sensitivity, has been associated with neural reward system responses to an appetitive (food) stimulus (Beaver *et al.*, 2006) The insula, a limbic cortical region involved in perception of somatovisceral states (Craig, 2009) and reward (Naqvi and Bechara, 2009), shows baseline hypo-activity in persons ranking highly in the personality construct Sensation Seeking, for whom intense sensation may then provide compensatory activation (Straube *et al.*, 2010). In contrast to reward and approach behavior, the trait measure Behavioral Inhibition (BIS; Carver and White, 1994) has been associated with activity of dorsal ACC regions (Beaver *et al.*, 2008) associated with expression of fear (Milad *et al.*, 2007) and aggressive personality traits (Denson *et al.*, 2009). It should be noted that brain reward networks broadly overlap anatomically with limbic regions involved in negative emotion and its regulation (Peters *et al.*, 2009).

F. The Limbic System in REM Sleep Dreaming

All the above personality trait-related regions are part of a network of cortical and subcortical limbic structures that, during REM sleep, reactivate from a relatively quiescent state in NREM at the same time as much of the multimodal association cortex remains in a quiescent state (Braun *et al.*, 1997, 1998; Maquet *et al.*, 1996, 2005; Nofzinger *et al.*, 1997, 2004; Pace-Schott, 2010). Nofzinger and colleagues have termed the former areas the "anterior paralimbic REM activation area" and, during REM sleep, these regions can become as active or more active than during waking (Nofzinger *et al.*, 1997, 2004). Activity in this network during REM sleep dreaming has been widely hypothesized to be involved in the processing of emotional memories (Nishida *et al.*, 2009; Wagner *et al.*, 2001; Walker, 2009) as well as with an emotion regulatory function during dreaming (Levin and Nielsen, 2007; Nielsen and Levin, 2007; Walker and van der Helm, 2009). It is therefore likely that stable inter-individual differences in features of REM sleep and dreaming would correlate specifically with personality measures. Indeed, preliminary findings suggest that REM is quantitatively associated positively with traits reflecting emotional competence and negatively with those reflecting emotional difficulty (Pace-Schott and McNamara, under review).

Valence biases in attention and emotional memory have been hypothesized to mediate inter-individual differences in personality measures such as neuroticism and extraversion (Haas and Canli, 2008). For example, in an observational learning paradigm, neuroticism was associated with enhanced amygdala-hippocampal activity during encoding of objects paired with pictures of fearful facial expression; activity that, in turn, was associated with enhanced recall of those objects paired with those fearful expressions (Hooker *et al.*, 2008). Notably, sleep-dependent processes involving emotional memory systems have been suggested to be linked to affective traits such as vulnerability to mood disorder (Walker and Van Der Helm, 2009). Therefore, inter-individual differences in characteristic affective tone or themes in dreaming may share with waking personality traits a bias toward attending to or remembering one or the other pole in a spectrum from affectively negative to positive emotional experience.

IV. Conclusions

Despite there being trait-like aspects of general dream recall and of dream content, very few psychometrically assessed correlates for DRF and dream content have been found. More successful has been the investigation of correlates of frequency of particular types of dreams, such as nightmares and lucid dreams, and also of how waking-life experience is associated with dream content. There is also potential in establishing neurobiological correlates of individual differences in dream recall and dream content.

References

Agargun, M. Y., Kara, H., Ozer, O. A., Selvi, Y., Kiran, U., and Ozer, B. (2003). Clinical importance of nightmare disorder in patients with dissociative disorders. *Psychiatry Clin. Neurosci.* **57**, 575–579.

Armitage, R., Rochlen, A., Fitch, T., Trivedi, M., and Rush, A. J. (1995). Dream recall and major depression: a preliminary report. *Dreaming* **5**, 189–198.

Barrett, D., and Loeffler, M. (1992). Comparison of dream content of depressed vs nondepressed dreamers. *Psychol. Rep.* **70**, 403–406.

Beaulieu-Prévost, D., and Zadra, A. (2005). How dream recall frequency shapes people's beliefs about the content of their dreams. *North Am. J. Psychol.* **7**, 253–264.

Beaulieu-Prévost, D., and Zadra, A. (2007). Absorption, psychological boundaries and attitude towards dreams as correlates of dream recall: two decades of research seen through a meta-analysis. *J. Sleep Res.* **16**, 51–59.

Beaver, J. D., Lawrence, A. D., Passamonti, L., and Calder, A. J. (2008). Appetitive motivation predicts the neural response to facial signals of aggression. *J. Neurosci.* **28**, 2719–2725.

Beaver, J. D., Lawrence, A. D., Van Ditzhuijzen, J., Davis, M. H., Woods, A., and Calder, A. J. (2006). Individual differences in reward drive predict neural responses to images of food. *J. Neurosci.* **26**, 5160–5166.

Bernstein, D. M., and Belicki, K. (1995–1996). On the psychometric properties of retrospective dream content questionnaires. *Imagin. Cogn. Pers.* **15**, 351–364.

Bernstein, D. M., and Roberts, B. (1995). Assessing dreams through self-report questionnaires: relations with past research and personality. *Dreaming* **5**, 13–27.

Berquier, A., and Ashton, R. (1992). Characteristics of the frequent nightmare sufferer. *J. Abnorm. Psychol.* **101**, 246–250.

Blagrove, M. (2007). Dreaming and Personality. In: The New Science of Dreaming, Volume 2: Content, Recall, and Personality Correlates (D. Barrett and P. McNamara, eds.), Praeger Publishers, Westport, CT, Chapter 5, pp. 115–158.

Blagrove, M., and Akehurst, L. (2000). Personality and dream recall frequency: further negative findings. *Dreaming* **10**, 139–148.

Blagrove, M., Bell, E., and Wilkinson, A. (in press). Association of lucid dreaming frequency with Stroop task performance. *Dreaming*.

Blagrove, M., Farmer, L., and Williams, E. (2004). The relationship of nightmare frequency and nightmare distress to well-being. *J. Sleep Res.* **13**, 129–136.

Blagrove, M., and Fisher, S. (2009). Trait state interactions in the etiology of nightmares. *Dreaming* **19**, 65–74.

Blagrove, M., and Hartnell, S. J. (2000). Lucid dreaming: associations with internal locus of control, need for cognition and creativity. *Pers. Individ. Dif.* **28**, 41–47.

Blagrove, M., and Haywood, S. (2006). Evaluating the awakening criterion in the definition of nightmares: how certain are people in judging whether a nightmare woke them up? *J. Sleep Res.* **15**, 117–124.

Blagrove, M., and Tucker, M. (1994). Individual differences in locus of control and the reporting of lucid dreaming. *Pers. Individ. Dif.* **16**, 981–984.

Blagrove, M., and Wilkinson, A. (2010). Lucid dreaming frequency and change blindness performance. *Dreaming* **20**, 130–135.

Bone, R. N. (1968). Extroversion, neuroticism and dream recall. *Psychol. Rep.* **23** (3, Pt. 1), 922.

Bone, R. N., and Corlett, F. (1968). Brief report: frequency of dream recall, creativity, and a control for anxiety. *Psychol. Rep.* **22** (3, Pt. 2), 1355–1356.

Braun, A. R., Balkin, T. J., Wesensten, N. J., Carson, R. E., Varga, M., Baldwin, P., Selbie, S., Belenky, G., and Herscovitch, P. (1997). Regional cerebral blood flow throughout the sleep-wake cycle. An H2(15)O PET study. *Brain* **120** (Pt. 7), 1173–1197.

Braun, A. R., Balkin, T. J., Wesensten, N. J., Gwadry, F., Carson, R. E., Varga, M., Baldwin, P., Belenky, G., and Herscovitch, P. (1998). Dissociated pattern of activity in visual cortices and their projections during human rapid eye movement sleep. *Science* **279**, 91–95.

Brodsky, S. L., Esquerre, J., and Jackson, R. R. (1990–1991). Dream consciousness in problem solving. *Imagin. Cogn. Pers.* **10**, 353–360.

Bruni, O., Lo Reto, F., Recine, A., Ottaviano, S., and Guidetti, V. (1999). Development and validation of a dream content questionnaire for school age children. *Sleep. Hypn.* **1**, 41–46.

Butler, S. F., and Watson, R. (1985). Individual differences in memory for dreams: the role of cognitive skills. *Percept. Mot. Skills* **61**, 823–828.

Canli, T. (2004). Functional brain mapping of Extraversion and Neuroticism: learning from individual differences in emotion processing. *J. Pers.* **72**, 1105–1132.

Canli, T. (2008). Toward a neurogenetic theory of Neuroticism. *Ann. N. Y. Acad. Sci.* **1129**, 153–174.

Canli, T., Ferri, J., and Duman, E. A. (2009). Genetics of emotion regulation. *Neuroscience* **164**, 43–54.

Canli, T., Sivers, H., Whitfield, S. L., Gotlib, I. H., and Gabrieli, J. D. (2002). Amygdala response to happy faces as a function of extraversion. *Science* **296**, 2191.

Canli, T., Zhao, Z., Desmond, J. E., Kang, E., Gross, J., and Gabrieli, J. D. (2001). An fMRI study of personality influences on brain reactivity to emotional stimuli. *Behav. Neurosci.* **115**, 33–42.

Cann, D. R., and Donderi, D. C. (1986). Jungian personality typology and the recall of everyday and archetypal dreams. *J. Pers. Soc. Psychol.* **50**, 1021–1030.

Cartwright, R. D., and Wood, E. (1993). The contribution of dream masochism to the sex ratio difference in major depression. *Psychiatry Res.* **46**, 165–173.

Carver, C. S., and White, T. L. (1994). Behavioral inhibition, behavioral activation, and affective responses to impending reward and punishment: the BIS/BAS scales. *J. Pers. Soc. Psychol.* **67**, 319–333.

Cernovsky, Z. Z. (1984). Dream recall and attitude toward dreams. *Percept. Mot. Skills* **58**, 911–914.

Chivers, L., and Blagrove, M. (1999). Nightmare frequency, personality and acute psychopathology. *Pers. Individ. Dif.* **27**, 843–851.

Claridge, G., Clark, K., and Davis, C. (1997). Nightmares, dreams, and schizotypy. *Br. J. Clin. Psychol.* **36**, 377–386.

Cloninger, C. R., Przybeck, T. R., Svrakic, D. M., and Wetzel, R. D. (1994). The Temperament and Character Inventory (TCI): A Guide to its Development and Use. Center for Psychobiology of Personality, St. Louis, MO.

Cloninger, C. R., Svrakic, D. M., and Przybeck, T. R. (1993). A psychobiological model of temperament and character. *Arch. Gen. Psychiatry* **50**, 975–990.

Cohen, D. B. (1971). Dream recall and short-term memory. *Percept. Mot. Skills* **33** (3, Pt. 1), 867–871.

Cohen, D. B. (1974a). Toward a theory of dream recall. *Psychol. Bull.* **81**, 138–154.

Cohen, D. B. (1974b). Effect of personality and presleep mood on dream recall. *J. Abnorm. Psychol.* **83**, 151–156.

Cohen, D. B. (1979). Sleep and Dreaming: Origins, Nature and Functions. Pergamon Press, Oxford.

Cohen, D. B., and MacNeilage, P. F. (1974). A test of the salience hypothesis of dream recall. *J. Consult. Clin. Psychol.* **42**, 699–703.

Conduit, R., Crewther, S. G., and Coleman, G. (2004). Poor recall of eye-movement signals from Stage 2 compared to REM sleep Implications for models of dreaming. *Conscious. Cogn.* **13**, 484–500.

Congdon, E., and Canli, T. (2008). A neurogenetic approach to impulsivity. *J. Pers.* **76**, 1447–1484.

Connor, G. N., and Boblitt, W. E. (1970). Reported frequency of dream recall as a function of intelligence and various personality test factors. *J. Clin. Psychol.* **26**, 438–439.

Coolidge, F. L., and Bracken, D. D. (1984). The loss of teeth in dreams: an empirical investigation. *Psychol. Rep.* **54**, 931–935.

Cory, T. L., Ormiston, D. W., Simmel, E., and Dainoff, M. (1975). Predicting the frequency of dream recall. *J. Abnorm. Psychol.* **84**, 261–266.

Costa, P. T., and McCrae, R. R. (1992). Revised NEO Personality Inventory (NEO-PI-R) and NEO Five-Factor Inventory (NEO-FFI) Professional Manual. Psychological Assessment Resources, Inc, Odessa, FL.

Cowen, D., and Levin, R. (1995). The use of the Hartmann boundary questionnaire with an adolescent population. *Dreaming* **5**, 105–114.

Craig, A. D. (2009). How do you feel—now? The anterior insula and human awareness. *Nat. Rev. Neurosci.* **10**, 59–70.

Daoust, A. -M., Lusignan, F. -A., Braun, C. M. J., Mottron, L., and Godbout, R. (2008). EEG correlates of emotions in dream narratives from typical young adults and individuals with autistic spectrum disorders. *Psychophysiology* **45**, 299–308.

Deckersbach, T., Miller, K. K., Klibanski, A., Fischman, A., Dougherty, D. D., Blais, M. A., Herzog, D. B., and Rauch, S. L. (2006). Regional cerebral brain metabolism correlates of Neuroticism and Extraversion. *Depress. Anxiety* **23**, 133–138.

Delgado, M. R., Nearing, K. I., Ledoux, J. E., and Phelps, E. A. (2008). Neural circuitry underlying the regulation of conditioned fear and its relation to extinction. *Neuron* **59**, 829–838.

Dement, W. C., and Kleitman, N. (1957). The relation of eye movements during sleep to dream activity: an objective method for the study of dreaming. *J. Exp. Psychol.* **53**, 339–346.

Denson, T. F., Pedersen, W. C., Ronquillo, J., and Nandy, A. S. (2009). The angry brain: neural correlates of anger, angry rumination, and aggressive personality. *J. Cogn. Neurosci.* **21**, 734–744.

Depue, R. A., and Collins, P. F. (1999). Neurobiology of the structure of personality: dopamine, facilitation of incentive motivation, and Extraversion. *Behav. Brain Sci.* **22**, 491–517, discussion 18–69.

Deyoung, C. G., Hirsh, J. B., Shane, M. S., Papademetris, X., Rajeevan, N., and Gray, J. R. (2010). Testing predictions from personality neuroscience: brain structure and the big five. *Psychol. Sci.* **21**, 820–828.

Domhoff, B. (1968). An unsuccessful search for further correlates of everyday dream recall. *Psychophysiology* **4**, 386.

Domhoff, G. W. (1996). Finding Meaning in Dreams: A Quantitative Approach. Plenum Press, New York.

Domhoff, G. W. (2003). The Scientific Study of Dreams: Neural Networks, Cognitive Development, and Content Analysis. APA Press, Washington, DC.

Domino, G. (1976). Primary process thinking in dream reports as related to creative achievement. *J. Consult. Clin. Psychol.* **44**, 929–932.

Drevets, W. C., Savitz, J., and Trimble, M. (2008). The subgenual anterior cingulate cortex in mood disorders. *CNS Spectr.* **13**, 663–681.

Erlacher, D., Schredl, M., Watanabe, T., Yamana, J., and Gantzert, F. (2008). The incidence of lucid dreaming within a Japanese university student sample. *Int. J. Dream Res.* **1**, 39–43.

Evans, K. K., and Singer, J. A. (1994–1995). Studying intimacy through dream narratives: the relationship of dreams to self-report and projective measures of personality. *Imagin. Cogn. Pers.* **14**, 211–226.

Farley, F. H., Schmuller, J., and Fischbach, T. J. (1971). Dream recall and individual differences. *Percept. Mot. Skills* **33**, 379–384.

Fitch, T., and Armitage, R. (1989). Variations in cognitive style among high and low frequency dream recallers. *Pers. Individ. Dif.* **10**, 869–875.

Foulkes, D. (1999). Children's Dreaming and the Development of Consciousness. Harvard University Press, Cambridge.

Foulkes, D., and Rechtschaffen, A. (1964). Presleep determinants of dream content: effects of two films. *Percept. Mot. Skills* **19**, 983–1005.

Funkhouser, A. T., Wurmle, O., Comu, C. M., and Bahro, M. (2001). Boundary questionnaire results in the mentally healthy elderly. *Dreaming* **11** (83-), 88.

Gackenbach, J. I. (2009). Video game play and consciousness development: a replication and extension. *Int. J. Dream Res.* **2**, 3–11.

Gackenbach, J., Heilman, N., Boyt, S., and LaBerge, S. (1985). The relationship between field independence and lucid dreaming ability. *J. Ment. Imagery* **9**, 9–20.

Giesbrecht, T., and Merckelbach, H. (2006). Dreaming to reduce fantasy?—Fantasy proneness, dissociation, and subjective sleep experiences. *Pers. Individ. Dif.* **41**, 697–706.

Greicius, M. D., Flores, B. H., Menon, V., Glover, G. H., Solvason, H. B., Kenna, H., Reiss, A. L., and Schatzberg, A. F. (2007). Resting-state functional connectivity in major depression: abnormally increased contributions from subgenual cingulate cortex and thalamus. *Biol. Psychiatry* **62**, 429–437.

Gruber, R. E., Steffen, J. J., and Vonderhaar, S. P. (1995). Lucid dreaming, waking personality and cognitive development. *Dreaming* **5**, 1–12.

Guralnik, O., Levin, R., and Schmeidler, J. (1999). Dreams of personality disordered subjects. *J. Nerv. Ment. Dis.* **187**, 40–46.

Haas, B. W., and Canli, T. (2008). Emotional memory function, personality structure and psychopathology: a neural system approach to the identification of vulnerability markers. *Brain Res. Rev.* **58**, 71–84.

Haas, B. W., Constable, R. T., and Canli, T. (2008). Stop the sadness: neuroticism is associated with sustained medial prefrontal cortex response to emotional facial expressions. *NeuroImage* **42**, 385–392.

Haas, B. W., Omura, K., Constable, R. T., and Canli, T. (2007). Emotional conflict and Neuroticism: personality-dependent activation in the amygdala and subgenual anterior cingulate. *Behav. Neurosci.* **121**, 249–256.

Hakamata, Y., Iwase, M., Iwata, H., Kobayashi, T., Tamaki, T., Nishio, M., Matsuda, H., Ozaki, N., and Inada, T. (2009). Gender difference in relationship between anxiety-related personality traits and cerebral brain glucose metabolism. *Psychiatry Res.* **173**, 206–211.

Hall, C. S. (1947). Diagnosing personality by the analysis of dreams. *J. Abnorm. Soc. Psychol.* **42**, 68–79.

Hall, C. S. (1953). The Meaning of Dreams. Harper, Oxford, UK.

Hall, C. S., and Van de Castle, R. I. (1966). The Content Analysis of Dreams. Appleton Century Crofts, New York.

Hartmann, E. (1989). Boundaries of dreams, boundaries of dreamers: thin and thick boundaries as a new personality measure. *Psychiatr. J. Univ. Ott.* **14**, 557–560.

Hartmann, E. (1991). Boundaries in the Mind: A New Psychology of Personality. Basic Books, New York.

Hartmann, E., and Basile, R. (2003). Dream imagery becomes more intense after 9/11/01. *Dreaming* **13**, 61–66.

Hartmann, E., Elkin, R., and Garg, M. (1991). Personality and dreaming: the dreams of people with very thick and very thin boundaries. *Dreaming* **1**, 311–324.

Hartmann, E., Rosen, R., and Rand, W. (1998). Personality and dreaming: boundary structure and dream content. *Dreaming* **8**, 31–39.

Hartmann, E., Russ, D., Oldfield, M., Sivan, L., and Cooper, S. (1987). Who has nightmares? The personality of the lifelong nightmares sufferer. *Arch. Gen. Psychiatry* **44**, 49–56.

Hartmann, E., Russ, D., Van der Kolk, R., Falke, R., and Oldfield, M. (1981). A preliminary study of the personality of the nightmare sufferer: relationship to schizophrenia and creativity? *Am. J. Psychiatry* **138**, 794–797.

Hartmann, E., Zborowski, M., Rosen, R., and Grace, N. (2001). Contextualizing images in dreams: more intense after abuse and trauma. *Dreaming* **11**, 115–126.

Hauri, P. (1976). Dreams in patients remitted from reactive depression. *J. Abnorm. Psychol.* **85**, 1–10.

Hiscock, M., and Cohen, D. B. (1973). Visual imagery and dream recall. *J. Res. Pers.* **7**, 179–188.

Hobson, J. A., Pace-Schott, E. F., and Stickgold, R. (2000). Dreaming and the brain: toward a cognitive neuroscience of conscious states. *Behav. Brain Sci.* **23**, 793–842, disscussion 904–1121.

Holzinger, B., LaBerge, S., and Levitan, L. (2006). Psychophysiological correlates of lucid dreaming. *Dreaming* **16**, 88–95.

Hooker, C. I., Verosky, S. C., Miyakawa, A., Knight, R. T., and D'esposito, M. (2008). The influence of personality on neural mechanisms of observational fear and reward learning. *Neuropsychologia* **46**, 2709–2724.

Howarth, E. (1962). Extroversion and dream symbolism: an empirical study. *Psychol. Rep.* **10**, 211–214.

Hunt, H., Ruzycki-Hunt, K., Pariak, D., and Belicki, K. (1993). The relationship between dream bizarreness and imagination: artifact or essence? *Dreaming* **3**, 179–199.

Kagan, J., Snidman, N., and Arcus, D. (1995). The role of temperament in social development. *Ann. N. Y. Acad. Sci.* **771**, 485–490.

Kales, A., Soldates, C., Caldwell, A., Charney, D., Kales, J., Markel, D., and Cadieux, R. (1980). Nightmares: clinical characteristics and personality patterns. *Am. J. Psychiatry* **137**, 1197–1201.

Kim, S. H., Hwang, J. H., Park, H. S., and Kim, S. E. (2008). Resting brain metabolic correlates of Neuroticism and Extraversion in young men. *NeuroReport* **19**, 883–886.

Kim, M. J., and Whalen, P. J. (2009). The structural integrity of an amygdala-prefrontal pathway predicts trait anxiety. *J. Neurosci.* **29**, 11614–11618.

Koenigsberg, H. W., Siever, L. J., Lee, H., Pizzarello, S., New, A. S., Goodman, M., Cheng, H., Flory, J., and Prohovnik, I. (2009). Neural correlates of emotion processing in borderline personality disorder. *Psychiatry Res.* **172**, 192–199.

Kramer, M. (2007). The Dream Experience: A Systematic Exploration. Routledge, New York.

Lang, R. J., and O'Connor, K. P. (1984). Personality, dream content and dream coping style. *Pers. Individ. Dif.* **5**, 211–219.

Levin, R. (1998). Nightmares and schizotypy. *Psychiatry* **61**, 206–216.

Levin, R., and Fireman, G. (2001–2002). The relation of fantasy proneness, psychological absorption, and imaginative involvement to nightmare prevalence and nightmare distress. *Imagin. Cogn. Pers.* **21**, 111–129.

Levin, R., and Fireman, G. (2002). Nightmare prevalence, nightmare distress, and self-reported psychological disturbance. *Sleep* **25**, 205–212.

Levin, R., Fireman, G., and Rackley, C. (2003). Personality and dream recall frequency: still further negative findings. *Dreaming* **13**, 155–162.

Levin, R., Galin, J., and Zywiak, B. (1991). Nightmares, boundaries, and creativity. *Dreaming* **1**, 63–73.

Levin, R., Gilmartin, L., and Lamontanaro, L. (1998–1999). Cognitive style and perception: the relationship of boundary thinness to visual-spatial processing in dreaming and waking thought. *Imagin. Cogn. Pers.* **18**, 25–41.

Levin, R., and Nielsen, T. A. (2007). Disturbed dreaming, posttraumatic stress disorder, and affect distress: a review and neurocognitive model. *Psychol. Bull.* **133**, 482–528.

Levin, R., and Raulin, M. L. (1991). Preliminary evidence for the proposed relationship between frequent nightmares and schizotypal symptomatology. *J. Pers. Disord.* **5**, 8–14.

Levin, R., and Young, H. (2001–2002). The relation of waking fantasy to dreaming. *Imagin. Cogn. Pers.* **21**, 201–219.

Maquet, P., Peters, J., Aerts, J., Delfiore, G., Degueldre, C., Luxen, A., and Franck, G. (1996). Functional neuroanatomy of human rapid-eye-movement sleep and dreaming. *Nature* **383**, 163–166.

Maquet, P., Ruby, P., Maudoux, A., Albouy, G., Sterpenich, V., Dang-Vu, T., Desseilles, M., Boly, M., Perrin, F., Peigneux, P., and Laureys, S. (2005). Human cognition during REM sleep and the activity profile within frontal and parietal cortices: a reappraisal of functional neuroimaging data. *Prog. Brain Res.* **150**, 219–227.

Milad, M. R., Quinn, B. T., Pitman, R. K., Orr, S. P., Fischl, B., and Rauch, S. L. (2005). Thickness of ventromedial prefrontal cortex in humans is correlated with extinction memory. *Proc. Natl. Acad. Sci. U.S.A.* **102**, 10706–10711.

Milad, M. R., Quirk, G. J., Pitman, R. K., Orr, S. P., Fischl, B., and Rauch, S. L. (2007). A role for the human dorsal anterior cingulate cortex in fear expression. *Biol. Psychiatry* **62**, 1191–1194.

Milad, M. R., Rauch, S. L., Pitman, R. K., and Quirk, G. J. (2006). Fear extinction in rats: implications for human brain imaging and anxiety disorders. *Biol. Psychol.* **73**, 61–71.

Murzyn, E. (2008). Do we only dream in colour? A comparison of reported dream colour in younger and older adults with different experiences of Black and White media. *Conscious. Cogn.* **17**, 1228–1237.

Najam, N., Mansoor, A., Kanwal, R. H., and Naz, S. (2006). Dream content: reflections of the emotional and psychological states of earthquake survivors. *Dreaming* **16**, 237–245.

Naqvi, N. H., and Bechara, A. (2009). The hidden island of addiction: the insula. *Trends Neurosci.* **32**, 56–67.

Nielsen, T. A., Laberge, L., Paquet, J., Tremblay, R. E., Vitaro, F., and Montplaisir, J. (2000). Development of disturbing dreams during adolescence and their relation to anxiety symptoms. *Sleep* **23**, 727–736.

Nielsen, T., and Levin, R. (2007). Nightmares: a new neurocognitive model. *Sleep Med. Rev.* **11**, 295–310.

Nielsen, T., Paquette, T., Solomonova, E., Lara-Carrasco, J., Colombo, R., and Lanfranchi, P. (2010). Changes in cardiac variability after REM sleep deprivation in recurrent nightmares. *Sleep* **33**, 113–122.

Nikles, C. D.II, Brecht, D. L., Klinger, E., and Bursell, A. L. (1998). The effects of current-concern- and nonconcern-related waking suggestions on nocturnal dream content. *J. Pers. Soc. Psychol.* **75**, 242–255.

Nishida, M., Pearsall, J., Buckner, R. L., and Walker, M. P. (2009). REM sleep, prefrontal theta, and the consolidation of human emotional memory. *Cereb. Cortex* **19**, 1158–1166.

Nofzinger, E. A., Buysse, D. J., Germain, A., Carter, C., Luna, B., Price, J. C., Meltzer, C. C., Miewald, J. M., Reynolds, C. F. III, and Kupfer, D. J. (2004). Increased activation of anterior paralimbic and executive cortex from waking to rapid eye movement sleep in depression. *Arch. Gen. Psychiatry* **61**, 695–702.

Nofzinger, E. A., Mintun, M. A., Wiseman, M., Kupfer, D. J., and Moore, R. Y. (1997). Forebrain activation in REM sleep: an FDG PET study. *Brain Res.* **770**, 192–201.

Ochsner, K. N., and Gross, J. J. (2005). The cognitive control of emotion. *Trends Cogn. Sci. (Regul. Ed.)* **9**, 242–249.

Ongur, D., and Price, J. L. (2000). The organization of networks within the orbital and medial prefrontal cortex of rats, monkeys and humans. *Cereb. Cortex* **10**, 206–219.

Pace-Schott, E. F. (2010). The Neurobiology of dreaming. In: Principles and Practice of Sleep Medicine, 5th ed. (M. H. Kryger, T. Roth, and W. C. Dement eds.), Elsevier, Philadelphia.

Pace-Schott, E. F., and Mcnamara, P. Sleep architecture and emotional regulation in healthy adults under review.

Pagel, J. F., and Shockness, S. (2007). Dreaming and insomnia: polysomnographic correlates of reported dream frequency. *Dreaming* **17**, 140–151.

Patrick, A., and Durndell, A. (2004). Lucid dreaming and personality: a replication. *Dreaming* **14**, 234–239.

Peters, J., Kalivas, P. W., and Quirk, G. J. (2009). Extinction circuits for fear and addiction overlap in prefrontal cortex. *Learn. Mem.* **16**, 279–288.

Phelps, E. A., and Ledoux, J. E. (2005). Contributions of the amygdala to emotion processing: from animal models to human behavior. *Neuron* **48**, 175–187.

Propper, R. E., Stickgold, R., Keeley, R., and Christman, S. D. (2007). Is Television Traumatic? Dreams, Stress, and Media Exposure in the Aftermath of September 11, 2001. *Psychol. Sci.* **18**, 334–340.

Rauch, S. L., Milad, M. R., Orr, S. P., Quinn, B. T., Fischl, B., and Pitman, R. K. (2005). Orbitofrontal thickness, retention of fear extinction, and extraversion. *NeuroReport* **16**, 1909–1912.

Revonsuo, A. (2000). The reinterpretation of dreams: an evolutionary hypothesis of the function of dreaming. *Behav. Brain Sci.* **23**, 877–901, discussion 04–1121.

Rim, Y. (1986). Dream content and daytime coping styles. *Pers. Individ. Dif.* **7**, 259–261.

Rim, Y. (1988). Comparing coping styles: awake and asleep. *Pers. Individ. Dif.* **9**, 165–170.

Robbins, P. R., and Tanck, R. H. (1988–1989). Depressed mood, dream recall and contentless dreams. *Imagin. Cogn. Pers.* **8**, 165–174.

Robbins, P. R., Tanck, R. H., and Houshi, F. (1985). Anxiety and dream symbolism. *J. Pers.* **53**, 17–22.

Roberts, J., and Lennings, C. J. (2006). Personality, psychopathology and nightmares in young people. *Pers. Individ. Dif.* **41**, 733–744.

Rochlen, A., Hoffmann, R., and Armitage, R. (1998). EEG correlates of dream recall in depressed outpatients and healthy controls. *Dreaming* **8**, 109–123.

Schäfer, V., and Bader, K. (2009). The impact of early-life maltreatment on dreams of patients with insomnia. *Int. J. Dream Res.* **2**, 18–26.

Schechter, N., Schmeidler, G. R., and Staal, M. (1965). Dream reports and creative tendencies in students of the arts, sciences and engineering. *J. Consult. Psychol.* **29**, 415–421.

Schonbar, R. A. (1965). Differential dream recall frequency as a component of "life style". *J. Consult. Psychol.* **29**, 468–474.

Schredl, M. (2002). Dream recall frequency and openness to experience: a negative finding. *Pers. Individ. Dif.* **33**, 1285–1289.

Schredl, M. (2003). Effects of state and trait factors on nightmare frequency. *Eur. Arch. Psychiatr. Clin. Neurosci.* **253**, 241–247.

Schredl, M., Brenner, C., and Faul, C. (2002). Positive attitude toward dreams: reliability and stability of a ten-item scale. *North Am. J. Psychol.* **4**, 343–346.

Schredl, M., Ciric, P., Gotz, S., and Wittmann, L. (2003). Dream recall frequency, attitude towards dreams and openness to experience. *Dreaming* **13**, 145–153.

Schredl, M., and Engelhardt, H. (2001). Dreaming and psychopathology: dream recall and dream content of psychiatric inpatients. *Sleep. Hypn.* **3**, 44–54.

Schredl, M., and Erlacher, D. (2004). Lucid dreaming frequency and personality. *Pers. Individ. Dif.* **37**, 1463–1473.

Schredl, M., Jochum, S., and Souguenet, S. (1997). Dream recall, visual memory, and absorption in imaginings. *Pers. Individ. Dif.* **22**, 291–292.

Schredl, M., Kleinferchner, P., and Gell, T. (1996). Dreaming and personality: thick vs thin boundaries. *Dreaming* **6**, 219–223.

Schredl, M., and Montasser, A. (1996–1997a). Dream recall: state or trait variable? Part I: Model, theories, methodology and trait factors. *Imagin. Cogn. Pers.* **16**, 181–210.

Schredl, M., and Montasser, A. (1996–1997b). Dream recall: state or trait variable? Part II: State factors, investigations, and final conclusions. *Imagin. Cogn. Pers.* **16**, 231–261.

Schredl, M., and Reinhard, I. (2008). Dream recall, dream length, and sleep duration: state or trait factor. *Percept. Mot. Skills* **106**, 633–636.

Schredl, M., Schäfer, G., Hofmann, F., and Jacob, S. (1999). Dream content and personality: thick vs. thin boundaries. *Dreaming* **9**, 257–263.

Schredl, M., Wittmann, L., Ciric, P., and Götz, S. (2003). Factors of home dream recall: a structural equation model. *J. Sleep Res.* **12**, 133–141.

Schwartz, C. E., Kunwar, P. S., Greve, D. N., Moran, L. R., Viner, J. C., Covino, J. M., Kagan, J., Stewart, S. E., Snidman, N. C., Vangel, M. G., and Wallace, S. R. (2010). Structural differences in adult orbital and ventromedial prefrontal cortex predicted by infant temperament at 4 months of age. *Arch. Gen. Psychiatry* **67**, 78–84.

Schwartz, C. E., Snidman, N., and Kagan, J. (1999). Adolescent social anxiety as an outcome of inhibited temperament in childhood. *J. Am. Acad. Child Adolesc. Psychiatry* **38**, 1008–1015.

Schwartz, C. E., Wright, C. I., Shin, L. M., Kagan, J., Whalen, P. J., McMullin, K. G., and Rauch, S. L. (2003). Differential amygdala response to novel versus newly familiar neutral faces: a functional MRI probe developed for studying inhibited temperament. *Biol. Psychiatry* **53**, 854–862.

Semiz, U. B., Basoglu, C., Ebrinc, S., and Cetin, M. (2008). Nightmare disorder, dream anxiety, and subjective sleep quality in patients with borderline personality disorder. *Psychiatry Clin. Neurosci.* **62**, 48–55.

Sergerie, K., Chochol, C., and Armony, J. L. (2008). The role of the amygdala in emotional processing: a quantitative meta-analysis of functional neuroimaging studies. *Neurosci. Biobehav. Rev.* **32**, 811–830.

Simard, V., Nielsen, T. A., Tremblay, R. E., Boivin, M., and Montplaisir, J. Y. (2008). Longitudinal study of bad dreams in preschool-aged children: prevalence, demographic correlates, risk and protective factors. *Sleep* **31**, 62–70.

Snyder, J., and Gackenbach, J. (1988). Individual differences associated with lucid dreaming. In: Conscious Mind, Sleeping Brain—Perspectives on Lucid Dreaming (J. Gackenbach and S. LaBerge eds.), Plenum Press, New York, pp. 221–259.

Solms, M. (1997). The Neuropsychology of Dreams. Lawrence Erlbaum Associates, Mahwah, NJ.

Solms, M. (2000). Dreaming and REM sleep are controlled by different brain mechanisms. *Behav. Brain Sci.* **23**, 843–850, discussion 904–1121.

Spanos, N. P., Stam, H. J., Radtke, H. L., and Nightingale, M. E. (1980). Absorption in imaginings, sex-role orientation, and the recall of dreams by males and females. *J. Pers. Assess.* **44**, 277–282.

Spielberger, C. D., Gorsuch, R. L., and Lushene, R. E. (1990). Manual for the State-Trait Anxiety Inventory (Self-Evaluation Questionnaire). Consulting Psychologists Press, Palo Alto.

Straube, T., Preissler, S., Lipka, J., Hewig, J., Mentzel, H. J., and Miltner, W. H. (2010). Neural representation of anxiety and personality during exposure to anxiety-provoking and neutral scenes from scary movies. *Hum. Brain Mapp.* **31**, 36–47.

Stuart, K., and Conduit, R. (2009). Auditory inhibition of rapid eye movements and dream recall from REM sleep. *Sleep* **32**, 399–408.

Sylvia, W., Clark, P. M., and Monroe, L. J. (1978). Dream reports of subjects high and low in creative ability. *J. Gen. Psychol.* **99**, 205–211.

Takeuchi, T., Ogilvie, R. D., Ferrelli, A. V., Murphy, T. I., and Belicki, K. (2001). The dream property scale: an exploratory English version. *Conscious. Cogn.* **10**, 341–355.

Tart, C. T. (1962). Frequency of dream recall and some personality measures. *J. Consult. Psychol.* **26**, 467–470.

Tellegen, A., and Atkinson, G. (1974). Openness to absorbing and self-altering experiences ("absorption"), a trait related to hypnotic susceptibility. *J. Abnorm. Psychol.* **83**, 268–277.

Tonay, V. K. (1993). Personality correlates of dream recall: who remembers? *Dreaming* **3**, 1–8.

Velikonja, D., Warriner, E., and Brum, C. (2009). Profiles of emotional and behavioral sequelae following acquired brain injury: cluster analysis of the Personality Assessment Inventory. *J. Clin. Exp. Neuropsychol.* **21**, 1–12.

Volpe, N., and Levin, R. (1998). Attributional style, dreaming and depression. *Pers. Individ. Dif.* **25**, 1051–1061.

Voss, U., Holzmann, R., Tuin, I., and Hobson, J. A. (2009). Lucid dreaming: a state of consciousness with features of both waking and non-lucid dreaming. *Sleep* **32**, 1191–1200.

Wagner, U., Gais, S., and Born, J. (2001). Emotional memory formation is enhanced across sleep intervals with high amounts of rapid eye movement sleep. *Learn. Mem.* **8**, 112–119.

Walker, M. P. (2009). The role of sleep in cognition and emotion. *Ann. N. Y. Acad. Sci.* **1156**, 168–197.

Walker, M. P., and Van Der Helm, E. (2009). Overnight therapy? The role of sleep in emotional brain processing. *Psychol. Bull.* **135**, 731–748.

Wamsley, E. J., Perry, K., Djonlagic, I., Babkes Reaven, L., and Stickgold, R. (2010). Cognitive replay of visuomotor learning at sleep onset: temporal dynamics and relationship to task performance. *Sleep* **33**, 59–68.

Warriner, E. M., and Velikonja, D. (2006). Psychiatric disturbances after traumatic brain injury: neurobehavioral and personality changes. *Curr. Psychiatry Rep.* **8**, 73–80.

Watson, D. (2001). Dissociations of the night: individual differences in sleep-related experiences and their relation to dissociation and schizotypy. *J. Abnorm. Psychol.* **110**, 526–535.

Watson, D. (2003). To dream, perchance to remember: individual differences in dream recall. *Pers. Individ. Dif.* **34**, 1271–1286.

Webb, W. B., and Kersey, J. (1967). Recall of dreams and the probability of stage 1 ñ REM sleep. *Percept. Mot. Skills* **24**, 627–630.

Wegner, D. M., Wenzlaff, R. M., and Kozak, M. (2004). Dream rebound: the return of suppressed thoughts in dreams. *Psychol. Sci.* **15**, 232–236.

Woods, D. J., Cole, S., and Ferrandez, G. (1977). Dream reports and the test of emotional styles: a convergent-discriminant validity study. *J. Clin. Psychol.* **33**, 1021–1022.

Wood, J. M., Sebba, D., and Domino, G. (1989–1990). Do creative people have more bizarre dreams? A reconsideration. *Imagin. Cogn. Pers.* **9**, 3–16.

Yang, Y., and Raine, A. (2009). Prefrontal structural and functional brain imaging findings in antisocial, violent, and psychopathic individuals: a meta-analysis. *Psychiatry Res.* **174**, 81–88.

CONSCIOUSNESS IN DREAMS

David Kahn* and Tzivia Gover†

*Department of Psychiatry, Harvard Medical School, Boston, MA, USA
†Holyoke Community College, Holyoke, MA, USA

I. Introduction
II. How Does Dream Consciousness Come About?
III. What Characterizes Dream Consciousness?
IV. Characteristics of Dreams
V. Dream Consciousness and the Dream Body
VI. How Do Dream Consciousness and Lucidity Differ from Wake Consciousness?
VII. What We Can Learn from Dream Consciousness
   References

This chapter argues that dreaming is an important state of consciousness and that it has many features that complement consciousness in the wake state. The chapter discusses consciousness in dreams and how it comes about. It discusses the changes that occur in the neuromodulatory environment and in the neuronal connectivity of the brain as we fall asleep and begin our night journeys. Dreams evolve from internal sources though the dream may look different than any one of these since something entirely new may emerge through self-organizing processes. The chapter also explores characteristics of dreaming consciousness such as acceptance of implausibility and how that might lead to creative insight. Examples of studies, which have shown creativity in dream sleep, are provided to illustrate important characteristics of dreaming consciousness. The chapter also discusses the dream body and how it relates to our consciousness while dreaming. Differences and similarities between wake, lucid, non-lucid and day dreaming are explored and the chapter concludes with a discussion on what we can learn from each of these expressions of consciousness.

I. Introduction

Dreams have alternately been hailed as messages from the gods and dismissed as random hallucinations. The pendulum of popular opinion has swung from one extreme to the other throughout recorded history and between cultures and

181

camps, with scientists, psychologists, sages, and philosophers all weighing in. Aristotle (Gallop, 1996), for one, believed dreams were formed by the dreamer's impaired mind, and Plato (Talbot, 2009) argued that dreams represent a frightening breakdown of reason. In the Victorian era, some scientists posited that dreaming was pathological.

But rather than place dreaming on a mystical pedestal, or look at dreaming as a deficient form of consciousness, it is highly instructive to look at dreaming as an alternative form of consciousness and a different way of thinking.

Though Plato and Aristotle could not have proven it, today we know that the dreaming brain is, in a sense, differently abled. At least two important regions, the dorsal lateral prefrontal cortex (DLPFC) and the precuneus in the parietal lobe, are deactivated during rapid eye movement (REM) sleep, the period when most dreaming takes place.

Because of this, we lack the ability to fully exercise our short-term memory when we dream, both within the dream and upon awakening. This helps explain breaks in continuity during the dream and why it is difficult to recall dreams on waking. Also, we are unable to locate our physical body in space when asleep, which is why the dreamer does not realize her or his body is at home in bed during nocturnal adventures in familiar or fantastical landscapes. Making decisions or directing our will is likewise difficult while dreaming, because of these changes in brain activity during sleep.

Thoughts emanate from and generate attitudes, memories, and feelings from brain activity when awake. When dreaming, the same is true, but with altered brain activity. If these alterations are viewed as imperfections, or evidence that the brain is simply firing on too few cylinders, it is easy to dismiss dream content and write it off. If, on the other hand, dreaming is to be accepted as a different but valuable form of consciousness, there is much to learn, wonder at, and explore.

Thoughts while dreaming may start with neuronal impulses, as waking thoughts do, too. But awake, there are many distractions. Our external senses are alert and highly functioning during waking hours; sights, smells, bodily sensations, and needs externalize our focus. Impulses quickly attach themselves to the waking business at hand. A hunger pang in the stomach may trigger thoughts about what's in the refrigerator, for example.

Asleep and dreaming, the senses are muted if not shut down. Now the impulses are not directed into thoughts by external stimuli, but instead form and flow into an infinite stream of possible images, memories, emotions, and attitudes. Our minds might then react to the brain's output and look for connections between these thoughts, thus weaving the visually and emotionally intense scenarios we know as dreams.

The contents of much of our thinking when awake can fairly easily be attributed to physical and emotional well-being. Our most basic survival depends on our ability to plan, calculate, and even imagine. Waking brain function and thinking enables us to remember where we live, where we stored our food, and how to decide on a safe route of travel. But what about musing, fantasizing, brooding, or even doing Sudoku puzzles—could not we live well enough without these forms of thinking? Surely we could, but without the pleasures of art and play, not to mention doing really difficult puzzles.

And what then of dreams? Anyone who can recall their dreams knows that the detail, visual beauty, and complexity of some dreams are truly astounding. Also, the consistent and widespread experience of people gaining new insights and deeper self-knowledge by reflecting on their dreams is well known. Whether insight and knowledge are made during the dream or later upon awakening, the mind's persistence in searching for connection and making meaning from these images and fantasies seems to point in the direction of dreaming being an inherently valuable form of thinking, one that complements and adds dimension to the thinking achieved during waking hours.

II. How Does Dream Consciousness Come About?

Are you one self or many selves? What is it like to be you? How do you sense yourself in different external environments, in different internal environments? How does the self change when you are watching a sunset, are absorbed in a goal-oriented task such as balancing a checkbook, are listening to Mozart or chanting *kirtan*? How do you sense yourself in these different external environments? Consciousness is the You who you sense as yourself. Consciousness is the inner visual image, inner auditory image, feelings, assumptions, beliefs, opinions, attitudes, and combinations of thoughts and feelings (personal conversation [2010] with Linda Trichter Metcalf, author, with Tobin Simon, of *Writing the Mind Alive*, 2002). Moment to moment, this constellation tells you who you are and helps define your consciousness when awake.

Now consider what is the self when you are asleep and dreaming? In dreaming consciousness, the thoughts, actions, behaviors, beliefs, and associated feelings are often quite different than they are when awake. Hence, one of your selves is your dream self; the self who shows up as the protagonist. Or, more accurately, your dream self embodies one of your many selves.

One goal we set for ourselves is to review how changes in the brain during sleep and dreaming can account for changes in conscious experience. We explore this by noting specific changes that occur in the activation and chemistry of the brain as we fall asleep, and how these changes can account for the changes in the way that the mental and emotional contents of our dreams are expressed. We will see that many of these differences are accounted for by specific changes in the brain as we pass from the wake to the sleep state.

What happens in the brain when we dream? Brain data findings using electro-encephalogram, magnetoencephalogram, positron emission tomography, and functional magnetic resonance imaging suggest how the characteristics of dreaming consciousness come about as a result of changes in brain activation and chemistry.

As we fall asleep, a change in brain activity and brain chemistry begins to occur in parallel with the appearance of dream-like mental activity. Consciousness is already beginning to change at sleep onset. As sleep deepens, the cerebral energy metabolism and blood flow associated with neuronal activity begins to decrease. This is especially so during the slow-wave, also called non-REM (NREM), deep sleep stages 3 and 4. (Even in the deep sleep stages, brain activity remains at 80% of wake levels. The brain, asleep or not, is always active.) The connectivity between brain regions is also reduced in these deep sleep stages. The long-range connectivity between distant brain regions when awake gives way to only local connectivity between adjacent brain regions in deep sleep. This is similar to what occurs under anesthesia having the effect of reducing awareness of sensation and external stimuli on our physical body (Ferrarelli et al., 2010). Long-range connectivity between cortical brain regions is necessary for the integration of information and for communication between distant brain regions in the wake and sleep states.

As we move out of deep sleep, the large, slow brain waves characteristic of deep sleep begin to change, resembling more and more the waves typical of the brain when awake. Appropriately, this wake-like state has been called "paradoxical sleep," because its brain wave structure so much resembles that of the wake brain. This paradoxical sleep stage usually appears about 90 min after falling asleep. This sleep stage has been aptly named the REM stage of sleep, as it is accompanied by REMs and is associated with dreaming. In this REM stage, many of the brain regions whose neural activity diminished as sleep ensued become reactivated (Braun et al., 1997, 1998; Maquet et al. 1996; Nofzinger et al. 1997). Not all brain regions, however, become reactivated as we move from the deep sleep stages 3 and 4. Instead, there is a *selective* reactivation of neural activity as we move into the REM stage of sleep. This selective re- and deactivation affects dreaming and dream content by changing our short-term autobiographical memory recall, and by changing how we integrate previously learned information. For example, we may combine events from different times in our lives and we may combine different people we know into a blend of one or more people. Physiologically, there is reactivation of the limbic, paralimbic, and amygdala regions, the areas important for producing emotion. The medial prefrontal cortex (MPFC) is also reactivated in REM. It is believed that the MPFC allows for internally motivated behavior, as occurs in our wake lives and in our dreams. An example of internally motivated behavior is one that arises from thoughts, or feelings about ourselves or about some real or imagined behavior of others.

On the other hand, the brain circuit consisting of regions that are important for executive functions, autobiographical memory recall, and goal selection (DLPFC) and for body location (precuneus) is selectively deactivated. Further, during REM sleep dreaming, there is a diminished coupling between frontal, prefrontal, and parietal regions of the brain in the gamma frequency range (Desmedt and Tonberg, 1994; Perez-Garci *et al.*, 2001). This decoupling or dissociation between the prefrontal and parietal regions of the brain contributes to the lack of executive control over the unfolding dream scenario. Furthermore, areas that are important for visual processing and emotions are selectively activated while areas needed for full volitional control are not. Thus, we have little volitional control over the feelings and visual imagery that are internally generated while dreaming as the (non-lucid) dream unfolds. However, internally motivated and strongly emotional feelings and imagery are alive and well.

It is perhaps no coincidence that some of the areas that are selectively activated and deactivated during dreaming are involved in the spontaneous generation of music, for example, the generation of improvised jazz. Jazz improvisation has been found to be characterized by deactivation of the DLPFC together with activation of the MPFC (Limb and Braun, 2008). As Limb and Braun (2008) state in their article on the neural substrates of jazz improvisation, "... creative intuition may operate when an attenuated DLPFC no longer regulates the contents of consciousness, allowing unfiltered, unconscious, or random thoughts and sensations to emerge." This is what we believe accounts for some of the innovation and creativity in dreams where the DLPFC is also deactivated. Further, in both dreaming and jazz improvisation, there is a dissociation between the MPFC and the DLPFC, the former being activated, the latter not. With deactivation of the DLPFC, there is minimal focused attention allowing "free-floating spontaneous unplanned associations and sudden insights."

Additionally, while all neuromodulatory brainstem systems are available when awake and engaged in goal-directed behavior, when we fall asleep some of the neuromodulatory brainstem systems shut off. The main neuromodulators in the brain include serotonin, norepinephrine, dopamine, and acetylcholine. These neuromodulators affect cognitive function, mood, attention, the ability to retrieve memories, and the ability to pay attention. When we reach REM, two of these systems completely shut off, specifically the locus coeruleus (LC) neurons and the dorso raphe nucleus neurons, which house norepinephrine and serotonin, respectively. This causes a change in our brain chemistry that affects how our minds process information during the REM stage of dreaming. The LC neurons, for example, play a key role in maintaining vigilance, attention, and decision making. The extreme change in brain chemistry whereby the aminergic system completely shuts down while the cholinergic system remains high contributes to the occurrence of hallucinatory images during dreaming and the reduced ability to recognize implausibility within the dream, for example, encountering talking animals, as well as to a reduced ability to stay focused, such as the difficulty of dialing phone numbers or reading text in dreams. Because the inhibitory aminergic system is shut down and the cholinergic system is not, activity in the dreaming brain during REM sleep is more prone to "errors," that is, more likely to undergo unexpected twists and turns, for example, finding oneself suddenly in another city, and more likely to make uncommon associations within the unfolding dream narrative (Mamelak and Hobson, 1989), such as finding that fish are using snorkels to breathe air while walking on land.

III. What Characterizes Dream Consciousness?

Dreaming consciousness evolves without the intrusion of sensory input. Dreams are clearly emanations from our brain/mind as they do not require external stimulation. Dreams, then, are "all in our heads" (Metzinger, 2009). They take what is in our heads and create stories from it. The contents of our conscious minds are filtered through an altered sleeping brain. So dreams are both a distillation of what our lives consist of without the distractions of the external world and, at times, also entirely new creations put together by self-organizing processes in the dreaming mind without input or direction from external sources. Self-organizing processes are those in which the elements of a system themselves "decide" what comes next. The behavior that emerges comes about through the exchange of information between the interacting elements. In the dreaming brain, reciprocal and re-entrant interactions between neurons lead to new neural activity from which novel neural patterns can emerge. From this ongoing activity, dream images, thoughts, and feelings emerge. These are influenced not only by internal memories, conflicts, and desires but also, importantly, by ongoing wisps of information exchange, for example, the appearance of an unexpected person or animal that pushes the ongoing dream into unexpected directions (Kahn and Hobson, 1993). The dream, in a sense, has a mind of its own.

IV. Characteristics of Dreams

An important characteristic of dreams is that they are hyperassociative, that is, there is a change in the kind of associations that become linked during dreaming. One study explored this through the use of a semantic priming task that measured the reaction time to find a word when it is preceded by a word that is associated with it. The authors of the study used weak and strong associations. For example, a weak association is dog–elephant or hot–temperature. A strong association is dog–cat or hot–cold. The study found that semantic priming was state dependent. Subjects awakened from REM sleep showed greater priming by weak primes than by strong ones during the moments following awakening from REM, the REM carryover period. In other words, primes that were weakly associated with the word that followed it were more rapidly associated than the strong primes during an REM carryover period (Stickgold et al., 1999). Subjects tested during normal wake hours showed greater priming by strong primes, as expected. The authors of the study speculate that this change in the kind of associations that occurred in the REM carryover period is because of the absence of norepinephrine during REM. Norepinephrine is needed to be able to pick out the signal from the noise, in this case to disregard the minimally relevant words. Without the presence of norepinephrine, minimally relevant information became relevant. Hence, unexpected and enhanced associations are likely to occur in the dreaming brain.

Additionally, there is evidence that flexible and creative processing occurs without focused effort in REM sleep dreams. As the dream unfolds, it often contains unexpected associations between dream characters and between dream events. This creativity has been shown to carry over into the wake state where creative problem solving improved if sleep contained REM. Subjects who had achieved REM sleep during a nap did better on a remote associates task (RAT) than those who only had NREM or had no nap at all. In a RAT, subjects are asked to produce a word that is associated with test words. For example, the authors primed the test word "sweet" before sleep by showing the subjects the words "heart," "sixteen", and "cookie." When subjects who were so primed were asked to find a word that is associated with a preceding word in a different RAT, they found the correct word (sweet) most often if they had a nap that contained REM. "REM enhances the integration of unassociated information for creative problem solving" (Cai et al., 2009).

Several other studies, in fact, have demonstrated that after a night's sleep there is an enhanced ability to solve anagram problems and an enhanced ability to find a hidden rule or obtain an insight that helps solve a difficult problem (Wagner et al., 2004; Walker, 2009a; Walker et al., 2002).

Another important characteristic of dreaming consciousness is that in dreams the dreamer often may engage in logical thought as good as that used when awake, while at the same time the dreamer is uncritical of illogical events and behaviors that are happening in the dream. In dreaming consciousness, therefore, implausible events often go unrecognized by the dreamer during the dream, and remain so until the dreamer awakens (Kahn and Hobson, 2004). Here is an excerpt from a recent dream from one of the authors, which illustrates this dual component to thinking in dreams:

> I am in a plane; my wife is also in the plane, seated farther back. The plane lands at Logan Airport. Then the plane continues to go on before I can get

off. But now it is a train. I wonder how far the next stop is from Logan. The train, which now is more like a subway, stops, but I miss getting off. I try going from car to car but find that I am in a sort of large empty supply-like room on the train. I wonder if my wife is worried. I finally get off and I am about to call her on my cell phone when I see that there is a call coming in that requires me to hit the answer button.

In this dream, logic is retained in wondering how far it is to get back to the airport, in wondering if my wife is worried, and in wanting to use the cell phone.

But I accept that the plane becomes a train, then a subway car, and then I accept that it no longer resembles a subway car. And I accept that I have a wife (which I do not).

Another feature of dreaming consciousness may be its ability to help in emotion regulation (Walker, 2009b). When subjects were shown photos of faces that expressed different emotions, positive and negative, the subjects who had taken a nap and reached REM sleep were better able to identify positive emotions than subjects who did not reach REM sleep in their naps. In fact, those subjects who did not nap or reach REM sleep reported seeing more negative emotions. This indicates that REM might help in processing negative emotions.

Yet another characteristic of dreaming consciousness is the existence of a theory of mind (ToM), that is, an ability to make an informed guess as to what a person is thinking based on the person's actions (Kahn and Hobson, 2005), for example, the realization that if a colleague is opening a file drawer, that colleague is probably looking for a file. Studies have shown that dream characters interact with each other and have thoughts, feelings, and intentions among themselves and with the dreamer. In one excerpt, a subject reported:

I was winning in a game of ping pong with my boyfriend, I knew he was thinking that my winning would make me feel good.

This excerpt is an example of the presence of ToM in the dream; the dreamer is able to think about what someone else is thinking about.

Another characteristic of consciousness in the dream state is that people think more about social interactions while dreaming than they do when awake, and that those interactions are often charged with aggression. On the average a dreamer reports almost four characters per dream, five if the dreamer includes him or herself. This points to a wide range of interactions with others in dreams. In a study in which wake reports were elicited by randomly beeping subjects to find out what they were thinking, researchers found that social interactions were reported less in these wake reports than in dream reports (McNamara et al., 2005). Looking at these social interactions occurring in dreams in more detail, the study found that dreamer-initiated aggressions were found to occur only in the

REM and not in the NREM stage of sleep. On the other hand, dreamer-initiated friendliness was reported twice as often in NREM dream reports. Apparently, not only are dreams more vivid in the REM stage but they are also more aggressive. While it is not entirely clear why dreamer-initiated aggressive social interactions are more likely to occur in the REM stage of dreaming, a possible explanation might lie in the selective activation that occurs in REM versus NREM. In REM several affect-laden limbic areas reactivate while they remain deactivated in NREM stages 3 and 4. These affect-laden limbic areas that reactivate in REM but remain deactivated are the medial prefrontal, anterior cingulate, insula, and temporal pole.

V. Dream Consciousness and the Dream Body

Despite the fact that our physical eyes are closed, our senses "asleep," and our body in bed is paralyzed, our dream body can see, hear, taste, smell, move, and feel.

Why do we have a body at all when we are dreaming? After all, the dream is a form of thinking; it is constructed of images, memories, thoughts, and attitudes. Yet not only do we have a body in our dreams, but we have a highly attuned body that can feel texture, taste, run, and make love. It functions very much like our awake body and in some ways even surpasses it. For instance, we can fly in our dreams, fall from tall buildings without getting hurt, disarm powerful opponents, and squeeze through narrow passageways or pedal bicycles up near-vertical inclines. It seems that in dreaming we project our consciousness into the dream body and fully inhabit it.

But how does the dreamer within the dream know where his or her dream body is within the dreamscape? When awake, a person knows where he or she is with the activation of brain regions such as the precuneus in the parietal lobe. In REM sleep dreaming, the precuneus is deactivated. Are there other areas that take over, because almost every area that is active when awake is also active in REM sleep? Does the dreaming brain show us that the brain is more flexible than we think, that there are other ways of getting things done? Or is the explanation that we are so involved in the experience of the dream that our dream body becomes as real as our physical body?

In a number of experiments that induced illusory ownership of fake body parts, we see that consciousness can be tricked into believing that a fake limb is real. In the so-called rubber hand illusion (Slater *et al.*, 2009), the tapping on a person's hidden hand and synchronously on a fake rubber hand led to a feeling of ownership of the fake hand. Similar experiments have shown that this effect is not limited to the hand. When a person's back was stroked while viewing their own body through a mirror display, the person located him or herself to the position of the virtual body being viewed. Further, actual physical stroking is not necessary for this mistaken belief. One need only take the image as one's own, that is, take the image into one's consciousness, to believe it to be happening to oneself and not to someone outside of oneself.

In dreams, something similar appears to be taking place. Images and our place within these images are taken into our consciousness—or we project our consciousness into the images, particularly into the image of the dream body. We do not see images of ourselves—we experience ourselves. We know where we are in space without using the precuneus because we experience the image of ourselves as ourselves. The precuneus is necessary only when we need to know where our physical body is in space. In dreams, the image takes the place of the physical body; it becomes the physical body, at least as far as our dreaming consciousness is concerned. Hence, just as in the rubber hand illusion, we totally believe that the image of the self is the self.

VI. How Do Dream Consciousness and Lucidity Differ from Wake Consciousness?

Lucid dreaming, when the dreamer is conscious that he or she is dreaming, offers a hybrid of wake and dream consciousness. The lucid state is not as stable as the wake or REM states and hence often moves back into either non-lucid dreaming or into waking.

In studying the physiological correlates of lucid dreaming (Voss *et al.*, 2009), it was found that there is a shift of brain activity in the direction of waking as one becomes lucid within REM. A degree of self-reflective awareness occurs such that the dreamer becomes aware that he or she is lying in bed dreaming. In order for this to happen, several brain regions that had become inactive during the REM stage of sleep, such as the DLPFC and the precuneus, reactivate when lucidity emerges in REM. Additionally, during lucid dreaming, compared with non-lucid dreaming, there is an increase in overall cortical connectivity and greater activity in the gamma band of frequencies. The gamma band, around 40 Hz, is a frequency band known to be associated with conscious processing. Its predominance during lucidity indicates that conscious processing is taking place. This processing leads the dreamer to the awareness that he or she is, in fact, dreaming. In short, self-reflective awareness and volitional control re-emerge in the dreamer when there is a reactivation of the DLPFC, an increase in gamma frequency power, and an increase in global cortical connectivity.

Lucid dreaming is unique in that it is perhaps the only time we are fully engaged in dreaming while, at the same time, having access to our awake consciousness. When awake the brain is creating reality, but we do not know it. That is, we are not aware that our brains are actually creating a representation of the physical world. Likewise, in non-lucid REM, we do not know we are dreaming. In non-lucid dreaming, the dreamer believes that he or she is seeing and moving through a real physical space and the dreamer is unaware that this space does not exist, or that the scene emanates from the dreamer's own mind. This belief does not diminish even if some very strange behaviors and events occur. The dreamer is not aware that he or she is lying in bed asleep (unless the dreamer becomes lucid).

This unawareness rarely happens during focused wake behavior, though it may emerge during the wake state when the mind is wandering or when fantasizing or when engaging in a guided visualization.

Consciousness when awake proceeds within a known world. We predict what is going to happen next and develop over time a model of the world from sensory input and learned experience. We know what to expect. In fact, we become so certain of our predictions that if the unexpected happens we may not even notice it. This is especially so if we are fully engaged in a task, as shown in the study in which viewers were attentive to the number of times members of two teams passed a basketball in a basketball game. The viewers were unaware of a person dressed in a gorilla costume walking on the basketball court between the players (Simons, 2010).

In those cases when, in fact, we do notice something unexpected, we are surprised or even startled by this unexpected event. Often we will try to find a reason for its occurrence, for example, this is an opera and unusual things happen in an opera. In dream consciousness, on the contrary, we do not take much notice of unusual events.

As the philosopher and scientist Thomas Metzinger says, consciousness situates us in the world: in the wake world when we are awake and in the dream world when we dream. When we dream we no longer have a model that is entirely based on learned knowledge of what is and what is not expected to occur. Further, sensory perceptions are not brought in from the wake world to help orient us. Importantly, even though the brain makes the dream, it is not recognized as a model but as a reality (Windt and Metzinger, 2008). Sometimes the reality is questioned because of its bizarreness, but the dreamer may concoct an equally bizarre explanation for rationalizing it away, thus remaining in the dream.

At other times, however, when the dreamer becomes lucid, he or she recognizes the dream as a dream, that is, as a model of reality. There are different degrees of lucidity, ranging from the recognition that one is dreaming to directing the course of the dream, that is, to having volitional control over the narrative of the dream.

In the following dream report it is the bizarre elements in a dream that cue the dreamer into the fact that she is dreaming:

My sister and I are standing outside my house, in a quiet suburban neighborhood. As we stand talking to one another, a subway train zooms on a perpendicular trajectory through the street and up into the sky. I look at the train, then turn to my sister and say, "Wow, that's weird!" I add, "Have you ever seen a subway car shoot up vertically like that?" She doesn't answer, but clearly she has never seen such a thing. I can tell that she's trying to come up with some kind of logical explanation for what we've just seen. Then I say, "There's only one explanation! It's a dream." But my sister is no longer listening to me. I get more excited. "Joanne, this is your opportunity to wake up! Wake up within the dream!" But she just shakes her head, rejecting my suggestion. I tell her to look into my eyes. "We're in this dream together!"

For the rest of the night I have dreams in which I find my sister and try to get her to remember the dream we were in together.

Another lucid dream experience is catalyzed by a false awakening, in which the dreamer believes she has woken up, but then realizes that she is in fact still sleeping:

I wake and look through my open bedroom door into my study. There I see a stranger standing in the room. I am frightened, but then realize I am not awake after all, but am still dreaming. I see that the stranger has his back to me, and since I now know I'm dreaming, I decide who this person will be. I decide it will be my brother. When the man turns around it is indeed my brother. I say, "Come on, let's fly. It's easy." So my brother and I fly down the stairs to the front door. I suggest we fly through the wall, instead. I instruct my brother: "Just imagine that it's a door and go through it." Together we fly around the front lawn, over the shrubs and so on. Now I wonder whether this is in fact a lucid dream or an Out of Body Experience (OBE). If it is an OBE, I wonder if the neighbors would see us flying, if they should look outside in our direction.

In this dream, the false awakening prompts lucidity within the dream. Once the dream becomes lucid, the dreamer contemplates yet another possibility, that of an OBE.

Importantly, the forms of consciousness we have explored here —wake, lucid, and non-lucid—complement each other in the sense that each offers a unique perspective on life's experiences. Goal-oriented, purposeful behaviors are best done during wake consciousness. Experiencing the impossible—flying, time travel, morphing into animal form, and engaging with the deceased—is generally possible only in dreaming consciousness. Lucidity allows us to peek behind the veil that separates waking and dreaming consciousness—and to experience two forms of consciousness simultaneously.

VII. What We Can Learn from Dream Consciousness

Science gives us a great deal of information about the brain's neural activity and chemistry when dreaming. It can explain why dreams are heavy on emotion and light on logic, for example. But science is still groping for an understanding of why we dream and what, if anything, can be made of dream content. Comparing the brain's activity and function, and the resulting states of consciousness when awake, in normal REM sleep, and when lucid dreaming, we find similarities and differences in brain activity and function in each of these states that help provide us with a richer understanding of consciousness.

While dreaming we are not constrained by what we know is possible. Even if there were no carryover into wake behavior, while dreaming we benefit by thinking the unthinkable and, importantly, believing it and experiencing it.

And sometimes this exposure to an "unreal" world may lead to an infusion of fresh ideas, as has been shown in many cases where scientific and artistic inspiration came through dreams. Even if the highly rational perspective were true, and dreams have no inherent meaning, people often gain insight into themselves by looking at the dream after waking, just as after we read a book, recite a poem, or look at a painting we gain insight into ourselves and the world.

Dreaming consciousness, especially lucid dreaming consciousness, provides another way for the brain-mind to understand itself. Unlike the wake brain-mind that relies on external sensory input and on the activation of specific brain regions to locate itself in space, these are not all available to the brain-mind while asleep and dreaming. Yet, we certainly are aware of where our dream body is in dream space (you are in your childhood home, outdoors planting a garden, or in your ninth-grade history classroom). What in dream consciousness is letting you know where your dream body is? What do these alternative ways of proprioceptively locating the dream body tell us about consciousness? What does it tell us about how the brain-mind works? And when we lift the veil that guards the transparency of our representation of the physical and dream worlds by becoming lucid, how does this lucidity inform us about the underlying brain basis for consciousness? Surely, addressing these questions will help provide a way for the brain-mind to better understand itself.

In each of these expressions of consciousness, its brain basis is clear, even if not yet fully elucidated. The brain is a dynamic pulsating living organ whose cells mutually interact within a neurochemical milieu that affects the expression of consciousness, as do the neuronal networks that are and are not engaged. For example, when awake, a network consisting of the fronto-parietal-hippocampal regions becomes engaged for performance on a task that requires recall and learning; when awake and the mind is wandering, the medial parietal and medial prefrontal cortices become engaged for stimulus-independent mind wandering; when in REM sleep, the DLPFC and pre-cuneus regions do not reactivate even though most other brain regions do. Further, during REM, the brain switches from cholinergic–aminergic to purely cholinergic. If lucidity occurs during this REM dream, there is partial reactivation of the prefrontal and parietal regions to give yet another expression of consciousness.

So, as Aristotle and Plato believed, the brain of the sleeping dreamer is, in fact, different, though as we have seen, it is not disabled. It is differently abled by virtue of its changing dynamics and neurochemistry. Dreaming, it can be argued, is, in fact, a no-holds-barred form of thinking that is often visually rich, emotionally charged, creative, associative, and seemingly boundless in its content and creative configurations. In particular, dreaming consciousness affords us the ability to be a part of and experience a world unconstrained by the realities of the wake physical world—and, when awake, to use the images and stories we have created while dreaming. Although easy to dismiss as meaningless hallucination, and although easily forgotten, dreams allow us to experience things beyond our abilities in waking reality, and beyond the laws of physical science and nature. Dreams give us the opportunity to bring into the physical world the insights and the creative and original perspectives contained within them.

References

Braun, A. R., Balkin, T. J., Wesensten, N. J., Carson, R. E., Varga, M., Baldwin, P., Selbie, S., Belenky, G., and Herscovitch, P. (1997). Regional cerebral blood flow throughout the sleep–wake cycle. *Brain* **120**, 1173–1197.

Braun, A. R., Balkin, T. J., Wesensten, N. J., Gwadry, F., Carson, R. E., Varga, M., Baldwin, P., Belenky, G., and Herscovitch, P. (1998). Dissociated pattern of activity in visual cortices and their projections during human rapid eye-movement sleep. *Science* **279**, 91–95.

Cai, D. J., Mednick, S. A., Harrison, E. M., Kanady, J. C., and Mednick, S. C. (2009). REM, not incubation improves creativity by priming associative networks. *Proc. Natl. Acad. Sci. U.S.A.* **106**(25), 10130–10134.

Simons, D. J. (2010). DVD Gorilla movie. http://viscog.beckman.illinois.edu/flashmovie/15.php

Desmedt, J. E., and Tonberg, C. (1994). Transient phase locking of 40 Hz electrical oscillations in prefrontal and parietal human cortex reflects the process of conscious somatic perception. *Neurosci. Lett.* **168**, 126–129.

Ferrarelli, F., Massimini, M., Sarasso, S., Casali, A., Riedner, B. A., Angelini, G., Tononi, G., and Pearce, R. A. (2010). Breakdown in cortical effective connectivity during midazolam-induced loss of consciousness. *Proc. Natl. Acad. Sci. U.S.A.* **107**, 2681–2686.

Gallop, D. (1996). Aristotle: On Sleep and Dreams. Aris & Phillips LTD, Warminster, UK.

Kahn, D., and Hobson, J. A. (1993). Self-organization theory of dreaming. *Dreaming* **3**, 151–178.

Kahn, D., and Hobson, J. A. (2004). State-dependent thinking: a comparison of waking and dreaming thought. *Conscious Cogn.* **14**, 429–439.

Kahn, D., and Hobson, A. (2005). Theory of mind in dreaming: awareness of feelings and thoughts of others in dreams. *Dreaming* **15**(1), 48–57.

Limb, C. H., and Braun, A. R. (2008). Neural substrates of spontaneous musical performance: an fMRI study of jazz improvisation. *PLoS ONE* **3**(2), 1–9.

Mamelak, A. N., and Hobson, J. A. (1989). Dream bizarreness as the cognitive correlate of altered neuronal behavior in REM sleep. *J. Cognit. Neurosci.* **1**, 201–222.

Maquet, P., Peteres, J. M., Aerts, J., Delfiore, G., Degueldre, C., Luxen, A., and Franck, G. (1996). Functional neuroanatomy of human rapid-eye-movement sleep and dreaming. *Nature* **383**, 163.

McNamara, P., McLaren, D., Smith, D., Bowen, A., and Stickgold, R. (2005). A "Jekyll and Hyde" within: aggressive versus friendly interactions in REM and non-REM dreams. *Psychol. Sci.* **16**(2), 130–136.

Metzinger, T. (2009). The Ego Tunnel. Basic Books, New York.

Nofzinger, E. A., Mintun, M. A., Wiseman, M. B., Kupfer, D. J., and Moore, R. Y. (1997). Forebrain activation in REM sleep: an FDG PET study. *Brain Res.* **770**, 192–201.

Perez-Garci, E., del-Rio-Portilla, Y., Guevara, A.M.A., Arch, C., and Corsi-Cabera, M. (2001). Paradoxical sleep is characterized by uncoupled gamma activity between frontal and perceptual cortical regions. *Sleep* **24**(1), 118–126.

Metcalf, L. T., and Simon, T. (2002). Writing the Mind Alive: The Proprioceptive Method for Finding Your Authentic Voice. Ballantine Books, New York.

Slater, M., Perez-Marcos, D., Ehresson, J. H., and Sanchez-Vives, M. (2009). Inducing illusory ownership of a virtual body. *Front. Neurosci.* **3**(2), 214–220.

Stickgold, R., Scott, L., and Rittenhouse, C. (1999). Sleep-induced changes in associative memory. *J. Cognit. Neurosci.* **11**(2), 182–193.

Talbot, M. (2009). Nightmare scenario: learning to rewrite our bad dreams. The New Yorker November 16, 43.

Voss, U., Holzmann, R., Tuin, I., and Hobson, J. A. (2009). Lucid dreaming: a state of consciousness with features of both waking and non-lucid dreaming. *Sleep* **32**(9), 1191–1200.

Wagner, U., Gais, S., Halder, H., Verleger, R., and Born, J. (2004). Sleep inspires insight. *Nature* **427**, 352–355.

Walker, M. (2009a). Sleep dependent memory integration. *Front. Neurosci.* **3**(3), 418–419.

Walker, M. (2009b). REM, dreams and emotional brain homeostasis. *Front. Neurosci.* **3**(3), 442–443.

Walker, M. P., Liston, C., Hobson, J. A., and Stickgold, R. (2002). Cognitive flexibility across the sleep–wake cycle: REM-sleep enhancement of anagram problem solving. *Brain Res. Cogn. Brain Res.* **14**, 317–324.

Windt, J. M., and Metzinger, T. (2008). The philosophy of dreaming and self-consciousness: what happens to the experiential subject during the dream state? In: New Science of Dreaming, vol. 3 (B. Deirdre and N. McPatrick, eds.), Greenwood Press, London.

THE UNDERLYING EMOTION AND THE DREAM: RELATING DREAM IMAGERY TO THE DREAMER'S UNDERLYING EMOTION CAN HELP ELUCIDATE THE NATURE OF DREAMING

Ernest Hartmann

Department of Psychiatry, Tufts University School of Medicine, Boston, MA, USA

There is a widespread consensus that emotion is important in dreams, deriving from both biological and psychological studies. However, the emphasis on examining emotions explicitly mentioned in dreams is misplaced. The dream is basically made of imagery. The focus of our group has been on relating the dream imagery to the dreamer's underlying emotion. What is most important is the underlying emotion—the emotion of the dreamer, not the emotion in the dream. This chapter discusses many studies relating the dream—especially the central image of the dream—to the dreamer's underlying emotion. Focusing on the underlying emotion leads to a coherent and testable view of the nature of dreaming. It also helps to clarify some important puzzling features of the literature on dreams, such as why the clinical literature is different in so many ways from the experimental literature, especially the laboratory-based experimental literature. Based on central image intensity and the associated underlying emotion, we can identify a hierarchy of dreams, from the highest-intensity, "big dreams," to the lowest-intensity dreams from laboratory awakenings.

INTERNATIONAL REVIEW OF
NEUROBIOLOGY, VOL. 92
DOI: 10.1016/S0074-7742(10)92010-2

197

I. Emotion and Dreaming: Introduction

There has been a great deal of interest in emotions and dreams, but the conclusions overall are not clear. The most common approach has been to study the incidence of emotions mentioned in dreams, and there are many such studies. Early studies showed that fewer than half of dream reports contained a mention of emotion (Snyder, 1970). However, when dreamers are asked about emotion in their dreams, the proportion is much higher (Nielsen *et al.*, 1991; Strauch and Meier, 1996). Also, the original finding of predominantly negative emotions in dreams (Hall and Van de Castle, 1966) has been questioned. It now appears that positive emotions are much more frequent than anticipated (Nielsen *et al.*, 1991; Strauch and Meier, 1996).

The scales most commonly used for the detailed analysis of dream content—the Hall–van de Castle scales—pay limited attention to emotion. The scorer is only allowed to score emotion specifically mentioned in the dream report, and only five emotions are considered: anger, apprehension, sadness, confusion, and happiness. This limited scoring of emotions was instituted to insure good reliability of scoring, and because of the authors' findings that "it is surprising how seldom emotional states are mentioned in dream reports unless they are asked for explicitly, and even then they are not always present" (p 20, Domhoff, 1996). Indeed these content analysis studies have demonstrated the frequent lack of mentioned emotions and have reported few results on emotion.

Inge Strauch and her collaborators come to somewhat different conclusions about the presence of emotion in dreams. They have found a large variety of positive as well as negative emotions in their large dream sample. Interestingly, this group finds "joy" to be the most common emotion among the 12 emotions they consider. However, if their negative emotions—"anger," "fear," "stress"—are combined, it would lead to a much larger total than "joy." The apparent differences from Hall and Van de Castle's and Domhoff's results may be due to the fact that Strauch's group asked each dreamer, after obtaining the complete dream report, "how did you feel during the dream?" (summarized in Strauch and Meier, 1996, p 27).

Thus there is only incomplete agreement about the actual presence and frequency of various emotions in dreams. In many studies by my collaborators and myself (see below), we have also found that explicitly mentioned emotion is absent in many dreams, even when there is obviously strong emotion in the dreamer. This has led us to study dream imagery in detail, and to investigate the relationship of the imagery to underlying emotion. After all, what stands out in dreams, especially in memorable or "big" dreams, is the imagery (Hartmann, 2008). What we really need to understand then is the relation of this imagery to the underlying emotion of the dreamer. Of course the emotion of the dreamer is not as easy to study or score: there is no standard "dreamer report" to be scored.

We have approached the problem by starting with dreams at times of strong emotional arousal—after a severe trauma, after abuse, and at times of stress, such as after 9/11. I will review these studies below and add some relevant previously unpublished material.

First, a brief review of previous approaches to studying the effect of the dreamer's emotion on dreams is presented. Early on there were numerous attempts to induce an emotion in the dreamer—sometimes by presenting emotional material of different kinds before (or during) sleep—and then to look at the effect on the night's dreams. For instance, there was a well-designed sleep laboratory study by Cartwright *et al.* (1969) investigating the effect on dreams of watching a pornographic film in the evening. Dreams were collected from each rapid eye movement (REM) period over five nights from each of ten young men—on nights with and nights without the film. The results were not dramatic: only a few significant effects, out of many variables studied. Scorers found more direct incorporation of the sleep laboratory setting than of material from the film; the number of dream characters per dream was reduced; the judges also concluded that the amount of "symbolic representation" of the film content (but not direct representation) was higher than expected on the film nights. There were no significant changes on most important measures.

Arkin and Antrobus (1991) reviewing this field state: "The results have told us relatively little about how dreams are constructed. The findings seem rather meager next to the rich, elaborate causal links that one may construct post-hoc" (p 306). Similar conclusions are expressed by Kramer (2007) and Nielsen (2007). Nielsen suggests that the lack of clear results may be due to the difficulty of imposing an emotion. The results should be better, if one use the dreamer's own emotions.

Along these more naturalistic lines, there have been attempts to determine whether emotionally important material is incorporated into dreams. Do the dreamer's emotional concerns enter into or determine the dream? Thus, Nikles *et al.* (1998) showed that when participants were asked to dream about concern-related themes and other themes, the concern-related themes influenced the dreams more, especially the central imagery of the dream.

Piccione *et al.* (1977) reported on a study in which subjects identified a great many of their daytime events and rated them as to emotionality. Judges then independently rated which daytime events appeared to be incorporated into the subjects' dreams. Results showed that the events judged to be incorporated into dreams had been rated significantly more emotional than daytime events not incorporated.

Similar results were found by Whitman *et al.* (1963). They asked ten psychiatric patients and their psychiatrists (therapists) to sleep in the lab for one night, just before the psychiatrist was to present his/her patient's case to a supervisory conference. Dreams were collected from all REM periods. It turned out that the patients' dreams included some reference to their psychiatrists over 40% of the time. The psychiatrists dreamt of their patients only 12% of the time, while they dreamt of the supervisory conference 38% of the time. Obviously both groups dreamt predominantly about their worries and emotional concerns. Thus, there have been some positive results relating dreams to emotionally important material.

There are also several studies in which students simply list their emotional concerns in the evening and their dreams are examined for possible relationships to these concerns. This seems a reasonable idea, but the results have often shown little effect (Rados and Cartwright, 1982; Roussy *et al.*, 1996). Overall, the results have been equivocal, showing no clear relationship between waking concerns and dream content or at best a small, barely significant, relationship.

These studies appear to contradict our view of the importance of underlying emotion and major tenets of the contemporary theory (below). However, there is no real contradiction. I believe that important underlying emotions and emotional concerns clearly influence our dreams, but these concerns may not be evident in the small day-to-day concerns we jot down on a list. The day-to-day concerns are readily accessible, so people mention them in interviews or questionnaires. In the studies that failed to find a relationship between waking concerns and dreams in college students, the concerns mentioned were such things as "What shall I do this weekend, or this vacation? Shall I take a trip with this boyfriend? What courses will I take next semester?" These concerns were seldom evident in the dreams. However, I have several times been able to look over such a series of dreams and discuss them with the student a few years later. From this later viewpoint the dreams seemed much more obvious. The student would say something like, "Oh yes, of course, it's so clear from the dreams. What I was really concerned about was separating from Mother and establishing my own identity, etc." But of course this was not a simple conscious concern, which s/he could have mentioned on a questionnaire at the time. So I would say that the studies of day-to-day waking concerns were not able to determine the true emotional concerns of the dreamer.

II. Dreams and Emotional Arousal: Starting with Trauma and Stress

I believe that the clearest results have come from studies starting with times of emotional arousal, or heightened emotion: for instance after trauma, at very stressful times, at times of loss.

There is a vast clinical literature (in psychoanalytic, psychiatric, and psychological journals) on dreams in traumatized and stressed individuals. Such reports are hard to evaluate systematically, but it is hard to escape the overall conclusion that dreams in the clinical literature, which often involve emotional arousal, are far more exciting and dramatic and have more powerful central images (see below) than dreams from the research literature, especially the sleep laboratory literature. This will be discussed again below.

In terms of research, I believe dreams after trauma provide a good starting point, because trauma obviously produces emotional arousal and it is often clear what emotion is being experienced. Some of us living in relatively peaceful developed world countries may think of trauma as a rare, exceptional event, but in much of the world, there is almost constant trauma. And it is likely that our ancestors—thousands of years ago when we were developing the relevant fore-brain structures—lived far more traumatic lives than we do now.

Starting with the most extreme situations, we can consider populations living in conditions of severe trauma. Punamäki and her collaborators have reported on several severely traumatized populations (Punamäki, 2008; Punamäki *et al.*, 2005; Valli, 2006). Her first impression of these populations was that everyone seemed to sleep poorly and to scream at night. Obviously nightmares were common. Punamäki concentrated especially on children's and adolescents' dreams. Using a variety of questionnaires and interviews, she reports a very high incidence of disturbed dreams involving screaming at night, and dreams involving blood, soldiers, and violence with the dreamer almost always as the victim. Her findings are impressive, though hard to summarize. She compares the traumatized population with non-traumatized or less-traumatized groups and finds clear differences on all these nightmare-like features. The dreams she reports definitely include powerful imagery, though they have not been scored specifically for central images.

Consistent with these findings, Winget *et al.* (1972), working with a Western-world population, found in a citywide sample in Cincinnati that the lower socio-economic class was associated with more troubled and disturbed dreams on a number of measures, and it is well recognized that lower socioeconomic class is associated with more trauma.

We have studied various traumas in individuals who have not led traumatic lives overall but have experienced a single major trauma. Also, we have studied the relationship of abuse to dreams, and the effect of 9/11 on dreams—making the assumption that we all suffered a trauma, or at least a stressful state, a state of emotion arousal, in the period following 9/11.

A. Dreams after a Traumatic Event: The Tidal Wave Dream

I was walking along a beach with a friend, I'm not sure who, when *suddenly a huge wave, 30 feet tall swept us away*. I struggled and struggled in the water. I'm not sure whether I made it out. Then I woke up.

This dream, or something like it, is very common in people who have recently experienced a trauma of any kind (Hartmann, 1998/2001; Hartmann *et al.*, 2001c). We have heard it from victims of rape or attempted rape, from victims of attacks, from people whose close relatives or friends were killed or attacked, and from people who have barely escaped from a burning house.

We consider this dream especially important, in fact paradigmatic, because it lets us see so clearly what is going on. The dream does not picture the actual traumatic experience—the burning house or the rape. It pictures the powerful emotion of the dreamer—"I am terrified. I am overwhelmed." Similar tidal wave dreams have been reported after a major fire by Siegel (1996). The image is not always literally a tidal wave. We have many examples, from people who have experienced a severe trauma, of images such as being swept away by a whirlwind, being tortured, or being chased off a cliff.

The simple picturing of an emotional state seems to occur most when there is a single powerful emotion present, as in someone who has just been traumatized. Terror is perhaps the most straightforward emotion in these situations, but there are others, which are also pictured in dreams. For instance, vulnerability is often pictured:

I dreamt of a small animal lying in the road bleeding.
Several of us were wandering around on a huge plain. There was no shelter. There was rain beating down on us. We had no place to go. We were all lost and helpless.

Guilt, especially survivor guilt, is often pictured in dreams. For instance, a man who escaped from a burning house in which his brother died dreamt:

There's a fire somewhere, in a house very different from ours. In the dream my brother and everyone else escaped, but I was still in the house getting burned when I woke up.

Sadness is also frequently portrayed very clearly. Here are dreams from two different women in the week after their mothers' deaths:

There was an empty house, empty and barren, the furniture all gone. All the doors and windows were open and the wind was blowing through.
A huge tree has fallen down right in front of our house. We're all stunned.

In all these cases, the central imagery of the dream seems to be picturing, very clearly, though metaphorically, the emotions of the dreamer (Hartmann, 1996, 1998/2001).

All the above is of course "anecdotal" or "clinical" evidence, illustrating rather than demonstrating its point. Therefore we went on to see whether we could develop actual research evidence for this view of dreams

B. The Central Image of the Dream

We first called the tidal wave image and similar powerful central images the contextualizing image (CI), as it appeared to provide a context, a picture-context, for the emotion of the dreamer (Hartmann, 1996; Hartmann *et al.*, 1997). However, this term was found unwieldy and confusing by some, so the image is now called simply the central image, keeping the abbreviation CI. A scoring sheet for the CI has been developed (Fig. 1) which can be used on any written or recorded dream report. This scoring has now been used in about 50 different research studies.

The scorer, who knows nothing about the dreamer or the circumstances surrounding the dream, looks at a dream report and first decides whether or not there is a scorable CI. If there is (this turns out to be the case in about 60% of the dreams scored), the scorer jots down a few words describing the image, and then scores the intensity of the image on a 7-point scale (0, 0.5, 1.0, 1.5, 2.0, 2.5, 3.0) based on how powerful, vivid, bizarre, and detailed the image seems ("0" means no CI and "3" means about as powerful an image as you have seen in dreams). S/he then tries to guess what emotion or emotions, from a list of emotions provided, might be pictured by this image. The CI intensity turns out to be an especially important measure. Although it is of course a subjective judgment by the scorer, there is good agreement between scorers—an inter-rater reliability of $r = 0.70$–0.90 (Hartmann *et al.*, 1998b, 2001a).

As there are 18 emotions to choose from, it has been more difficult to obtain good inter-rater reliability on the individual emotion pictured by the dream imagery. However there is quite good agreement between raters when emotions are grouped into three categories: (1) fear/terror and helplessness/vulnerability, (2) other negative emotions (#3–10 on rating sheet), and (3) all positive emotions (#11–18) (Hartmann *et al.*, 2001b).

First, we did a number of studies to establish the characteristics of the CI. We showed that, on a blind basis, CI intensity is rated higher in dreams than in daydreams (Hartmann *et al.*, 1998a, 2001a), as expected. We also found, as expected, that CI intensity is higher in content from REM awakenings than from non–rapid eye movement (NREM) awakenings, which in turn score higher than material obtained from waking periods (Hartmann and Stickgold, 2000).

We then went on to look at whether CI intensity is high in "big" dreams—dreams that stand out in some way, presumably related to their emotional importance for the dreamer. In one study we found that CI intensity is

Definition: A Central Image (contextualizing image) is a striking, arresting, or compelling image—not simply a story but an image which stands out by virtue of being especially powerful, vivid, bizarre, or detailed.

List of emotions

1.	fear, terror	9.	shame, inadequacy
2.	helplessness, vulnerability, being trapped, being immobilized	10.	disgust, repulsion
		11.	power, mastery supremacy
3.	anxiety, vigilance	12.	awe, wonder, mystery
4.	despair, hopelessness (giving up)	13.	happiness, joy, excitement
5.	anger, frustration	14.	hope
6.	disturbing — cognitive dissonance, disorientation, weirdness	15.	peace, restfulness
		16.	longing
7.	guilt	17.	relief, safety
8.	grief, loss, sadness, abandonment, disappointment	18.	love (relationship)

If there is a second contextualizing image in a dream, score on a separate line

Dream ID#	1. CI? (Y/N)	2. What is it?	3. Intensity (rate 0 – 3)	4. What emotion?	5. Second emotion?

FIG. 1. Scoring dreams for the central image.

rated higher in "dreams that stand out" than in "recent dreams" from the same persons (Hartmann *et al.*, 2001a). Likewise, CI intensity is scored higher in dreams characterized as "the earliest dream you can remember" than in "recent dreams" (Hartmann and Kunzendorf, 2006–2007). Thus, CI intensity appears to be high in dreams that are remembered and are presumably emotionally important.

In one study, we specifically examined dreams considered "important" by the dreamer. A group of 57 persons each sent us a recent dream they considered "important" and a dream they considered "unimportant" or less important. CI intensity was significantly higher in the "important" dreams (mean for important dreams $= 1.193$, mean for unimportant dreams $= 0.807$; difference $= 0.386$, SD $= 1.048$; $t = 2.78$, $p < 0.007$) (Hartmann, 2008).

We also studied one group of "especially significant" dreams. A group of 23 students very interested in their dreams each reported one "especially significant" dream (Knudson, 2001). The mean CI intensity in these 23 dreams was 2.617 (means of two experienced raters). This is the highest mean CI intensity score of any group we have seen, much higher than means of recent dreams in various groups. These students did not supply a "nonsignificant" dream for a direct comparison. However, comparing these "highly significant" dreams with our largest group of recent dreams, from 286 students, we found a highly significant difference (mean for significant dreams $= 2.62$, SD $= 0.48$; mean for recent dreams $= 0.75$, SD $= 1.03$; $t = 16.0$, $p < 0.0001$) (Hartmann et al., 2006). These studies establish that high CI intensity is related to dreams considered emotionally important—even though in many cases no emotion is mentioned in the dream.

C. THE CI IN DREAMS AFTER TRAUMA; CIS IN REPORTED ABUSE

The studies above demonstrated that CIs are prominent and CI intensity is high in various types of important or "big dreams." We went on to examine dreams after a traumatic event, as such times usually involve strong—and mainly negative—emotions. We were able to obtain long dream series from ten different persons who had recently experienced serious traumas. All of these dreams (451 dreams) were scored for CIs, on a blind basis, as above. In each of the ten trauma cases, the mean CI score was higher than the mean CI score of the student group (Hartmann et al., 2001c). Overall, the mean CI intensity for the dreams of the people experiencing trauma (a total of 451 dreams) was 1.43 ± 0.49, compared to a student group (286 dreams from 286 students) which was 0.75 ± 1.03, $t = 4.1$, $p < 0.001$. In four of the ten cases, a series of dreams before as well as immediately after trauma were available. In all four of these, the CI score was higher after trauma than before.

As the student group differed greatly from the trauma group, a subgroup of 30 students was formed carefully matched with the trauma group for age and sex. Comparing the trauma group with these matched control students again showed much higher scores in the trauma group (1.43 ± 0.49 vs 0.57 ± 0.92, $t = 3.6$, $p < 0.001$) (Hartmann et al., 2001c). The "emotion pictured" also showed a difference. The trauma group had more negative emotions, and especially more of emotions 1 and 2. The two cases who experienced the worst trauma (one violent rape and one case of torture in a Central American country) had the highest CI intensity scores, and also had "emotions pictured" scored as almost exclusively emotions 1 (fear/terror) or 2 (helplessness/vulnerability).

In another study we looked at the effects of reported abuse. Three hundred and six students filled out several forms and questionnaires as part of a study on dreams and personality. One page was headed simply "Please write down the most recent dream you can remember." Thus, 306 dream reports were available from this group. On another page, each student was asked to check off a number of items of demographic data. Here they were asked to answer "yes" or "no" to the following six questions. Students were asked about two types of abuse at each of three time periods:

1   Have you experienced any physical abuse ... in childhood?, in adolescence?, more recently?
2   Have you experienced any sexual abuse ... in childhood?, in adolescence?, more recently?

Relatively small numbers of students checked yes on any single question. However, we formed a sizeable group ($N = 52$) consisting of all students who answered "yes" to any one or more of the six questions. Scoring the dreams on a blind basis for CI intensity showed that the 52 students reporting any abuse had a mean score of 1.12 ± 1.2, whereas the students reporting no abuse had a mean CI intensity of 0.65 ± 0.97. This is a significant difference ($t = 2.63$, $p < 0.02$). "Emotions pictured" in the "abuse" group showed higher levels of emotions 1 and 2 (fear/terror and helplessness/vulnerability.) We did not interview these students, so we have no way of knowing exactly what abuse they had experienced. Nonetheless, the CI intensity of a recent dream was sensitive to this report of abuse (Hartmann *et al.*, 2001c).

D. Dreams before and after 9/11

Trauma and abuse are difficult to study systematically, as the trauma is different in each person, and the methods of dream collection differ as well. Therefore we did a more systematic study, using 9/11 as a day that we considered was traumatic or at least very stressful for everyone in the United States. We found a number of people who had been recording all their remembered dreams for years, and were willing to send us 20 dreams—the last ten they had recorded before 9/11 and the first ten dreams after 9/11. Thus, we examined 880 dreams from 44 persons on a blind basis. We found that the "after" dreams had a significantly higher CI intensity than the "before" dreams ($p < .002$). Somewhat to our surprise, the before and after dreams did not differ on length, "dream-likeness," "vividness," or presence of towers, airplanes, or attacks. There was also a slight but significant increase in content involving "attacks." CI intensity was definitely the measure that most clearly differentiated the after dreams versus the before dreams (Hartmann and Brezler, 2008). These results confirm our earlier findings of higher CI intensity at times of stress or emotional arousal.

In all these results after trauma, abuse, or 9/11, there was, in addition to the higher CI intensity, a shift in the ratings of "emotion pictured by the CI" toward fear/terror and helplessness/vulnerability. All these studies show that when we can know or estimate the power of the dreamer's emotion, the power of the CI of the dream appears to correspond—increasing in situations of increased emotion. Also, after trauma or stress, the negative emotions, especially fear/terror and helplessness/vulnerability, were the ones rated most as being pictured by the CI.

Clearly the underlying emotion had an effect on dreams in these various stressful conditions. The CIs were more intense, and they tended to picture negative emotions such as terror and vulnerability, almost always in a metaphoric way.

E. FOLLOW-UP OF 9/11 STUDY: THE EXCEPTION THAT PROVES THE RULE

A large study such as the one of dreams before and after 9/11 inevitably includes some noise (random, individual variability). This is unavoidable, but sometimes it is possible to look behind the noise and examine the individual participants. In this study, after completing the initial analysis, we attempted, a few years later, to get some follow-up information from the participants. We did in fact obtain usable info from 34 of the 44 participants. We asked them to answer several questions including:

> Aside from the terrorist attacks of 2001, was there anything important, unusual or disturbing going on in your life—either negative or positive—around the time in 2001 covered by the dreams that you sent? Was your life changing in any way? If so, please describe.

Most did not recall anything specific happening in 2001. A few said the years around 2001 were a period of change in their lives, but they described no specific events or time periods.

Only one participant did describe something specific. This was a woman who reported having serious gynecological surgery 3 months before 9/11 followed by many complications that summer, just before 9/11. So for this woman, the pre-9/11 dreams clearly did not come from a peaceful period that was then disturbed or shattered by the events of 9/11. On examining the dream data, it turned out that this woman was the one participant who had the most pronounced shift in CI intensity in the "opposite" direction from most participants: in other words her dreams had more intense powerful imagery before 9/11 compared to after. In this sense, she constitutes "the exception that proves the rule."

Analyzing the data without this one participant results in an even more clear-cut increase in CI intensity after 9/11 and also a more clear-cut shift in emotions pictured toward fear/terror and helplessness/vulnerability.

F. STRESSFUL SITUATIONS

Breger *et al.* (1971) studied dreams intensively under two specific stressful situations: having serious surgery and being the focus person in an intense group therapy experience. Extremely powerful dreams were found in both situations, and the dreams could clearly be seen as dealing with the emotional concerns of the dreamers. For instance, the first dream Breger reports from a man about to have a major cardiovascular surgery includes the following:

> ... this quarter of beef had been delivered apparently and we were talking about cutting it up ... to preserve it you know. This ex-boss of mine she come in the picture. She discussed with me and with my daughter too how this meat should be cut up. There was a kind of heated argument there about how the meat was gonna be cut.

The first dream reported from the group therapy experience was from a young man who had just been the focus of the group and felt intensely criticized:

> I was in a swimming pool ... all the water dripped out ... I was swimming along ... then there was nothing left ... everything disappeared ... I was alone in the pool ... no lifeguard ... no nothing. Everything just dropped out right underneath me. [...] I was just wallowing on the bottom crying out.

CI scoring was not available at the time, but I have retrospectively scored the dreams of Breger's study and found very high CI intensities.

My group has not specifically studied situations labeled as "stressful"; however, the dreams after 9/11 could be considered to have occurred in a stressful situation. Also, the students who reported one or another kind of abuse may well be considered to be in a situation of stress, or at least emotional arousal. The results are all consistent in showing more CI intensity at times of stress or arousal.

G. Personality and the Central Image

There are certain people—those characterized by thin boundaries—who are very sensitive and in whom everything seems to "get through" more easily than in most people. Many studies have demonstrated that persons with "thin boundaries" remember more dreams, are more interested in their dreams, and are more likely to have dramatic or "big" dreams than those with "thick boundaries" (Hartmann, 1990, 1991; Hartmann and Kunzendorf, 2006–2007; Hartmann et al., 1991, 1998b). Those with thin boundaries have in fact been called "dream people," as opposed to the thick-boundaried "thought-people" (Hartmann, 1998/2001, 1999; Hartmann et al., 1998b).

Thus, it is of interest to examine whether the dreams of people with thin boundaries are also characterized by higher CI intensity scores. One study found a significant correlation between thin boundary score (SumBound) and CI intensity in the recent dreams of a group of 286 students (Zborowski et al., 1998). In another study, CI intensity was found to be significantly higher in the recent dreams of a group of students selected for having thin boundaries than in students with thick boundaries (Hartmann and Kunzendorf, 2005–2006).

We showed that a small group of persons characterized by very thin boundaries had dreams that were scored extremely high on CI intensity, much higher than the dreams of those with thick boundaries (Hartmann, 2008; Hartmann et al., 1991). In a larger group ($N=80$), we showed a significant correlation between SumBound (indicating thinness of boundaries) and CI intensity in dreams (Hartmann, 2008; Hartmann et al., 1998b). Also, we have recently obtained follow-up information including the boundary questionnaire from most of the participants in the 9/11 study. Both "thin" and "thick" participants showed an increase in CI intensity after 9/11, and there was a trend toward higher CI intensity in participants with thin boundaries.

Thus, people with thin boundaries—those whose emotions are more prominent and "get through" easily and who are more influenced by their emotions (Hartmann, 1991)—have more intense CIs in their dreams.

III. The Contemporary Theory of Dreaming

All these studies of the effect of underlying emotion on dreaming play a major part in what has been called the contemporary theory of dreaming.

1. Dreaming is a form of mental functioning. It is one end of a continuum of mental functioning (and cerebral cortical functioning), which runs from

focused waking thought at one end, through reverie and daydreaming, to dreaming at the other end.

2. Dreaming is hyperconnective. At the dreaming end of the continuum, connections are made more easily than in waking, and connections are made more broadly and loosely. Dreaming is always creation, not replay.

3. The connections are not made randomly. They are guided by the emotions of the dreamer.

4. The dream, and especially the CI of the dream, pictures or expresses the dreamer's emotion. The CI is a measure of the power of the emotion: the more powerful the emotion, the more powerful (intense) is the CI.

5. This making of broad connections guided by emotion probably has an adaptive function, which we can conceptualize as "weaving in" new material—in other words taking new experiences and gradually connecting them, multiply connecting them, into existing memory systems.

6. In addition to this basic function of dreaming, the entire focused waking thought-to-dreaming continuum has an adaptive function. It is useful for us to be able to think in clear, focused, serial fashion at certain times, and at other times to associate more broadly, and loosely—in other words to daydream and to dream.

The 9/11 study, along with the other studies on trauma and stress, supports several important aspects of the contemporary theory of dreaming. First is the importance of emotion in driving or guiding the dream. This is at the center (tenets 3 and 4) of the theory.

The data also are important for tenet 2: dreams are hyperconnective, always making new connections; dreams are creation, not replay (Hartmann, 2010a, 2010b). Thus, in the 9/11 study, there was not a single dream of planes hitting towers, or anything similar, even though all the participants had seen this scene many times on TV. Likewise, in our other studies involving over 1000 dreams after traumatic events, there was not a single dream simply picturing the event itself. This may seem surprising, since PTSD patients often report experiencing the war scenes, or traumatic scenes, as replays: "just the way it was." But I have reviewed in detail elsewhere (Hartmann, 2010a) a great deal of data on repetitive dreams, including PTSD dreams, that are sometimes thought to be "replays," and I have found that even in those situations the dream includes a change. The dream scene is not "the way it was."

Finally, the studies are related to tenet 5 of the theory on the functions of dreaming. The exact functions of dreaming are admittedly still unknown. But if, as we have suggested, dreaming has a role in the integration of new material into memory, the importance of underlying emotion is crucial. Our memory systems should be organized according to what is emotionally important. The cerebral cortex specializes in producing imagery, and the imagery appears to be involved in the cortex's memory storage systems. Thus, it is important for the imagery to be closely tied to the underlying emotion. However, the material itself and its associations constitute what is to be remembered—not the emotion itself. The emotion is a basic driving force or guiding force for the imagery and for memory—indicating or labeling what is to be remembered. In this context it is not surprising that some of our important dreams have intense CIs, but no explicitly mentioned emotions.

IV. The Clinical Literature versus the Research Literature on Dreams: A Hierarchy of Emotional Intensity

Focusing on the underlying emotion also allows us to understand some major intriguing findings in the overall dream literature. For instance, one prominent finding that has struck me and many others is that the clinical literature, involving dreams reported in psychotherapy or psychoanalysis, etc., seems to be full of much more exciting, vivid dreams, compared to the research literature, which consists of relatively dull material. And within the research literature, the dreams from laboratory awakenings seem even less exciting than the dreams reported at home (for instance "the most recent dreams"). By and large, this is true not only of NREM but even of REM awakenings.

As mentioned previously, we performed one study of dream material and other material obtained from students wearing a "nightcap" device (Hartmann and Stickgold, 2000). They were repeatedly signaled—during REM, NREM sleep, sleep onset, and awake periods—and asked to dictate a report as to the ongoing mental activity. CI scoring, on a blind basis, revealed that CIs were more common and CI intensity was significantly higher in material from REM sleep than in the other conditions. However, all the CI intensity scores were relatively low. Even the CI intensity scores from REM awakenings were lower than the scores we have found in "most recent" dreams, and much lower than the CI intensity scores in "memorable" dreams, "earliest" dreams (Hartmann and Kunzendorf, 2005–2006), "important" dreams, and "significant" dreams (Hartmann, 2008).

These recent studies using CI scoring confirm earlier studies using a variety of scoring systems, available as early as the 1950s and 1960s. Domhoff summarized these early results in a paper with an apt and memorable subtitle: "Lab dreams and home dreams: Home dreams are better" (1969). "Better" meant chiefly more interesting, eventful, and emotional.

These results are in part due to the fact that emotional intensity increases during the course of an REM period, so dreams from awakenings after 25–30 min of REM are more emotional than dreams after shorter periods of

REM (Kramer and Roth, 1979). "Home dreams," remembered in the morning, presumably contain material from longer or more completed REM periods, while the typical "lab dream" comes from a scheduled awakening usually 5 or 10 min into a REM period.

Another factor to consider is the effect of the laboratory situation itself. One might think that sleeping in a laboratory, with a bunch of wires attached, might be quite stressful, and thus produce memorable dreams with high CI intensity. However, the laboratory situation actually seems to be peaceful or calming place for most participants. I reported many years ago, surveying several thousand dream reports from laboratory awakenings, that these included almost no nightmares (Hartmann, 1970). Interviews revealed that most people did not consider the lab stressful or anxiety provoking. Several mentioned that they felt unusually safe in the lab. One said, "I feel safe. Someone's keeping an eye on me so nothing can happen." Another said, "It's like having a guardian angel."

Most important, I believe, is the factor of selection. While awakenings during sleep presumably capture unselected dream material, as it is happening, the "most recent dreams" or other home dreams already involve some selection: only a fraction of the dreams of a night are ever remembered, and evidence suggests that more emotionally intense dreams are better remembered (Cohen, 1970; Trinder and Kramer, 1971). Thus, a typical sample of "most recent dreams" already involves a selection process, compared with REM and NREM laboratory dreams. Then there is a further selection among remembered dreams when we consider various kinds of "big" dreams—"memorable" dreams, "important" dreams, "significant" dreams, etc. (Hartmann, 2008).

Thus, we can consider a hierarchy of dreams based on their powerful imagery (CI intensity). The highest scores are found in "big" dreams—especially those chosen as "highly significant," and also dreams after trauma—followed by dreams labeled as "memorable" or "important," and "earliest" dreams, then "most recent dreams," finally REM dreams, and lowest of all, NREM or other dreams obtained from laboratory awakenings. Based on our discussion above, this also represents a hierarchy of emotional importance or emotional power.

This hierarchy of image intensity and underlying emotional power is also relevant to the "meaningfulness" of dreams. Someone examining a collection of lab-awakening dreams, or a collection of "most recent dreams" from unknown dreamers, may initially come to the conclusion that dreams are random nonsense. However, no one looking at a powerful dream, such as the "tidal wave dream," knowing it came from a man whose house had recently burned down, would be likely to call the dream random nonsense. Such a dream is obviously meaningful in relation to the underlying emotion.

References

Arkin, A., and Antrobus, J. (1991). The effects of external stimuli applied prior to and during sleep on sleep experience. In: The Mind in Sleep: Psychology and Psychophysiology, 2nd ed. (S. J. Ellman and J. S. Antrobus, eds.), John Wiley & Sons, Oxford, UK, pp. 265–307.

Breger, L., Hunter, I., and Lane, R. (1971). The Effect of Stress on Dreams. International Universities Press, New York.

Cartwright, R. D., Bernick, N., and Borowitz, G. (1969). Effect of an erotic movie on the sleep and dreams of young men. *Arch. Gen. Psychiatr.* **20**, 262–271.

Cohen, D. B. (1970). Current research in the frequency of dream recall. *Psychol. Bull.* **73**, 433–440.

Domhoff, W. G. (1969). Home dreams versus laboratory dreams: home dreams are better. In: Dream Psychology and the New Biology of Dreaming (M. Kramer, ed.), Charles C. Thomas, Springfield, IL, pp. 199–217.

Domhoff, W. G. (1996). Finding Meaning in Dreams: A Quantitative Approach. Plenum, New York.

Hall, C. S., and Van de Castle, R. (1966). The Content Analysis of Dreams. Meredith, New York.

Hartmann, E. (1970). A note on the nightmare. In: Sleep and Dreaming (E. Hartmann, ed.), Little, Brown & Co, Boston, MA, pp. 192–197.

Hartmann, E. (1990). Non-dreamers have "thick boundaries"; good dream-recallers "thin boundaries." *Sleep Res.* **19**, 135.

Hartmann, E. (1991). Boundaries in the Mind: A New Psychology of Personality. Basic Books, New York.

Hartmann, E. (1996). Outline for a theory on the nature and functions of dreaming. *Dreaming* **6**, 147–170.

Hartmann, E. (1998/2001). Dreams and Nightmares: The New Theory on the Origin and Meaning of Dreams. Plenum Press, New York. Revised paperback edition, 2001, Perseus Books, New York.

Hartmann, E. (1999). Thought people and dream people: individual differences on the waking to dreaming continuum. In: Individual Differences and Conscious Experience (R. Kunzendorf and D. Wallace, eds.), John Benjamins, Amsterdam, pp. 251–268.

Hartmann, E. (2008). The central image (CI) makes "big" dreams big: the central image is the emotional heart of the dream. *Dreaming* **18**, 44–57.

Hartmann, E. (2010a). The dream always makes new connections: the dream is a creation, not a replay. *Sleep Med. Clin.* **5**, 241–248.

Hartmann, E. (2010b). The Nature and Function of Dreaming. Oxford University Press, New York.

Hartmann, E., and Brezler, T. (2008). A systematic change in dreams after 9/11/01. *Sleep* **31**, 213–218.

Hartmann, E., Elkin, R., and Garg, M. (1991). Personality and dreaming: the dreams of people with very thick or very thin boundaries. *Dreaming* **1**, 311–324.

Hartmann, E., and Kunzendorf, R. (2005–2006). The central image (CI) in recent dreams, dreams that stand out, and earliest dreams: relationship to boundaries. *Imagin. Cogn. Pers.* **25**, 383–392.

Hartmann, E., and Kunzendorf, R. (2006–2007). Boundaries and dreams. *Imagin. Cogn. Pers.* **26**, 101–115.

Hartmann, E., Kunzendorf, R., Rosen, R., and Grace, N. (2001a). Contextualizing images in dreams and daydreams. *Dreaming* **11**, 97–104.

Hartmann, E., Rosen, R., Gazells, N., and Moulton, H. (1997). Contextualizing images in dreams—images that picture provide a context for an emotion. *Sleep Res.* **26**, 274.

Hartmann, E., Rosen, R., and Grace, N. (1998a). Contextualizing images in dreams: more frequent and more intense after trauma. *Sleep* **21S**, 284.

Hartmann, E., Rosen, R., and Rand, W. (1998b). Personality and dreaming: boundary structure and dreams. *Dreaming* **8**, 31–40.

Hartmann, E., and Stickgold, R. (2000). Contextualizing images in content obtained from different sleep and waking states. *Sleep* **23S**, A172.

Hartmann, E., Zborowski, M., and Kunzendorf, R. (2001b). The emotion pictured by a dream: an examination of emotions contextualized in dreams. *Sleep. Hypn.* **3**, 33–43.

Hartmann, E., Zborowski, M., Rosen, R., and Grace, N. (2001c). Contextualizing images in dreams: more intense after abuse and trauma. *Dreaming* **11**, 115–126.

Knudson, R. (2001). Significant dreams: bizarre or beautiful? *Dreaming* **11**, 167–177.

Kramer, M. (2007). The Dream Experience: A Systematic Exploration. Routledge/Taylor & Francis Group, New York.

Kramer, M., and Roth, T. (1979). The stability and variability of dreaming. *Sleep* **1**, 319-325.

Nielsen, T., Deslauriers, D., and Baylor, G. W. (1991). Emotions in dream and waking event reports. *Dreaming* **1**, 287–300.

Nielsen, T., and Lara-Carrasco, J. (2007). Nightmares, dreams, and emotion regulation: a review. In: The New Science of Dreaming (D. Barrett and P. McNamara, eds.), Vol. II, Praeger, Westport, CT, pp. 253–284.

Nikles, C. D., Brecht, D. L., Klinger, E., and Bursell, A. L. (1998). The effects of current concern- and nonconcern-related waking suggestions on nocturnal dream content. *J. Pers. Soc. Psychol.* **75**, 242–255.

Piccione, P., Jacobs, G., Kramer, M., and Roth, T. (1977). The relationship between daily activities, emotions and dream content. *Sleep Res.* **6**, 133.

Punamäki, R. (2008). Trauma and dreaming: trauma impact on dream recall, content, and patterns, and the mental health function of dreams. In: The New Science of Dreaming (D.Barrett and P. McNamara, eds.), Vol. II, Praeger, Westport, CT, pp. 211–251.

Punamäki, R., Ali, K. J., Ismahil, K. H., and Nuutinen, J. (2005). Trauma, dreaming, and psychological distress among Kurdish children. *Dreaming* **15**, 178–194.

Rados, R., and Cartwright, R. D. (1982). Where do dreams come from? A comparison of presleep and REM sleep thematic content. *J. Abnorm. Psychol.* **91**, 433–436.

Roussy, F., Camirand, C., Foulkes, D., De Koninck, J., Loftis, M., and Kerr, N. (1996). Does early-night REM dream content reliably reflect presleep state of mind? *Dreaming* **6**, 121–130.

Siegel, A. (1996). Dreams of firestorm survivors. In: Trauma and Dreams (D. Barrett , ed.), Harvard University Press, Cambridge, MA, pp. 159–176.

Snyder, F. (1970). The phenomenology of dreaming. In: The Psychodynamic Implications of the Psychological Study of Dreams (H. Madow and L. H. Snow, eds.), Charles C. Thomas, Springfield, IL, pp. 124–151.

Strauch, I., and Meier, B. (1996). In Search of Dreams: Results of Experimental Dream Research. State University of New York Press, Albany, NY.

Trinder, J., and Kramer, M. (1971). Dream recall. *Am. J. Psychiatr.* **128**, 296–301.

Valli, K., Revonsuo, A., Pälkäs, O., and Punamäki, R. (2006). The effect of trauma on dream content—A field study of Palestinian children. *Dreaming* **16**, 63–87.

Whitman, R., Kramer, M., and Baldridge, B. (1963). Experimental study of supervision of psychotherapy. *Arch. Gen. Psychiatr.* **106**, 529–535.

Winget, C., Kramer, M., and Whitman, R. (1972). Dreams and demography. *Can. Psychiatr. Assoc. J.* **17**(Suppl. 2), SS203–SS208.

Zborowski, M., McNamara, P., Hartmann, E., Murphy, M., and Mattle, L. (1998). Boundary structure related to sleep measures and to dream content. *Sleep* **21S**, 284.

DREAMING, HANDEDNESS, AND SLEEP ARCHITECTURE: INTERHEMISPHERIC MECHANISMS

Stephen D. Christman* and Ruth E. Propper[†]

*Department of Psychology, University of Toledo, Toledo, OH, USA
[†]Psychology Department, Merrimack College, North Andover, MA, USA

Research on individual differences in sleep as a function of handedness is reviewed. Special emphasis is placed on a new way of approaching handedness in terms of degree (strong/consistent versus mixed/inconsistent), as opposed to the traditional focus on direction (right versus left). Handedness differences in sleep architecture reflect increased time in REM sleep and decreased time in NREM sleep in consistent right-handers. A framework relating increased interhemispheric interaction in mixed-handers, increased interhemispheric interaction during episodic retrieval, increased interhemispheric interaction during NREM sleep, and the role of NREM sleep in memory consolidation is discussed.

Of all the myriad human behaviors, sleep is one of the most ubiquitous and odd. People spend approximately 6–8 hours out of every 24 prone, immobile, and without conscious awareness of the external world. The fact that all humans, and most mammals and birds, experience sleep has been used as an argument for the importance of this state both evolutionarily and functionally, and much effort has been directed toward determining its origins and purposes. A very brief list of proposed functions for sleep includes: increased immune system functioning (e.g., Spiegel *et al.*, 2002), regulation of carbohydrate metabolism and endocrine function

---

[‡] Both authors contributed equally to this manuscript.

INTERNATIONAL REVIEW OF
NEUROBIOLOGY, VOL. 92
DOI: 10.1016/S0074-7742(10)92011-4

215

(e.g., Spiegel *et al.*, 1999), elimination of "parasitic" modes of thought (Crick and Mitchison, 1983), reinstatement of neuronal connections fostering intra-species individual variability (Jouvet, 1999), increased connectivity between divergent concepts leading to enhanced problem solving (e.g., Walker *et al.*, 2002), transformation of episodic memories into semantic ones (Payne *et al.*, 2009), memory consolidation (Stickgold, 2005), and enhanced neuronal plasticity (e.g., Frank *et al.*, 2001).

Given that sleep may be necessary for both physical and mental well-being, an examination of individual differences affecting sleep variables is important. Such investigations could help to shed light on the cerebral processes involved in sleep generally, as well as on factors contributing to physical impairment following sleep loss and on sleep disturbance components involved in mental illnesses. There are in fact stable individual differences in sleep, in cognitive impairment following sleep loss, and in neurobiological markers of sleep disturbance (Van Dongen *et al.*, 2005), with the factors responsible for these individual differences being unknown, although work suggests a genetic component (Viola *et al.*, 2007).

In this chapter, we focus on one potential individual difference that affects sleep variables: hand preference. We first define "hand preference," and discuss how individual differences in this measure are related to cortical organization. We follow this with a discussion of individual differences in handedness effects on sleep quantity and quality. We then discuss individual differences in handedness on dream characteristics. Last, we speculate on how individual differences in handedness effects on sleep and dreaming may shed light on communication between the cerebral hemispheres during sleep, as well as on the functions of sleep.

I. Hand Preference: Definition, Measurement, and Neurophysiology

Handedness can be conceptualized as varying along two dimensions: *degree* (strong/consistent vs mixed/inconsistent) and *direction* (left vs right), with these two dimensions being represented by different cortical areas (Dassonville *et al.*, 1997). Although the distinction between left- and right-handedness has long been noted, there is growing evidence that degree of handedness is at least as important, as behavioral research has shown that the consistently left-handed (CLH) and consistently right-handed (CRH) are more similar to each other than either is to the inconsistently handed (ICH) (e.g., Barnett and Corballis, 2002; Christman, 1993, Christman *et al.*, 2007; Kempe *et al.*, 2009; Niebauer *et al.*, 2002; Propper *et al.*, 2005). Converging support comes from physiological studies, as ICH are more likely to demonstrate bihemispheric language relative to both CRH and CLH (Khedr *et al.*, 2002), and are more likely to demonstrate symmetrical white matter tract volume, while both the CRH and CLH demonstrate left hemisphere asymmetry in white matter structure (Propper *et al.*, 2010). Thus, comparisons between "left-handers" and "right-handers" should control for *degree* of hand preference, as well as for *direction*, as studies that do not distinguish between degree versus direction of handedness lose statistical power by combining consistent and inconsistent right-handers (Schacter, 1994).

Before discussing the implications of these handedness dimensions for neuroanatomy and for individual differences in sleep, a few words about how handedness is measured and classified are in order. The handedness inventory (Oldfield, 1971) used in our studies asks about hand preference for ten common activities, with scores ranging from –100 (perfectly left-handed) to +100 (perfectly right-handed). The median of absolute values of scores on this handedness inventory is used to define the cutoff point between classification as inconsistent-versus consistent-handed. The median score on the handedness inventory is typically +80. By this criterion, simply doing one or two of the activities consistently with the non-dominant hand (and the remaining eight or nine with the dominant hand) is sufficient to be classified as ICH (that is, inconsistent handedness is not necessarily the same as ambidexterity).

While this practice of referring to someone who does nine things always with their right hand and only one with their left as being "inconsistent-handed" may seem counterintuitive, we argue that the use of a median split divides the sample into two natural, non-arbitrary groups: one comprised of people who perform virtually *all* actions always with their right hand, and one comprised of people who display *any* degree of inconsistent hand preference. Individuals who consistently use their left hand for most or all activities are rare, comprising only about 2% of the population (Lansky *et al.*, 1988). Owing to logistic difficulties associated with obtaining large samples of CLH, most of the studies reviewed in this chapter primarily focus on comparisons between ICH and CRH. It is worth noting that direction and degree of handedness are related, as about 60% of right-handers are consistent-handed, while about 75% of left-handers are inconsistent-handed (Christman, 2005). Thus, comparisons between left- and right-handers can also be thought of as "noisy" comparisons between inconsistent- and consistent-handers.

From a neurophysiological perspective, Propper, Christman, and colleagues have hypothesized that a key difference between the handedness groups involves inter-hemispheric interaction, with consistent handedness being associated with decreased interaction between the left and right cerebral hemispheres relative to inconsistent handedness. This hypothesis is based on both neural and behavioral findings.

First, there is evidence that consistent handedness is associated with a smaller corpus callosum. Witelson and Goldsmith (1991) examined a sample composed solely of right-handers and found that the correlation between strength of right-handedness and corpus callosum size was -0.69, meaning that almost half of the interindividual variation in callosal size was associated with degree of handedness. Similar findings have been reported by Clarke and Zaidel (1994), Denenberg *et al.* (1991), and Habib *et al.* (1991). The smaller callosal size in CRH is supportive of decreased interhemispheric interaction in such individuals.

Second, from a behavioral perspective, there is growing evidence that cognitive processes known to be functionally lateralized to opposite hemispheres show decreased interaction in consistent right-handers. Studies have shown decreased interaction in CRH between left-hemisphere (LH) and right hemisphere (RH) motor processes (Christman, 1993), LH-based word reading and RH-based color naming (Christman, 2001), LH-based processing of local form and RH-based processing of global form (Christman, 2001), LH-based episodic memory encoding and RH-based episodic retrieval (Christman *et al.*, 2004, 2006; Propper and Christman, 2004; Propper *et al.*, 2005), and LH-based belief maintenance processes and RH-based belief updating processes (Christman *et al.*, 2008, 2009; Jasper and Christman, 2005; Niebauer *et al.*, 2004). In addition, evidence indicates that CRH is associated with decreased access (presumably mediated by the corpus callosum) to RH-based processing of risk (Christman *et al.*, 2007) and RH-based representation of body image (Christman *et al.*, 2006). This decreased interaction between the LH and RH and decreased access to RH processing in CRH is hypothesized to reflect decreased interhemispheric interaction in CRH.

To summarize, we suggest that (1) degree of hand preference is at least as important as direction in studies that include hand preference as a variable and (2) consistent-handedness is associated with smaller corpus callosum size, and decreased interhemispheric interaction, relative to inconsistent-handedness. These two hypotheses have implications for the interpretation of handedness effects on sleep variables and for understanding the "functions" of sleep.

II. Handedness and Sleep

A. BEHAVIORAL MEASURES: SLEEP QUANTITY

The pattern of results in the literature examining the effects of individual differences in handedness on sleep strongly suggests differences between consistently and inconsistently handed individuals. Self-report questionnaires (Coren and Searleman, 1987; Hicks *et al.*, 1979, 1999; Propper, 2000, 2004; Propper and Simon, 2002), sleep diary measures (Violani *et al.*, 1988), laboratory-based investigations (Murri *et al.*, 1984; Nielsen *et al.*, 1990; Serafetinides, 1991), and naturalistic home-sleep monitoring (Propper *et al.*, 2004, 2007) have all demonstrated handedness differences in measures of either sleep quantity or quality.

With regard to sleep quantity, there is suggestive evidence that the ICH have shorter sleep durations than the consistently handed (CH) (Hicks *et al.*, 1979;

Propper, 2000; Propper and Simon, 2002). For example, Hicks *et al.* reported that in individuals satisfied with their amount of sleep, the ICH had nominally shorter sleep durations than CH. Propper and Simon found decreased self-reported sleep durations in the ICH versus CH in a population of undergraduate college students (who were not queried as to their sleep satisfaction). Among US Air Force (USAF) recruits, Propper (2004) reported that although the ICH and CH did not differ in their self-reported obtained sleep (presumably as a result of government intervention), the handedness groups did differ in how much sleep they desired obtaining, with ICH desiring less sleep than the CH. Propper and Simon (2002) also reported decreased desired sleep in the IC compared to the ICH.

Although this research is suggestive of shorter sleep durations in the ICH, relative to the CH, other sleep duration estimate studies have not found such differences. Violani *et al.* (1988) examined the sleep of ICH, CLH, and CRH. Rather than generally estimating their sleep duration on a questionnaire, the participants in this study were required to keep a sleep diary for 7 days wherein they wrote, within 30 minutes of waking, their sleep duration, their time in bed following their morning waking, and their dreams. Using sleep diary methodology, Violani *et al.* found no handedness differences in sleep duration.

Differences between handedness groups in sleep duration variability as measured by sleep duration deviation from a standard norm have also been reported, although again the direction of this difference has been conflicting. Hicks *et al.* (1979) using a questionnaire reported greater deviation from a standard norm in the ICH versus the CH, while Violani *et al.* (1988) using a sleep diary reported the opposite, with significantly greater sleep duration variability in the CRH relative to the ICH.

Differences in methodologies (i.e., questionnaire vs sleep diary) may have contributed to the conflicting reports. For example, it is unclear if questionnaire estimates or sleep diary reports of sleep durations accurately reflect sleep habits (Frederickson *et al.*, 2001). Similarly, it is unclear how well individuals are able to accurately estimate the sleep variables of quantity or quality of sleep either in response to a general questionnaire or in a sleep diary (i.e., Frederickson *et al.*, 2001). For example, Killgore *et al.* (2009) contrasted handedness differences in sleep using (1) self-report estimates, sleep diaries, and wrist activity monitors during an uncontrolled 7-day at-home phase versus (2) observations during a controlled overnight stay in a sleep laboratory. They found no handedness differences in the at-home phase, but found that right-handedness was associated with longer sleep duration and greater sleep efficiency under controlled laboratory conditions.

Based on the findings of self-report studies, we examined home sleep as a function of handedness. In two studies, we found that CRH and consistent left-handers exhibited longer sleep latency and greater percentage of sleep period spent awake (Propper *et al.*, 2007), compared to ICH (Propper *et al.*, 2004).

To summarize, on questionnaires, ICH individuals self-report decreased sleep duration and decreased sleep needs. However, objective home-based measures reveal either increased sleep duration and decreased sleep latency in these individuals or no handedness effects. Finally, in the only sleep laboratory assessment of sleep duration as a function of handedness of which we are aware, right-handedness was associated with increased sleep length. Clearly, more research is needed to untangle the effects of methodology on sleep duration assessment. Research investigating the relationship between hand preference and sleep duration could help in this regard.

B. BEHAVIORAL MEASURES: SLEEP QUALITY

On questionnaires, non-right-handed individuals self-report increased symptoms of insomnia, such as difficulty falling asleep, difficulty maintaining sleep and difficulty falling back to sleep following awaking during the night (Coren and Searleman, 1987; Hicks et al., 1999). Hicks et al. (1999) found that ICH reported increased sleep latency, increased night awakenings, and increased difficulty returning to sleep following a nocturnal awakening, compared to the strongly handed. Coren and Searleman (1987) also found greater difficulty falling asleep and maintaining sleep in the non–strongly right-handed using the self-report questionnaire method. Laboratory examination of sleep difficulties has suggested increased severity of sleep apnea in the left-handed relative to right-handed individuals (Hoffstein et al., 1993), supporting the self-report data of decreased sleep quality in the non-right-handed.

In contrast, Violani et al. (1988) reported no handedness differences in either sleep comfort or in time spent in bed following a morning waking (a measure of sleep termination insomnia) using sleep diary data. Similarly, in a sample of older adults, Porac and Searleman (2006) reported no differences between CH and ICH in terms of getting enough sleep.

Interestingly, we found no handedness effects on such symptoms of insomnia in our studies of objectively measured home sleep (Propper et al., 2004, 2007). The disagreement in the literature about possible handedness differences in sleep problems suggests that such handedness differences either are small and/or unreliable or are influenced by methodological factors. If the latter, examination of individual differences in sleep may help to elucidate the mechanisms involved in sleep-related pathologies and in mental illnesses that have sleep disturbance components. For example, pseudoinsomnia, occurring in approximately 10–12% of individuals, is the self-reporting of insomnia, but the demonstration of normal objectively measured sleep (Kelly, 1991). Given the neuroanatomical differences between the handedness groups, research investigating the relationship between hand preference and pseudoinsomnia could help to determine the brain mechanisms involved in the self-perception of sleep.

In any event, results suggest that questionnaire estimates of sleep variables do not reflect actual sleep, but rather some other cognitive process, and that this cognitive process is influenced by participant handedness. Future research could explicitly examine the relationship between questionnaire estimates of sleep variables, their relationship to actual recorded home sleep, and handedness.

C. Physiological Measures: Interhemispheric Interaction

The neurochemical and neuro-activational changes associated with sleep generally, and with the different sleep stages in particular, have been well described (e.g., see Hobson *et al.*, 1998; Pace-Schott and Hobson, 2002). Rather than focusing on individual differences effects of handedness at the micro- or neurochemical level (a level which, incidentally, has never to our knowledge been investigated), the focus of this section is on the role of corpus callosum–mediated hemispheric interaction during sleep, and on how individual differences in handedness may interact with sleep stage and allow for inferences to be drawn about both the neural activity taking place during sleep as well as the function of sleep.

A typical method by which corpus callosum activity and hemispheric communication during sleep has been examined is via interhemispheric electroencephalographic (EEG) coherence, which compares the relationship between EEG signals from (usually) homologous sites in the two hemispheres as a function of the signals' frequencies. Interhemispheric EEG coherence is thought to reflect corpus callosum–mediated communication between the two cerebral hemispheres (Montplaisir *et al.*, 1990; Nielsen *et al.*, 1993). Increased levels of coherence are believed to reflect increased callosal activity and thus indicate increased hemispheric connectivity, while decreased levels of coherence are thought to reflect the opposite (e.g., Montplaisir *et al.*, 1990; Nielsen *et al.*, 1990). With very few exceptions (i.e., Banquet, 1983), there appears to be a general pattern in the literature of an overall increase in interhemispheric EEG coherence during all sleep stages compared to wake (Achermann and Borbély, 1998; Dumermuth and Lehmann, 1981;Dumermuth *et al.*, 1983; Nielsen *et al.*, 1990).

A specific examination of interhemispheric EEG coherence effects as a function of sleep stage demonstrates that rapid eye movement (REM) sleep tends to display increased interhemispheric EEG coherence relative to wake or to non–rapid eye movement (NREM) sleep, particularly at parietal sites or in low- or high-frequency ranges (Achermann and Borbély, 1998; Dumermuth and Lehmann, 1981;Dumermuth *et al.*, 1983; Nielsen *et al.*, 1990). In general, NREM (stages 2, 3, and 4) tends to display interhemispheric EEG levels intermediate between those of wake and REM. In some low- and middle-frequency ranges, NREM displays greater interhemispheric EEG coherence at anterior and occipital sites compared to REM or to wake (Achermann and Borbély, 1998; Dumermuth and Lehmann, 1981; Dumermuth *et al.*, 1983), although this increase in NREM coherence relative to wake or REM at frontal sites has not always been found (Nielsen *et al.*, 1990).

To summarize, overall, the literature supports a general increase in interhemispheric interaction, and inferred corpus callosum activity, as measured via interhemispheric EEG coherence, from wake to NREM to REM sleep. We suggest that these results support the hypothesis that the organization of the sleep stages reflects corpus callosum activity and hemispheric interaction, and that corpus callosum activity and hemispheric communication change across the different sleep stages.

Only three laboratory-based examinations of handedness differences in sleep variables have examined individual differences in neurophysiological sleep measures (Murri *et al.*, 1984; Nielsen *et al.*, 1990; Serafetinides, 1991). Nielsen *et al.* (1990) reported increased EEG coherence in left-handed versus right-handed individuals during wake, stage 2, and REM sleep. Murri *et al.* (1984) found increased left- versus right-hemispheric activation in left- versus right-handed individuals in REM sleep, respectively. Serafetinides (1991) reported differences in EEG amplitude as a function of both handedness and sleep time (first 4 hours vs last 4 hours of sleep), with left-handed individuals demonstrating no change in amplitude as a function of sleep time, and right-handed individuals demonstrating decreased amplitude during the last 4 hours of sleep in both the left and right hemispheres.

We would like to point out that, while each of the above laboratory studies states a comparison was made between left- and right-handers, in all of the studies no information is given concerning the actual calculation of handedness, and, in addition, Nielson *et al*, confounded personal with familial sinistrality. Because strongly left-handed individuals constitute only about 2–3% of the population (Lansky *et al.*, 1988), it is likely that many of the left-handers who participated in these studies were, in fact, mixed-handed.

The importance of distinguishing between both degree and direction of handedness is reflected in studies from our lab wherein we explicitly compared CRH, ICH, and CLH individuals (Propper *et al.*, 2004). With regard to the time spent in NREM sleep, CLH spent the most time ($M = 76\%$), ICH spent the second most time ($M = 72\%$), and CRH the least ($M = 62\%$). The opposite pattern was obtained for REM sleep: CLH, $M = 20\%$; ICH, $M = 24\%$; and CRH, $M = 31\%$. The number of discrete REM episodes, however, showed a different story, with ICH having the most REM episodes ($M = 4.7$), CLH the second most ($M = 3.8$), and CRH the least ($M = 3.2$). Given the pattern of the literature, these results suggest (1) the CRH and the CLH should not be conceptualized as a single group in sleep studies; (2) CRH differ in their sleep architecture from both the CLH and the ICH; and (3) individual differences in sleep architecture are influenced by both degree (i.e., inconsistent vs consistent) and direction (i.e., left vs right) of handedness.

The presence of differences in physiological and behavioral indicators of interhemispheric interaction as a function of both degree and direction of handedness (e.g., Clarke and Zaidel, 1994; Schacter, 1994), and the fact different sleep stages are associated with different degrees of interhemispheric interaction (e.g.; Achermann and Borbély, 1998;Nielsen *et al.*, 1990), directly implies that the handedness differences in sleep architecture reported here and in previous work from our lab (i.e., Propper *et al.*, 2004) may be related to handedness differences in interhemispheric interaction during sleep.

One interpretation for these findings is based on REM sleep: If during REM sleep some behavioral or physiological processes involves increased hemispheric communication relative to NREM or to wake, then individuals who tend to have greater overall interhemispheric communication regardless of sleep/wake stage (i.e., inconsistent-handers) may be more efficient at fulfilling this REM function. That is, inconsistent-handers may be able to perform REM-dependent processes more efficiently and quickly and/or during other wake/sleep stages, resulting in less need for REM sleep relative to consistent-handers.

A second possibility is based on NREM, and involves the relationship between this sleep stage, interhemispheric interaction, episodic memory ability, and individual differences in handedness. Increased interhemispheric interaction has been related to enhanced episodic, but not semantic or implicit, memory retrieval (Christman and Propper, 2001;Christman *et al.*, 2003; Propper *et al.*, 2005). The increased interhemispheric communication in inconsistent-handers has been proposed as the mechanism by which this handedness group outperforms consistent-handers on tasks accessing episodic memories (e.g., Christman and Propper, 2010; Propper and Christman, 2004; Propper *et al.*, 2005). Interestingly, NREM sleep, particularly stages 3 and 4, has been implicated in episodic memory consolidation (Plihal and Born, 1997). If increased corpus callosum–mediated interhemispheric interaction is involved in both episodic memory and NREM sleep, with episodic memory being enhanced by this sleep stage, then it is not surprising that inconsistent-handers, individuals who display both increased interhemispheric interaction and increased episodic memory, demonstrated increased NREM in the present study. The increased NREM of the inconsistent-handed may in part underlie this handedness group's superior episodic memory ability. Future research could attempt to determine which of the above hypotheses is more accurate.

D. Physiological Measures: Hemispheric Activation

The topic of hemispheric activation during sleep is a broad and varied matter, and, as such, is beyond the scope of the current chapter. We will, however, briefly review the small number of studies of handedness differences in hemispheric activation during sleep. The general finding overall is that, relative to NREM sleep, REM sleep is associated with increased right hemisphere activation in right-handers (e.g., Angeleri *et al.*, 1984; Boldu *et al.*, 2003; Gordon *et al.*, 1982).

In contrast, non-right-handedness is associated with increased left hemisphere activation during REM sleep, although data for non-right-handers were also more variable (Murri *et al.*, 1984). Additionally, while right-handers exhibit different patterns of hemispheric activation during REM versus NREM sleep, left-hemispheric activation appears to exist across all sleep stages in non-right-handers (e.g., Lavie, 1986; Murri *et al.*, 1984; Serafetinides, 1991).

Unfortunately, what has not been explicitly addressed in prior research is the question of absolute levels of hemispheric activation as a function of handedness. For example, behavioral research on inconsistent handedness suggests that it is associated with increased access to right hemisphere processing, which seems potentially inconsistent with the finding of greater left hemisphere activation during sleep in non-right-handers. However, it might be the case that, for instance, absolute levels of right hemisphere activation during REM sleep are equivalent in right- and non-right-handers, with the left hemisphere being relatively less versus more activated in right- versus non-right-handers, respectively.

E. Handedness and Dreaming: Recall

A small number of studies have looked at handedness differences in dream recall. Violani *et al.* (1988) reported increased self-reported dream recall in right- relative to left- and mixed-handers. Van Nuys (1984) reported no handedness differences in dream recall, although an association was found between increased dream recall and leftward shifts of gaze. Interestingly, leftward shifts of gaze selectively activate right frontal regions and there is tentative evidence that they may enhance recall in laboratory settings (Christman, 2006). Hicks *et al.* (1999) reported greater recall of vivid dreams in left- relative to mixed- and right-handers. Finally, Christman (2007) reported increased dream recall in inconsistent- relative to consistent right-handers. Thus, there is no clear pattern of handedness differences in dream recall.

However, given the existence of methodological complications in the measurement of dream recall (e.g., Schredl, 2002), the nature and existence of handedness differences in dream recall remains an open question. It is worth noting that increased left hemisphere activation may be associated with enhanced dream recall (Bertini and Violani, 1984), but this is controversial. Given evidence reviewed above that non-right-handers display increased left hemisphere activation during sleep, this in turn suggests that non-right-handers may display overall higher levels of dream recall. Consistent with this, individual with lesser degrees of functional cerebral asymmetry report higher levels of dream recall (Doricchi *et al.*, 1993), and non-right-handedness is associated with lesser cerebral asymmetry (Hellige, 2001).

F. Handedness and Dreaming: Content and Vividness

Again, a small number of studies have looked at handedness differences in dream content. Cohen (1977) found that, among right-handers, dream content across a night's sleep progressively shifted toward greater left hemisphere involvement (assessed both behaviorally and electrophysiologically) in right- but not left- handers. Hicks *et al.* (1999) reported greater dream vividness among left- relative to mixed- and right-handers. Similarly, McNamara *et al.* (1998) reported that the dreams of left-handers were characterized by increased imagery and affective content. Left-handers also reported that their dreams were less likely to reflect everyday life experiences. Finally, Nielsen and Chénier (1999) reported that increased interhemispheric EEG coherence was associated with an increased proportion of dream characters whose faces were explicitly represented in the dream. Given evidence reviewed above that non-right-handedness is associated with increased interhemispheric EEG coherence, this raises the possibility of fine-grained handedness differences in dream content.

III. Summary

The research reviewed in this chapter documents a wide range of individual differences in both behavioral and physiological measures of sleep and dreaming as a function of both degree (e.g., consistent/strong versus inconsistent/mixed) and direction (e.g., right vs left) of handedness. In general, compared to inconsistent and/or consistent left-handedness, (consistent) right-handedness is tentatively associated with (1) increased total sleep duration (although this measure appears to be strongly influenced by methodological variations), (2) a possible decreased tendency toward various sleep problems, (3) decreased interhemispheric EEG coherence during sleep, (4) decreased time spent in NREM sleep, (5) increased time spent in REM sleep, (6) fewer REM episodes per night, and (7) increased right hemisphere activation during sleep. Individual differences in handedness effects on dream recall and dream content also appear to exist, but the direction of these differences is currently unclear.

It must be noted, however, that very few studies to date have explicitly compared both dimensions of handedness within the same paradigm, the notable exception being a pair of papers from our lab (Propper *et al.*, 2004a, 2004b). These studies found that direction of handedness was the best predictor of time spent in REM versus NREM sleep, with increasing right-handedness associated with more REM time and less NREM time, while degree of handedness was the best predictor of the number of REM episodes, with inconsistent-handers having more REM episodes than consistent left- or right-handers. Thus, it is imperative that future research on handedness differences in sleep explicitly assesses and compares both degree and direction of handedness.

Given the existence of handedness differences in the architecture and physiology of sleep, what are some of the implications of these differences? Perhaps the most intriguing involves the interrelations between handedness, interhemispheric interaction, episodic memory, and sleep. An association between inconsistent handedness and superior episodic retrieval has been robustly demonstrated (Christman *et al.*, 2004, 2006; Lyle *et al.*, 2008a, 2008b; Propper and Christman, 2004; Propper *et al.*, 2005). These findings have been interpreted in terms of the facts that (1) the retrieval of episodic memories is critically dependent on efficient interhemispheric interaction and (2) inconsistent handedness is associated with increased callosally mediated interhemispheric interaction.

Given evidence reviewed above that non-right-handedness is also associated with increased interhemispheric interaction during sleep and that sleep plays an important role in the consolidation of memory, the possibility is raised that the handedness differences in memory and in sleep are directly related.

Evidence for a differential reliance of episodic/explicit memory on early night NREM (specifically slow-wave) sleep, and procedural/semantic/implicit memory on late night REM or possibly stage 2 sleep (e.g., Plihal and Born, 1997, 1999), in conjunction with research reviewed above demonstrating an increase in interhemispheric interaction from wake, to NREM, to REM (e.g., Achermann and Borbély, 1998; Dumermuth and Lehmann, 1981; Dumermuth *et al.*, 1983; Nielsen *et al.*, 1990), suggests a possible causal relationship between interhemispheric interaction during sleep and some forms of memory consolidation, with the exact nature of that relationship unclear. For example, it may be that lesser interhemispheric interaction in NREM early in the night benefits episodic/explicit memory, while increased interhemispheric interaction late in the night benefits procedural/semantic/implicit memory. This is consistent with the findings that inconsistent- versus consistent-handers spend more time in NREM versus REM sleep, respectively.

As mentioned earlier, perhaps the superior episodic memory performance among inconsistent-handers reflects the fact that they spend more time in NREM sleep, thus allowing greater consolidation of episodic memories. Future research should consider measuring and manipulating both sleep and memory variables in the same study in order to more directly examine the interrelated roles of handedness, memory, and sleep. Taken together, at the very least these results suggest that both sleep and some forms of memory consolidation may rely on similar corpus callosum–mediated pathways, and that investigation of the relationship between memory, interhemispheric interaction, and sleep can inform theories of both sleep and memory.

The picture regarding dreams is much less clear and has received less attention. Although inconsistent-handers spend less time in REM sleep, they also appear to have more REM episodes. At present, it is not clear whether handedness differences in dream recall exist, and, if so, whether they simply reflect handedness differences in memory overall or whether they reflect a sleep-specific process (e.g., a handedness difference in the number and/or vividness of dreams).

In any case, there is another intriguing reason why handedness differences in interhemispheric processing during REM sleep are worth further study. The majority of eye movements during REM sleep involve horizontal, bilateral saccades (Hansotia et al., 1990), and a growing body of research shows that the same bilateral saccadic eye movements during waking states lead to increased interhemispheric interaction (e.g., Bruny et al., 2009; Christman et al., 2003, 2004, 2006; Lyle et al., 2008; Parker and Dagnall, 2007; Parker et al., 2008). Similarly, there is evidence that bilateral saccadic eye movements have systematic effects on gamma oscillations during both waking (Propper et al., 2007) and REM sleep (Hong and Harris, 2009).

As REM sleep, relative to NREM sleep, is associated with an overall increase in interhemispheric EEG coherence (e.g., Guevara et al., 1995), these findings suggest that the bilateral saccadic eye movements observed during REM sleep may play a key role in fostering increased interhemispheric interaction and integration. Critically, given evidence that bilateral saccadic eye movements have different effects on consistent- versus inconsistent-handers (Hanaver-Torrez and Lyle, 2010; Lyle et al., 2008), the possibility is raised that handedness differences exist in the functions and/or mechanisms underlying REM sleep.

In conclusion, it is evident that there are systematic differences in sleep and dreaming as a function of both degree and direction of handedness, and that these differences are related to mechanisms of interhemispheric interaction during NREM and REM sleep. Future research on the neural and functional mechanisms of sleep would be well advised to include strength of handedness as a variable. Even if one is not interested in handedness and individual differences *per se*, including handedness as a factor in analyses would likely move variability from the error term to an effect term, thereby allowing increased power in the analyses of other effects.

References

Achermann, P., and Borbély, A. A. (1998). Coherence analysis of the human sleep electroencephalogram. *Neuroscience* **85**(4), 1195–1208.

Angeleri, F., Scarpino, O., and Signorino, M. (1984). Information processing and hemispheric specialization: electrophysiological study during wakefulness, stage 2 and stage REM sleep. *Res. Commun. Psychol. Psychiatr. Behav.* **9**(1), 121–138.

Banquet, J. P. (1983). Inter- and intrahemispheric relationships of the EEG activity during sleep in man. *Electroencephalogr. Clin. Neurophysiol.* **55**(1), 51–59.

Barnett, K. J., and Corballis, M. C. (2002). Ambidexterity and magical ideation. *Laterality* **7**(1), 75–84.

Bertini, M., and Violani, C. (1984). Cerebral hemispheres, REM sleep, and dream recall. *Res. Commun. Psychol. Psychiat. Behav.* **9**(1), 3–14.

Boldu, C., Daous, A. -M., Daous, E., Braun, C. M. J., and Godbout, R. (2003). Hemispheric lateralization of the EEG during wakefulness and REM sleep in young healthy adults. *Brain Cogn.* **53**(2), 193–196.

Bruny, T. T., Mahoney, C. R., Augustyn, J. S., and Taylor, H. A. (2009). Horizontal saccadic eye movements enhance the retrieval of landmark shape and location information. *Brain Cogn.* **70**(3), 279–288.

Christman, S. (1993). Handedness in musicians: bimanual constraints on performance. *Brain Cogn.* **22**(2), 266–272.

Christman, S. D. (2001). Individual differences in Stroop and local–global processing: a possible role of interhemispheric interaction. *Brain Cogn.* **45**(1), 97–118.

Christman, S. D. (2005). *It's mixed versus strong, not left versus right: Handedness as a wide-ranging dimension of individual difference.* Presented at the Annual Meeting of the Midwestern Psychological Association, Chicago, May.

Christman, S. D. (2006). *Left hemisphere activation enhances encoding of episodic memories in females but not males.* Presented at the 47th Annual Meeting of the Psychonomic Society, Houston.

Christman, S. D. (2007). *Individual differences in déjà vu and jamais vu experiences: Degree of handedness and access to the right hemisphere.* Presented at the 19th Annual Meeting of the Association for Psychological Science, Washington, DC.

Christman, S. D., Bentle, M., and Niebauer, C. L. (2007). Handedness differences in body image distortion and eating disorder symptomatology. *Int. J. Eat. Disord.* **40**(3), 247–256.

Christman, S. S., Garvey, J. K., Propper, R. E., and Phaneuf, K. (2003). Bilateral eye movements enhance the retrieval of episodic memories. *Neuropsychology* **17**(2), 221–229.

Christman, S. D., Henning, B., Geers, A. L., Propper, R. E., and Niebauer, C. L. (2008). Mixed-handed persons are more easily persuaded and are more gullible: interhemispheric interaction and belief updating. *Laterality* **13**(5), 403–426.

Christman, S. D., Jasper, J. D., Sontam, V., and Cooil, B. (2007). Individual differences in risk perception versus risk taking: handedness and interhemispheric interaction. *Brain Cogn.* **63**(1), 51–58.

Christman, S. D., and Propper, R. E. (2001). Superior episodic memory is associated with interhemispheric processing. *Neuropsychology* **15**(4), 607–616.

Christman, S. D., and Propper, R. E. (2010). The interhemispheric basis for the retrieval of episodic memories: effects of handedness and bilateral eye movements. In: Current Issues in Applied Memory (G. Davies and D. Wright, eds.), Psychology Press, London.

Christman, S. D., Propper, R. E., and Brown, T. J. (2006). Increased interhemispheric interaction is associated with earlier offset of childhood amnesia. *Neuropsychology* **20**(3), 336–345.

Christman, S. D., Propper, R. E., and Dion, A. (2004). Increased interhemispheric interaction is associated with decreased false memories in a verbal converging semantic associates paradigm. *Brain Cogn.* **56**(3), 313–319.

Christman, S. D., Sontam, V., and Jasper, J. D. (2009). Individual differences in ambiguous figure perception: degree of handedness and interhemispheric interaction. *Perception* **38**(8), 1183–1198.

Clarke, J. M., and Zaidel, E. (1994). Anatomical-behavioral relationships: corpus callosum morphometry and hemispheric specialization. *Behav. Brain Res.* **64**(1–2), 185–202.

Cohen, D. B. (1977). Changes in REM dream content during the night: implications for a hypothesis about changes in cerebral dominance across REM periods. *Percept. Mot. Skills* **44**(3), 1267–1277.

Coren, S., and Searleman, A. (1987). Left sidedness and sleep difficulty: the alinormal syndrome. *Brain Cogn.* **6**(2), 184–192.

Crick, F., and Mitchison, G. (1983). The function of dream sleep. *Nature* **304**(14), 111–114.

Dassonville, P., Zhu, X. -H., Ugurbil, K., Kim, S. -G., and Ashe, J. (1997). Functional activation in motor cortex reflects the direction and the degree of handedness. *Proc. Natl. Acad. Sci. USA* **94**(25), 14015–14018.

Denenberg, V. H., Kertesz, A., and Cowell, P. E. (1991). A factor analysis of the human's corpus callosum. *Brain Res.* **548**(1–2), 126–132.

Doricchi, F., Milana, I., and Violani, C. (1993). Patterns of hemispheric lateralization in dream recallers and non-dream recallers. *Int. J. Neurosci.* **69**(1–4), 105–117.

Dumermuth, G., Lange, B., Lehmann, D., Meier, C. A., and Dinkelmann, R. (1983). Spectral analysis of all-night sleep EEG in healthy adults. *Eur. J. Neurol.* **22**(2), 322–339.

Dumermuth, G., and Lehmann, D. (1981). EEG power and coherence during non-REM and REM phases in humans in all night sleep analyses. *Eur. J. Neurol.* **20**(6), 429–434.

Frank, M. G., Issa, N. P., and Stryker, M. P. (2001). Sleep enhances plasticity in the developing visual cortex. *Neuron* **30**(1), 275–287.

Frederickson, J. D., Naponelli, S., and Helder, M. (2001). *Sleep and college students: Are they getting more than they think?* Poster presented at the American Psychological Society, Miami, FL.

Gordon, H. W., Frooman, B., and Lavie, P. (1982). Shift in cognitive asymmetries between wakings from REM and NREM sleep. *Neuropsychologia* **20**(1), 99–103.

Guevara, M. A., Lorenzo, I., Arce, C., and Ramos, J. (1995). Inter- and intrahemispheric EEG correlation during sleep and wakefulness. *Sleep* **18**(4), 257–265.

Habib, M., Gayraud, D., Oliva, A., Regis, J., Salamon, G., and Khalal, R. (1991). Effects of handedness and sex on the morphology of the corpus callosum: a study with brain magnetic resonance imaging. *Brain Cogn.* **16**(1), 41–61.

Hanaver-Torrez, S. D., and Lyle, K. B. (2010). *Can repetitive saccadic eye movements enhance mental rotation?* Presented at the Annual Meeting of the Midwestern Psychological Association, Chicago, May.

Hansotia, P., Broste, S., So, E., Ruggles, K., Wall, R., and Friske, M. (1990). Eye movement patterns in REM sleep. *Electroen. Clin. Neuro.* **76**(5), 388–399.

Hellige, J. B. (2001). Hemispheric Asymmetry: What's Right and What's Left. Harvard University Press, Cambridge.

Hicks, R. A., Bautista, J., and Hicks, G. J. (1999). Handedness and the vividness of dreams. *Dreaming* **9**(4), 265–269.

Hicks, R. A., DeHaro, D., Inman, G., and Hicks, G. J. (1999). Consistency of hand use and sleep problems. *Percept. Mot. Skills* **89**(1), 49–56.

Hicks, R. A., Pellegrini, R. J., and Hawkins, J. (1979). Handedness and sleep duration. *Cortex* **15**(2), 327–229.

Hobson, J. A., Stickgold, R., and Pace-Schott, E. F. (1998). The neuropsychology of REM sleep dreaming. *NeuroReport* **9**(3), R1–R14.

Hoffstein, V., Chan, C. K., and Slutsky, A. S. (1993). Handedness and sleep apnea. *Chest* **103**(6), 1860–1862.

Hong, C. C. -H., and Harris, J. C. (2009). Study of neural correlates of rapid eye movements in dreaming sleep using video camera for timing of REMs and functional MRI: it's implications. *Sleep Hypn.* **11**(1), 1–4.

Jasper, J. D., and Christman, S. D. (2005). A neuropsychological dimension for anchoring effects. *J. Behav. Decis. Mak.* **18**(5), 343–369.

Jouvet, M. (1999). The Paradox of Sleep: The Story of Dreaming. Translated by Laurence Garey. MIT Press, Cambridge, MA, xiii + 211 pp.

Kelly, D. D. (1991). Disorders of sleep and consciousness. In: Principles of Neural Science (E. R. Kandel, J. H. Schwartz, and T. M. Jessell, eds.), Appleton and Lange, Norwalk, pp. 805–819.

Kempe, V., Brooks, P. J., and Christman, S. D. (2009). Inconsistent handedness is linked to more successful foreign language vocabulary learning. *Psychon. Bull. Rev.* **16**(3), 480–485.

Khedr, E. M., Hamed, E., Said, A., and Basahi, J. (2002). Handedness and language cerebral lateralization. *Eur. J. Appl. Physiol.* **87**(4–5), 469–472.

Killgore, W. D., Lipizzi, E. L., Grugle, N. L., Killgore, D. B., and Balkin, T. J. (2009). Handedness correlates with actigraphically measured sleep in a controlled environment. *Percept. Mot. Skills* **109**(2), 395–400.

Lansky, L. M., Feinstein, H., and Peterson, J. M. (1988). Demography of handedness in two samples of randomly selected adults (N=2083). *Neuropsychologia* **26**(3), 465–477.

Lavie, P. (1986). Differential effects of awakening from REM and NONREM sleep on dichotic listening performance as a function of handedness. *Int. J. Neurosci.* **30**(1–2), 37–42.

Lyle, K. B., Logan, J., and Roediger, H. L. (2008a). Eye movements enhance memory for individuals who are strongly right-handed and harm it for individuals who are not. *Psychonom. Bull. Rev.* **15**(3), 315–320.

Lyle, K. B., McCabe, D. P., and Roediger, H. L. (2008b). Handedness is related to memory via interhemispheric interaction: evidence from paired associate recall and source memory tasks. *Neuropsychology* **22**(4), 523–530.

McNamara, P., Clark, J., and Hartmann, E. (1998). Handedness and dream content. *Dreaming* **8**(1), 15–22.

Montplaisir, J., Nielsen, T., Côté, J., Bolvin, D., Rouleau, I., and Lapierre, G. (1990). Interhemispheric EEG coherence before and after partial callosotomy. *Clin. Electroencephalogr.* **21**(1), 42–47.

Murri, L., Stefanini, A., Bonanni, E., Cei, G., Navona, C., and Denoth, F. (1984). Hemispheric EEG differences during REM sleep in dextrals and sinistrals. *Res. Commun. Psychol. Psychiat. Behav.* **9**(1), 109–120.

Niebauer, C. L., Aselage, J., and Schutte, C. (2002). Interhemispheric interaction and consciousness: degree of handedness predicts the intensity of a sensory illusion. *Laterality* **7**(1), 85–96.

Niebauer, C. L., Christman, S. D., Reid, S. A., and Garvey, K. (2004). Interhemispheric interaction and beliefs on our origin: degree of handedness predicts beliefs in creationism versus evolution. *Laterality* **9**(4), 433–447.

Nielsen, T., Abel, A., Lorrain, D., and Montplaisir, J. (1990). Interhemispheric EEG coherence during sleep and wakefulness in left- and right- handed subjects. *Brain Cogn.* **14**(1), 113–125.

Nielsen, T. A., and Chénier, V. (1999). Variations in EEG coherence as an index of the affective content of dreams from REM sleep: relationships with face imagery. *Brain Cogn.* **41**(2), 200–212.

Nielsen, T., Montplaisir, J., and Lassonde, M. (1993). Decreased interhemispheric EEG coherence during sleep in agenesis of the corpus callosum. *Eur. J. Neurol.* **33**(2), 173–176.

Oldfield, R. (1971). The assessment and analysis of handedness: the Edinburgh Inventory. *Neuropsychology* **9**(1), 97–113.

Pace-Schott, E. F., and Hobson, J. A. (2002). The neurobiology of sleep: genetics, cellular physiology and subcortical networks. *Nat. Rev. Neurosci.* **3**(8), 591–605.

Parker, A., and Dagnall, N. (2007). Effects of bilateral eye movements on gist based false recognition in the DRM paradigm. *Brain Cogn.* **63**(3), 221–225.

Parker, A., Relph, S., and Dagnall, N. (2008). Effects of bilateral eye movements on the retrieval of item, associative, and contextual information. *Neuropsychology* **22**(1), 136–145.

Payne, J. D., Schacter, D. L., Propper, R. E., Huang, L. W., Wamsley, E. J., Tucker, M. A., Walker, M. P., and Stickgold, R. (2009). The role of sleep in false memory formation. *Neurobiol. Learn. Mem.* **92**(3), 327–334.

Plihal, W., and Born, J. (1997). Effects of early and late nocturnal sleep on declarative and procedural memory. *J. Cogn. Neurosci.* **9**(4), 534–547.

Plihal, W., and Born, J. (1999). Effects of early and late nocturnal sleep on priming and spatial memory. *Psychophysiology.* **36**(5), 571–582.

Porac, C., and Searleman, A. (2006). The relationship between hand consistency, health, and accidents in a sample of adults over the age of 65 years. *Laterality* **11**(5), 405–414.

Propper, R. E. (2000). *Perceived sleep needs and feelings of alertness: Handedness and familial sinistrality effects.* Poster presented at the American Psychological Society, Miami, FL.

Propper, R. E. (2004). Handedness differences in self-assessment of sleep quantity: non-right versus strong-right-handers. *Sleep Biol. Rhythms* **2**(1), 99–101.

Propper, R. E., and Christman, S. D. (2004). Mixed- versus strong-handedness is associated with biases toward 'Remember' versus 'Know' judgments in recognition memory: role of interhemispheric interaction. *Memory* **12**(6), 707–714.

Propper, R. E., Christman, S. D., and Olejarz, S. (2007). Home-recorded sleep architecture as a function of handedness II: consistent right- versus consistent left-handers. *J. Neural Eng.* **195**(8), 689–692.

Propper, R. E., Christman, S. D., and Phaneuf, K. A. (2005). A mixed-handed advantage in episodic memory: a possible role of interhemispheric interaction. *Mem. Cogn.* **33**(4), 751–757.

Propper, R. E., Lawton, N., Przyborski, M., and Christman, S. D. (2004). An assessment of sleep architecture as a function of degree of handedness in college women using a home sleep monitor. *Brain Cogn.* **54**(3), 186–197.

Propper, R. E., O'Donnell, L. J., Whalen, S., Tie, Y., Norton, I. H., Suarez, R. O., Zollei, L., Radmanesh, A., and Golby, J. A. (2010). A combined fMRI and DTI examination of functional language lateralization and arcuate fasciculus structure: effects of degree versus direction of hand preference. *Brain Cogn.* **73**(2), 85–92. (Epub ahead of print): doi:10.1016/j.bandc.2010.03.004.

Propper, R. E., Pierce, J., Bellorado, N., Geisler, M. W., and Christman, S. D. (2007). Effects of bilateral eye movements on interhemispheric gamma EEG coherence: implications for EMDR therapy. *J. Neural Eng.* **95**(9), 785–788.

Propper, R. E., and Simon, B. *The weakly-handed need, and receive, less sleep than the strongly handed.* Poster presented at the meeting of the International Neuropsychological Society, 2002, February, Toronto, Canada.

Schacter, S. C. (1994). Ambilaterality: definition from handedness preference questionnaires and potential significance. *Int. J. Neurosci.* **77**(1–2), 47–51.

Schredl, M. (2002). Questionnaires and diaries as research instruments in dream research: methodological issues. *Dreaming* **12**(1), 17–26.

Serafetinides, E. A. (1991). Cerebral dominance and sleep: a comparison according to handedness and time of sleep. *Int. J. Neurosci.* **61**(1), 91–92.

Spiegel, K., Leproult, R., and Van Cauter, E. (1999). Impact of sleep debt on metabolic and endocrine function. *Lancet* **354**(9188), 1435–1439.

Spiegel, K., Sheridan, J. F., and Van Cauter, E. (2002). Effect of sleep deprivation on response to immunization. *J. Am. Med. Assoc.* **288**(12), 1471–1472.

Stickgold, R. (2005). Sleep-dependent memory consolidation. *Nature* **43**(7063), 1272–1278.

Van Dongen, H.P.A., Vitellaro, K. M., and Dinges, D. F. (2005). Individual differences in adult human sleep and wakefulness: leitmotif for a research agenda. *Sleep* **28**(4), 479–496.

Van Nuys, D. W. (1984). Lateral eye movement and dream recall: II. Sex differences and handedness. *Int. J. Psychosom.* **31**(3), 3–7.

Viola, A. U., Archer, S. N., James, L. M., Groeger, J. A., Lo, J. C. Y., Skene, D. J., Violani, C., Gennaro, De., and Solano, L. (1988a). Hemispheric differentiation and dream recall: subjective estimates of sleep and dreams in different handedness groups. *Int. J. Neurosci.* **39**(1–2), 9–14.

Viola, A. U., Archer, S. N., James, L. M., Groeger, J. A., Lo, J. C., Skene, D. J., von Schantz, M., and Dijk, D. J. (2007). PER3 polymorphism predicts sleep structure and waking perforance. *Curr. Biol.* **17**(7), 613–618.

Violani, C., De Gennaro, L., and Solano, L. (1988). Hemispheric differentiation and dream recall: subjective estimates of sleep and dreams in different handedness groups. *Int. J. Neurosci.* **39**(1–2), 9–14.

Walker, M. P., Liston, C., Hobson, A. J., and Stickgold, R. (2002). Cognitive flexibility across the sleep–wake cycle: REM-sleep enhancement of anagram problem solving. *Cogn. Brain Res* **14**(3), 317–324.

Witelson, S. F., and Goldsmith, C. H. (1991). The relationship of hand preference to anatomy of the corpus callosum in men. *Brain Res.* **545**(1–2), 175–182.

TO WHAT EXTENT DO NEUROBIOLOGICAL SLEEP-WAKING PROCESSES SUPPORT PSYCHOANALYSIS?

Claude Gottesmann*

Département de Biologie, Faculté des Sciences, Université de Nice-Sophia Antipolis, Nice, France

Sigmund Freud's thesis was that there is a censorship during waking that prevents memory of events, drives, wishes, and feelings from entering the consciousness because they would induce anxiety due to their emotional or ethical unacceptability. During dreaming, because the efficiency of censorship is decreased, latent thought contents can, after dream-work involving condensation and displacement, enter the dreamer's consciousness under the figurative form of manifest content. The quasi-closed dogma of psychoanalytic theory as related to unconscious processes is beginning to find neurobiological confirmation during waking. Indeed, there are active processes that suppress (repress) unwanted memories from entering consciousness. In contrast, it is more difficult to find neurobiological evidence supporting an organized dream-work that would induce meaningful symbolic content, since dream mentation most often only shows psychotic-like activities.

I. Introduction

The mind–body relationship, which has long occupied philosophers, is now beginning to give way to scientific explanation because of major current progress in neurobiology. One main field of investigation that remains

* Preliminary reports (Gottesmann, 2009, 2010b).

233

open to discussion concerns the mechanisms underlying conscious and unconscious processes.

The invention of psychoanalysis by Sigmund Freud was probably the most important advance in the psychological sciences in the last century. More important even than its discovery (Helmholtz, 1860), Freud's highlighting of the unconscious was a revolution which opened the way to the most crucial findings in fundamental psychology as well as in the fields of psychotherapy and psychiatry (Darcourt, 2006). The first major work devoted to describing and attempting to access the unconscious was, "The interpretation of dreams" (Freud, 1900a). This well-acknowledged monumental study of psychological function during sleep has been the direct or indirect source of numerous current clinical concepts, more or less related to the notion of censorship, even though it was first mentioned in relation to psychosis in a letter to Fliess in 1897 (Freud, 1897).

In Freud's theory, the higher integrated processes, i.e., waking consciousness, comprise self-awareness, attention, and reflection for adapted behavior. They support all the conscious, preconscious, and most importantly, unconscious processes. Indeed, during waking mind functioning, some psychological contents (thoughts, images, memory trace content) are rejected from consciousness and thus actively repressed because they are related to wishes or drives whose satisfaction would be incompatible with conscious well-being due to their potent emotional involvement (for example, culpability feelings, often because of morality standards resulting from education, e.g., parental or social super-ego). The occurrence of unconscious processes in waking consciousness is nevertheless shown by "parapraxes," the concept put forward by Freud (1901). As recalled by Laplanche and Pontalis (2006) (p 300–301) the German prefix "ver" is the common denominator of words showing involuntary parasitic intrusions during waking: das Vergessen (forgetting), das Versprechen (slip of the tongue), das Verlesen (misreading), das Verschreiben (slip of the pen), das Vergreifen (bungled action), and das verlieren (mislaying) [p 239 in Freud (1901)]. The described repression of traumatic memories, as seen in post-traumatic disease, is a well-known procedure for maintaining psychological security and well-being. However, repression is only one of the defense mechanisms identified by Freud. Regression is another, involving the possible return from the secondary mode of adult mental functioning (p 324, Freud, 1895) based on the reality principle (p 219, Freud, 1900b) to the primary mode of psychological functioning based on the pleasure principle. Projection, the process by which the subject expels and localizes in other(s) feelings and desires he refuses for himself, may also involve control processes to avoid personal implication. Reaction-formation, the adopted behavior psychologically opposite to a repressed desire, as well as sublimation of unconscious drives into artistic, intellectual, or political activities are other defense mechanisms described by Freud. Rationalization is also a common mental process that often acts as a defense mechanism intended to render logical or morally acceptable an action, idea, or feeling whose grounds are not entirely conscious.

For what concerns sleep mentation, the attempt to not only theorize dreaming but also interpret dream content in concrete terms was magnificently developed by Freud. His main assertion was that dreams are primarily the manifestation of unconscious wish-fulfilments: "We have accepted the idea that the reason why dreams are invariably wish-fulfilments is that they are products of the system unconscious, whose activity knows no other aim than the fulfilment of wishes and which has at his command no other forces than wishful impulses" (p 568, Freud, 1900a). However, to Freud, wishes, or drive representations, cannot enter the dreamer's consciousness in their crude form. Dreams have to be modified by a disguise-censorship to avoid anxiety, which would induce awakening because of often psychologically unacceptable feelings. Therefore, Freud thought that "dreams are the GUARDIANS of sleep and not its disturbers" (p 233, Freud, 1900a). Dream-work consistently occurs that transforms original latent dream content into acceptable manifest content that follows the primary pleasure principle, leading to a reduction of emotional tension. Indeed, although "an affect experienced in a dream is in no way inferior to one of equal intensity experienced in waking life" (p 460), and although "lively manifestations of affect can make their way into the dream itself" (p 467), "a dream is in general poorer in affect than the psychical material from the manipulation of which it has proceeded (p 467) . . . The inhibition of affect, accordingly, must be considered as the second consequence of the censorship of dreams, just as dream distortion is its first consequence" (p 468, Freud, 1900a).

Several psychological processes intervene in the dream-work to carry out this function. First, "compression" or "condensation" (p 595, Freud, 1911), which refers to the concentration, in a unique element or representation of several ideas or feelings. In the treatment course, identifying the intimate signification of a dream by returning from the manifest content to the latent content (the target of dream analysis) is made difficult because of this condensation of, generally, several constitutive thoughts. Even "thoughts which are mutually contradictory make no attempt to do away with each other, but persist side by side. They often combine to form condensations, just as though there was no contradiction between them, or arrive at compromises such as our conscious thoughts would never tolerate but such as are often admitted in our actions" (p 596, Freud, 1911). The dream-work process is the consequence of a censorship which prevents original unconscious wishes from reaching the conscious mind or the preconscious from which they are more easily accessible to the conscious.

The second mechanism by which unconscious ideas or feelings are allowed to generate dream manifest content is by displacement. As best told by my teacher of long ago (J.B. Pontalis), displacement is "the fact that an idea's emphasis, interest or intensity is liable to be detached from it and to pass on to other ideas, which were originally of little intensity but which are related to the first idea by a chain of associations" (p 121, Laplanche and Pontalis, 2006). Moreover, the transformation of the latent content into the manifest content has to attribute psychophysical, descriptive properties to the content and make it appear in the dream most often as a pictorial representation. Finally, a combination of condensation and displacement is encountered in the symbolization process that occurs in dreams. However, in many cases, through the second elaboration process related to the censorship work, "the rearrangement of a dream tends to present it in the form of a relatively consistent and comprehensible scenario" (p 412, Laplanche and Pontalis, 2006). Moreover, behind the manifest content of dreams, the latent content is often at first glance unintelligible since, in spite of general rules, "the symbols in dreams are peculiar to each dreamer" (p 209, Darcourt, 2006).

Today, although some psychoanalytic concepts, such as the existence of unconscious processes, either underlie or are largely taken into account in nearly all psychology-related clinical areas, their influence has decreased in recent decades because this extraordinarily rich model of mind functioning, a quasi-closed dogma, has not been extensively enriched by the experimental verification of its theoretical assertions (Kandel, 1999). However, with the development of the Neuro-Psychoanalysis approach, particularly supported by the neo-Freudian Mark Solms, there is a new generation of psychoanalysts who have a better knowledge of neurobiology and who are now taking a new look at the neurobiological basis of mind functioning as hypothesized by Sigmund Freud.

Indeed, after completing his medical studies, Freud carried out basic histological neuroscience research in Meynert's and Brücke's laboratories (Jones, 1953/1958) but was soon, as he was throughout his career, in search of the scientific underpinnings of psychology; this was shown early on by his "Project of a scientific psychology" (Freud, 1895). On several occasions in his writings, he underlined the necessity of ultimately understanding the biological basis of his findings: "we must recollect that all of our provisional ideas in psychology will presumably one day be based on an organic substructure" (p 144, Freud, 1914). However, at the culmination of his entire encyclopedic work, the neurosciences related to higher integrated processes were still either nonexistent or in their infancy. Following the uncertain and imprecise approach of psychosurgery in the late 1930s and 1940s (in those days the experimental support for the application of frontal leukotomies was almost completely lacking)—this preceded the beautiful (at first empirical) discovery of most psychotropic molecules (see Gottesmann and Gottesman, 2007)—the remarkable development of neuroscience techniques and the discoveries made over the last four to five decades have raised new hopes of identifying the reality of the relationship between mind and body.

The primary outstanding questions to be answered are how near is our current knowledge of neurobiology to that of mind functioning as posited by psychoanalysis and psychiatry. The issue that is particularly open to discussion concern the "biological bases of the unconscious with the implicit nagging question of possible naturalization of mind" (p 367, Buser, 2007), thus, to what extent do the neurobiological processes occurring during waking and rapid eye movement (REM) sleep support psychoanalysis? The study of sleep mentation, particularly dreaming, in comparison to waking mental processes represents a useful research tool with which to address such questions.

II. Results

A. WAKING NEUROBIOLOGICAL BASIS OF CONSCIOUSNESS

It was immediately obvious that behavior in all species requires invertebrate ganglion or vertebrate brain activation. The first electroencephalographic (EEG) sleep-waking recordings in humans (Berger, 1929) showed the presence of rapid, low-voltage cortical patterns during waking that are absent during sleep. This observation was later extended by animal research (Bremer, 1935, 1936); subsequently, these patterns were shown to be sustained by brainstem reticular processes (Moruzzi and Magoun, 1949; Steriade, 1996). Indeed, the isolated forebrain is able to generate low-voltage, waking-like EEG activities (Batsel, 1960, 1964; Belardetti et al., 1977; Villablanca, 1965, 1966), but to date not consciousness, as defined in humans as "self-observation, planning, prioritizing and decision-making abilities" (Muzur et al., 2002). The brainstem glutamatergic influences on the thalamus modulate thalamo-cortico-thalamic loops, particularly those involved in the newly discovered cortical gamma rhythm (Steriade et al., 1996); this rhythm was observed in humans during waking and shown to be decreased and disorganized in Alzheimer disease (Ribary et al., 1991). Today, it is acknowledged on the one hand that the role of the thalamus is to be the "functional interface between the arousal and the attentional system" (p 8988, Portas et al., 1998b), and on the other hand "it is the dialogue between the thalamus and the cortex that generates subjectivity" (p 532, Llinas and Paré, 1991). Brainstem glutamatergic influences also activate the basal forebrain nucleus (Meynert nucleus in humans), promoting cortical arousal through acetylcholine release (Buzsaki and Gage, 1989; Buzsaki et al., 1988; Jones, 2004; Kurosawa et al., 1989; Wigren et al., 2007). Acetylcholine has cortical facilitating influences (Bremer and Chatonnet, 1949; Szymusiak et al., 1990; Vanderwolf, 1988; Wikler, 1952), with its release being highest during waking (Celesia and Jasper, 1966; Cuculic and Himwich, 1968; Marrosu et al., 1995; Pepeu and Bartholini, 1968; Phillis and Chong, 1965; Szerb, 1967). This release mainly occurs through varicosities (Descarries et al., 1997)—meaning that acetylcholine may be viewed as a neuromodulator with long duration influences, rather than a true neurotransmitter acting only shortly at synaptic level. The cognitive functions of acetylcholine are now well established (Perry et al., 1999; Sarter and Bruno, 2000).

The high levels of the activating transmitter glutamate that are secreted (Léna et al., 2005) contribute to an awakened cortex. At the same time, the various monoamines also participate in cortex functioning. In addition to its cortical inhibitory influences (see below), dopamine induces waking activities at high doses (Bagetta et al., 1988; Di Chiara et al., 1976; Gessa et al., 1985; Kroft and Kuschinsky, 1991; Lavin and Grace, 2001; Monti et al., 1989). It is noteworthy that its release mainly occurs at the varicosity level, i.e., diffusely, although synaptic structures are present on numerous varicosities (Smiley and Goldman-Rakic, 1993). In addition to its more common cortical inhibitory properties, noradrenaline also has pontine (Cordeau et al., 1963; Fuxe et al., 1996; Lin et al., 1992), thalamic (Behrendt and Young, 2005; Moxon et al., 2007), and basolateral (Jones, 2004) activating influences. Finally, hypothalamic histaminergic neurons (Steinbusch and Mulder, 1984; Watanabe et al., 1984; Yamatodani et al., 1991), which project to several central structures, activate the basal forebrain nucleus (Espana et al., 2005; Khateb et al., 1995), increasing cortical acetylcholine release (Ceechi et al., 1998). Other histaminergic projections inhibit the preoptic sleep-inducing area (Lin et al., 1994) and therefore also promote waking (Jones, 2003).

Other recent developed criteria, such as blood flow, glucose uptake, and oxygen consumption, indicate that the forebrain, and especially the cortex, is activated during waking (Braun et al., 1997; Madsen et al., 1991; Maquet et al., 1996, 2004; Nofzinger et al., 1997).

Together, these activating processes support the ability of the forebrain structures, and mainly the cortex, to carry out their specific functions. Just like petrol allows engines to run, these processes allow cognitive functions to be carried out in the cortex during waking.

Nevertheless, in this description of basic cognitive function, the "driver" is missing. It most likely comes in the form of the inhibitory processes observed in the ganglia of "lower" species (Castellucci and Kandel, 1974), as well as in the mammalian forebrain, particularly in the cortex, which is often studied as a model for human brain function. Numerous such inhibitory processes have long been known (Adey et al., 1957; Hugelin and Bonvallet, 1957; Kaada, 1951; Meulders et al., 1963; Sloan and Jasper, 1950; Wall et al., 1951), starting with the original demonstration of Bubnoff and Heidenhain (1881) of this phenomenon in the brain. Indeed, even at this very early date these authors were able to inhibit cortical-induced motor activities in dogs using low-intensity peripheral and cortical stimulation. In humans, the first clinical observation of central inhibitory processes was provided by von Economo (1928) during the lethargic encephalitis epidemic in the First World War. He observed that, while the more common posterior hypothalamic lesions induce sleepiness, lesions of its anterior part—more precisely the preoptic area—lead to increased waking.

At the cortical neuron level, an electrophysiological demonstration of inhibitory processes was first provided by Creutzfeldt *et al.* (1956) and Krnjevic *et al.* (1966) through neuron firing arrest, as well as by Evarts (1960) and later others (Allison, 1968; Rossi *et al.*, 1965) by showing that evoked cortical potentials induced by radiation stimulation have long recovery cycles. Neurochemical studies confirmed these phenomena by showing the influence of gamma aminobutyric acid (GABA) on cortical neurons (Krnjevic *et al.*, 1966). Perhaps the first finding that revealed the importance of inhibitory phenomena, however, was the discovery that monoamines mainly inhibit cortical neurons while increasing the efficiency of neuron functioning. Noradrenaline mainly inhibits cortical neurons (Foote *et al.*, 1975; Frederickson *et al.*, 1971; Krnjevic and Phillis, 1963; Manunta and Edeline, 1999; Nelson *et al.*, 1973; Phillis *et al.*, 1973; Reader *et al.*, 1979). However, this neuromodulator also increases the signal-to-noise ratio of neuron functioning (Aston-Jones and Bloom, 1981b; Foote *et al.*, 1975; Warren and Dykes, 1996; Waterhouse *et al.*, 1990), thereby increasing neuron efficiency. Additional proof that noradrenaline has a regulatory function was provided by the observation that it transforms the bursting, irregular firing of cortical neurons into regular tonic firing (Wang and McCormick, 1993); this had previously been hypothesized by Evarts (1964) to be the consequence of an inhibitory control process. Moreover, both the maximal (but slow frequency) firing of locus coeruleus noradrenergic neurons (Aston-Jones and Bloom, 1981a; Hobson *et al.*, 1975) and the high level of cortical noradrenaline release during waking (Léna *et al.*, 2005) underline the significant role of this neuromodulator.

Today, many results point to the major role of noradrenaline in mental processes. Whatever the precise mechanism in aged primates, there is a parallel decrease in cortical catecholamines and performance (Arnsten and Goldman-Rakic, 1987; Goldman-Rakic and Brown, 1981). More generally, noradrenaline depletion increases error responses to irrelevant stimuli while decreasing responses to relevant stimuli (Milstein *et al.*, 2007; Selden *et al.*, 1990). Further, increases in cognitive performance are associated with increases in prefrontal noradrenaline release (Berridge *et al.*, 2006). In the rare cases of bilateral locus coeruleus stimulation that have been performed for therapeutic purposes, the observed consequence is "well-being (and) improved clarity of … thinking" (p 179, Libet and Gleason, 1994). Finally, as shown in our laboratory, noradrenaline release in the nucleus accumbens is also highest during waking (Léna *et al.*, 2005).

In the same way, serotonin, which mainly originates from the midbrain dorsal raphe nucleus but is also secreted by the medial nucleus (Kosofsky and Molliver, 1987), inhibits the majority of cortical neurons (Araneda and Andrade, 1991; Foote *et al.*, 1975; Krnjevic and Phillis, 1963; Nelson *et al.*, 1973; Phillis *et al.*, 1973; Reader *et al.*, 1979) following its release primarily at the varicosity level (Descarries *et al.*, 1975; Foote *et al.*, 1975; Kosofsky and Molliver, 1987). As with noradrenaline, neuron inhibition by serotonin increases neuron efficiency by enhancing the signal-to-noise ratio (Aston-Jones and Bloom, 1981b; Foote *et al.*, 1975; McCormick, 1992). The function of serotonergic neurons, which fire maximally but at low frequency during waking (McGinty and Harper, 1976; McGinty *et al.*, 1974; Rasmussen *et al.*, 1984; Trulson *et al.*, 1981), is well-established in higher integrated processes like the regulation of mood (Taylor *et al.*, 2006), personality, and well-being (Silver *et al.*, 2000; Van Hes *et al.*, 2003).

The third major monoamine involved in cognition is dopamine. The meso-cortical and mesolimbic tracts originating from the A_{10} ventral tegmental area release high levels of dopamine in the prefrontal cortex and nucleus accumbens during waking (Léna *et al.*, 2005). Cortical release occurs at the varicosity level, with or without the synapses in apposition (Smiley and Goldman-Rakic, 1993). As observed by Krnjevic and Phillis (1963), dopamine mainly inhibits glutamate-activated cortical neurons and induces long-term depression. In addition, the activation of mesocortical neurons inhibits cortical neurons (Tierney *et al.*, 2008). As with noradrenaline and serotonin, dopamine increases the efficiency of neu-ron functioning by increasing the signal-to-noise ratio "by enhancing the incom-ing neural signal to background noise or interference" (p 218, Luciana *et al.*, 1998). Behaviorally, dopamine sustains attention and working memory (Apud and Weinberger, 2007; Apud *et al.*, 2007; Rose *et al.*, 2010).

In some cases, acetylcholine is able to inhibit cortical neurons, particularly pyramidal cells (Giulledge and Stuart, 2005; Levy *et al.*, 2006; Nelson *et al.*, 1973). At the cellular level in monkeys, acetylcholine excites and inhibits almost the same number of neurons, with the "inhibitory responses to acetylcholine having shorter latency and faster recovery than the excitatory responses ... The thresh-old of both inhibitory and excitatory responses were almost the same. In some cases the responses to acetylcholine reversed from excitatory to inhibitory with increasing dose" (p 123, Nelson *et al.*, 1973).

B. Waking Consciousness

In the waking cognitive functioning, as more precisely considered in the psychoanalytic model, all defense mechanisms primarily involved in the repres-sion of mental content by censorship must be dissociated from memories that are transiently outside consciousness but able to spontaneously or voluntarily enter the conscious mind (unrepressed unconscious, Kandel, 1999). The fundamental rule by which the analysand "is asked to say what he thinks and feels, selecting nothing and omitting nothing from what comes into his mind" (p 178, Laplanche and Pontalis, 2006) intends to promote access to that which "falls into his mind" (Einfall) and which, in addition to this free association of ideas, could open the door to preconscious content and half-open the door to unconscious ones, thus somewhat overcoming the censorship. All of these processes involve the functioning of the phylogenetically most recently developed brain structures, in which strongly antagonistic but closely complementary activating and inhibitory influences necessarily cooperate. This concerns, of course, first the forebrain, predominantly the prefrontal cortex (Muzur *et al.*, 2002), in association with limbic (Maquet and Franck, 1997) and other related subcortical structures (Llinas and Paré, 1991). Going beyond earlier, highly significant but rather "passive" neurobiological observations, direct experimental research into the "suppression" (for cognitivists; "repression" in psychoanalytic terminology) of memories by think/no-think tasks has revealed "a suppression mechanism that pushes unwanted memories out of awareness, as posited by Freud . . . (These waking results—Fig. 1) provide a viable model of repression, and its potential evolution from an intentional to an unintentional process" (Anderson and Green, 2001). A later study by the same team showed (Fig. 2) that the brain structures involved in this process are the dorsolateral and ventrolateral prefrontal cortex, the anterior cingulate cortex, the dorsal premotor cortex, and the intraparietal sulcus, all of which are activated. In contrast, the "reduced hippocampal activation in the suppression condition indicates that subjects successfully stopped or reduced recollection of unwanted memories during scanning" (Anderson *et al.*, 2004). It is interesting that "increased activation in bilateral dorsolateral prefrontal cortex predicted increased memory inhibition." To conclude, "the current findings provide the first neurobiological model of the voluntary form of repression proposed by Freud" (Anderson *et al.*, 2004).

These suppression–repression phenomena evidenced in no-think conditions occurs with verbal as well as nonverbal cues and are more efficient with negative than with non-emotional information (Depue *et al.*, 2006) (Fig. 3). The same team further showed in a neuroimaging study (Fig. 4) that "the suppression of emotional memory involves at least two pathways with staggered phases of their modulatory influence. The first pathway involves cognitive control by right inferior frontal gyrus (rIFG) over sensory components of memory representation, as evidenced by reduced activity in fusiform gyrus and pulvinar . . . . A second pathway involves cognitive control by right middle frontal gyrus (rMFG) over memory processes and emotional components of memory representation via modulation of hippocampus and amygdala. The overall timing of these suppression effects appears to be orchestrated by a modulatory influence of BA_{10}, first over rIFG then over rMFG" (Depue et al., 2007). Indeed, the more the prefrontal

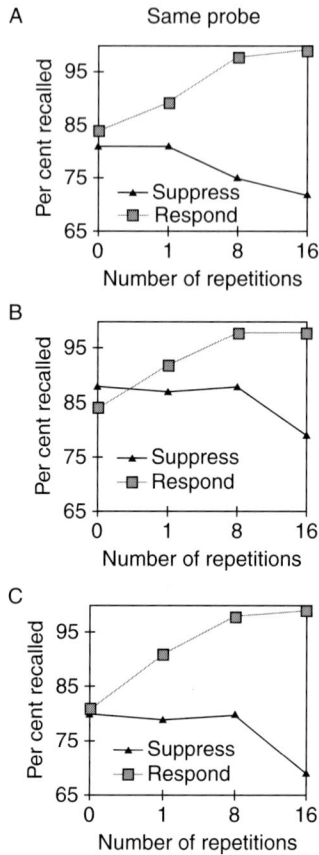

FIG. 1. Suppression processes during think/no-think tasks. (A) After learning pairs of unrelated words, the subjects were asked to "think about" or "not think about" the test word when the conditioning word was presented. With repetition, there was a progressive suppression of recall of the "no think" words. (B) The subjects were told to forget the previous instructions and to recall, with the incentive of a monetary reward, all the test words. There was no better recall of "no think" words. (C) The experimenter tried, without success, to induce parasitic intrusion of suppressed words (parapraxes). Modified from Anderson and Green (2001), Reprinted from *Nature*, with permission.

cortex is activated during no-think tasks, the more efficient suppression–repression is (Levy and Anderson, 2008) (Fig. 5). The particular function of the rIFG in inhibitory processes—and its support by noradrenaline—is also underlined by other teams (see Aston-Jones and Gold, 2009).

These inhibitory processes also involve other biochemical compounds, as the increase in centrally generated CaMKII not only disturbs immediate memory

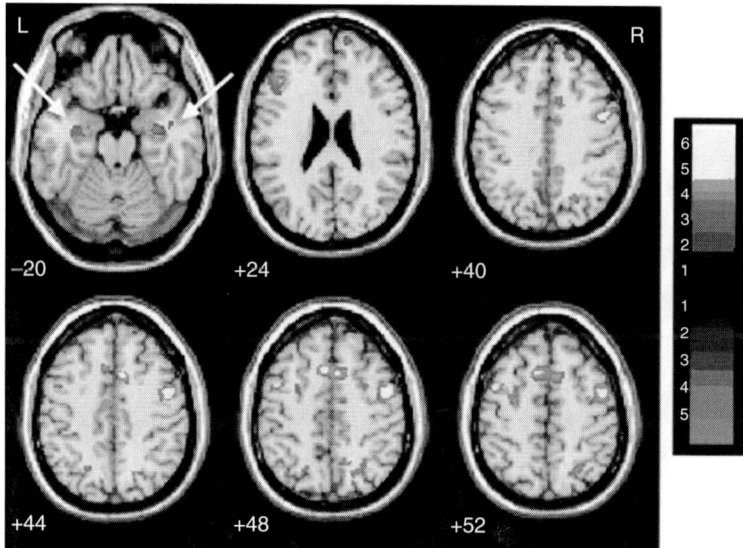

FIG. 2. Brain processes during think/no-think tasks. During suppression (repression) of unwanted memories, the dorsolateral and ventrolateral prefrontal cortex, the anterior cingulate cortex, the dorsal premotor cortex, and the intraparietal sulcus are activated (in light gray), while the hippocampus is deactivated (in deep gray). Reprinted from Anderson *et al.* (2004). *Science*, with permission.

fixation (Wang *et al.*, 2008) but also inhibits memory retrieval as well (Cao *et al.*, 2008). In the same way, propranolol, the β-adrenergic receptor antagonist, erases human fear responses and inhibits the return of fear (Kindt *et al.*, 2009).

Finally, "people who had experienced more traumatic events show enhanced memory inhibition abilities when compared to individuals who had experienced little or no trauma . . . (Moreover) older adults (between the age of 65–80) show a significantly reduced ability to inhibit compared to younger adults (aged 18–25)" (Levy and Anderson, 2008). All these processes of suppression–repression are reinforced by sleep (Rauchs *et al.*, 2008) and involve hippocampal consolidation prior to hippocampal–neocortical interactions (Gais *et al.*, 2008).

Nowadays, again, the main problem facing the new neuro-psychoanalysis approach is to know to what extent neurobiology can account for the basis of unconscious processes. During waking, unconsciousness-related events occur, as commonly evidenced by several kinds of parapraxes and the cognitive and neurobiological results described above. Moreover, it becomes obvious that subliminally shown, emotionally loaded stimuli can induce neurobiological modifications, as evidenced by event-related potentials within brain structures, like the amygdala, that are involved in emotion regulation (Naccache *et al.*, 2005). As underlined by Gaillard *et al.* (2006), at first glance it could be supposed, from a

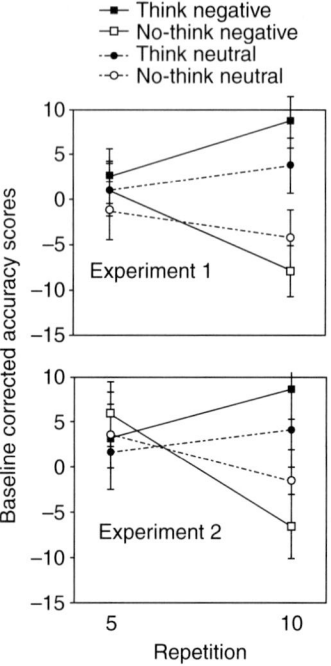

Fɪɢ. 3. Depue *et al.* (2006) applied the same "think-no think" paradigm with face-word (upper experiment) and face-picture pairs (bottom). In both cases there was a suppression of the "no think" tasks, with the suppression being better for negative words and pictures. Reprinted from *Psychological Science*, with permission.

psychoanalytic point of view, that repression would increase, rather than decrease, the threshold of negative stimuli identification. In fact, external—as well as interoceptive—perception must always be maximal to best induce adaptive processes, with the protective function of defense mechanisms serving only to repress the responses; this has been shown by experiments on taboo words presented by tachistoscope (Zajonc, 1962). More generally, unconscious activation of brain circuits has been strongly hypothesized in the case of visual stimuli that are critical to survival (Gaillard *et al.*, 2006). Thus, the initial opposition between cognitive and emotional unconsciousness (Buser, 2007) begins to be overcome, with the two being experimentally shown to be rather complementary. The fact that unconscious events are taken into account has also been demonstrated by face recognition experiments. Although there is a failure of face recognition in prosopagnosics, large galvanic skin responses are obtained with previously known faces, without consciousness of them (Tranel and Damasio, 1985). In the same way, modern neuroimaging has shown that the activation of

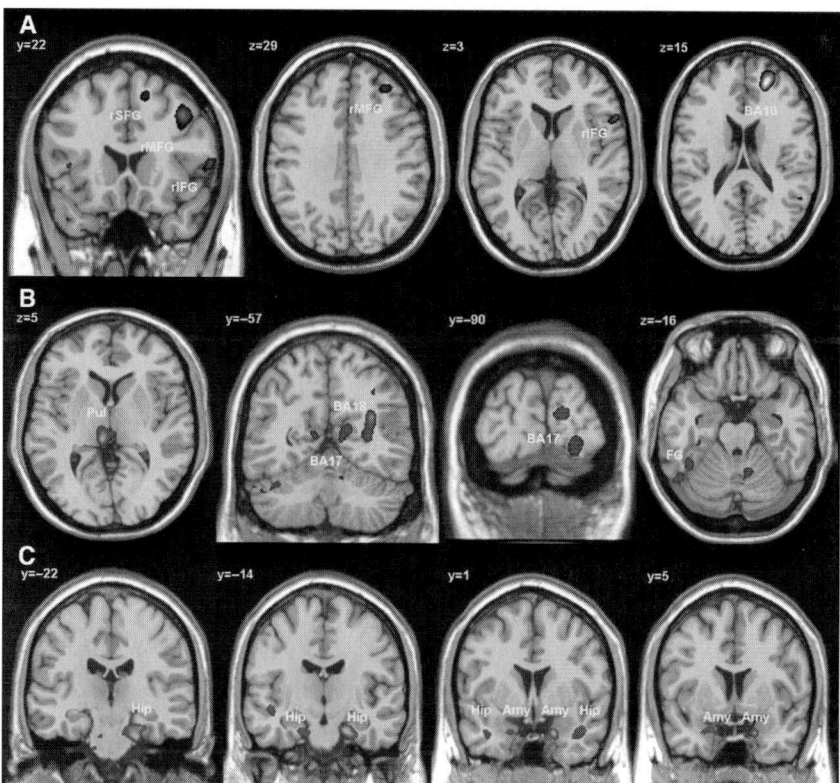

Fig. 4. Suppression of emotional memories. In a think/no-think (T/NT) paradigm, neuroimaging results show that in a first temporal step there is an inhibition by the right inferior frontal gyrus (A in red) of regions supporting sensory afferents of the memory representation (visual cortex, thalamus: B in blue), followed by right medial frontal gyrus control over regions supporting multimodal and emotional components of the memory representation (hippocampus, amygdala C, in blue); both types of control are influenced by fronto-polar regions. Red indicates greater activity for NT trials, blue indicates the reverse. The areas in blue are the more activated during T trials and the less activated for NT trials. Reprinted from Depue *et al.* (2007). *Science*, with permission.

the fusiform face area not only predicts whether a face will be later recognized or not but also reflects whether a face has been seen previously or not, independent of conscious memory, thus dissociating overt recognition from unconscious discrimination (Lehmann *et al.*, 2004). Finally, patients with right inferior parietal lesions show extinguished contralateral vision, but with right-side striate and extrastriate cortex activation, despite a lack of consciousness. This result implies that the activation of the primary visual cortex (and early extrastriate cortex) by visual stimuli is insufficient to evoke consciousness (Rees *et al.*, 2000), a conclusion

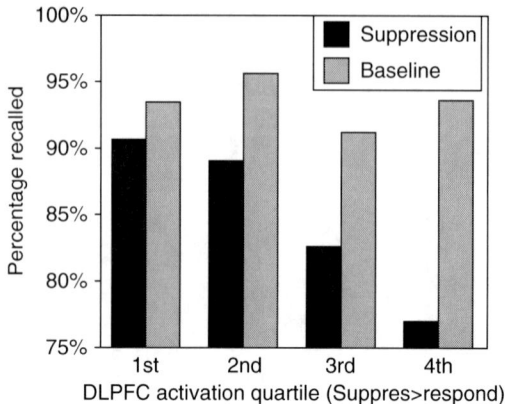

FIG. 5. Successful recruitment of the dorsolateral prefrontal cortex predicts behavioral inhibition. The subject group with the greatest prefrontal activation showed the highest suppression of recall. From Levy and Anderson (2008). Reprinted from *Acta Psychologica*, with permission.

which had previously been deduced from primate experiments (Crick and Koch, 1995).

Thus, we are beginning to have a better understanding of the neurobiological basis of cognitive functioning. Moreover, objective neurobiological evidence of unconscious processes has been demonstrated during waking, in particular a hypothetical prefrontal-originating censorship-like process situated in the BA_{10} area.

III. What Happens during Sleep

A. Slow Wave Sleep

During the night, two different sleep stages occur that are interspersed with each other. The first of these is slow wave sleep, which is mainly characterized by slow- and high-amplitude cortical waves (Berger, 1929; Blake and Gerard, 1937; Bremer, 1935; Loomis *et al.*, 1937). Slow wave sleep is divided into four stages, from light to deep slow wave sleep, as evidenced by slow wave EEG activity of progressively increasing amplitude. Let us examine now current knowledge about its neurobiological background and psychological activities and their possible psychoanalytic explanations.

1. *Neurobiological Findings*

The first major part of nighttime sleep comprises slow wave sleep. This sleep period is related to a decrease in brainstem (Moruzzi and Magoun, 1949; Steriade, 1996) and hypothalamic (Vanni-Mercier *et al.*, 1984, 2003) activating influences, the release of diencephalic sleep-inducing structures (Gallopin *et al.*, 2000; Maire and Patton, 1954; Manceau and Jorda, 1948; McGinty and Sterman, 1968; Nauta, 1946; von Economo, 1928), and medulla oblongata (Batini *et al.*, 1958, 1959; Bonvallet and Allen, 1963; Magnes *et al.*, 1961) EEG synchronizing influences. At the cortical level, the gamma rhythm decreases in amplitude (Gross and Gotman, 1999; Llinas and Ribary, 1993) and the blood flow is globally lower, except in the visual cortex and to a lesser extent in the secondary auditory cortex (Hofle *et al.*, 1997; Maquet *et al.*, 1997).

From a neurochemical standpoint, cortical acetylcholine release is decreased (Celesia and Jasper, 1966; Marrosu *et al.*, 1995). The cortical glutamate level is nearly unchanged, whereas it is decreased in the nucleus accumbens (Gottesmann, 2006a; Léna *et al.*, 2005). In contrast, noradrenergic neurons decrease their firing rate (Aston-Jones and Bloom, 1981a; Hobson *et al.*, 1975), and both cortical and nucleus accumbens noradrenaline release is decreased (Léna *et al.*, 2005). Serotonergic neurons also decrease their firing rate (McGinty and Harper, 1976; McGinty *et al.*, 1974; Rasmussen *et al.*, 1984), while histaminergic neurons become silent as early as during light slow wave sleep (Vanni-Mercier *et al.*, 1984, 2003). Finally, dopamine release is reduced both in the prefrontal cortex and in the nucleus accumbens (Gottesmann, 2006a; Léna *et al.*, 2005).

Concomitant with these neurochemical data showing a reduction in forebrain inhibitory influences, electrophysiological results, such as the observed decrease in cortical neuron firing (Evarts, 1964; Steriade, 1996), confirm a parallel lowering of both activating and inhibitory cortical processes. The decrease in organized cortical activity is confirmed by the near absence of gamma rhythm (Llinas and Ribary, 1993) and the loss of intracortical connectivity (Massimini *et al.*, 2005).

2. *Slow Wave Sleep Consciousness*

From a psychological standpoint, Foulkes (1962) was the first to describe mental activity during slow wave sleep. Indeed, he observed "thought-like" mental content that was mainly structured according to Freud's secondary mode based on the reality process. Often, the mental content is composed of the day's residue, and several studies have shown that waking cognitive processes are replicated during slow wave sleep (Euston *et al.*, 2007; Ji and Wilson, 2007). This mental content was different from dreams [also observed in some cases by other authors Bosinelli (1995); Bosinelli and Cigogna (2003); Tracy and Tracy

(1974)] because of "considerable and consistent qualitative differences" (p 22, Foulkes, 1962), making it easy to discriminate between them (Monroe *et al.*, 1965) (for details, see Gottesmann, 2005b). Nielsen (2003) speaks about "cognitive processes" rather than "dreaming." The specific "non-regressive" (Vogel *et al.*, 1966) mental activity of slow wave sleep is thus regulated similarly to during waking, but at a lower intensity level. The decrease in brainstem ascending activating influences (as evidenced by a reduction in cortical neuron firing Evarts, 1964; Steriade, 1996), concomitant with a parallel decrease in ascending inhibitory influences (as reflected by a decrement in monoamine—and acetylcholine—release and correlative disinhibition) could explain the occurrence of mental activity of a narrative rather than a descriptive nature, governed as during waking by the reality principle but to a lower cognitive degree (Fosse *et al.*, 2001). Thus, neurobiology is able to explain slow wave sleep mentation in possible accordance with psychoanalytic theory.

A particular case concerns arousal from night terrors (pavor nocturnus), which occur most often in teenagers during the deepest slow wave sleep (stage 4) (Gastaut and Broughton, 1964). As observed by Kales and Jacobson (1967), the subject "is usually unable to orient himself to his immediate environment or recall any specific dream other than the general feeling of fear" (p 88). More recent research has confirmed that "the individual is unaware of the fullness of their surroundings and is totally focused in their concern or activity" (p 599, Crisp, 1996). Night terrors are described as being "associated with limited or no mental content during the events and this feature is commonly used to distinguish them clinically from true nightmares" (p 246, Hartman *et al.*, 2001). It is of interest that in these episodes, characterized by a general "terrified scream accompanied by intense autonomic discharge" (p 245)—and often associated with sleep-talking "a history of major psychological trauma exists in only a minority of adult patients" (p 244). However, it could be a dissociative defense process for "excluding painful memories from awareness" (p 248, Hartman *et al.*, 2001). This is interesting, since these episodes occur during deep slow wave sleep, when forebrain activation is very low (Hofle *et al.*, 1997; Maquet *et al.*, 1997) and inhibitory processes are highly depressed, as shown by (1) a shorter recovery cycle of cortical evoked potentials (Allison, 1968; Demetrescu *et al.*, 1966; Rossi *et al.*, 1965) when compared to waking; (2) reductions in noradrenergic (Aston-Jones and Bloom, 1981a; Hobson *et al.*, 1975; Léna *et al.*, 2005), serotoner-gic (McGinty and Harper, 1976; McGinty *et al.*, 1974; Rasmussen *et al.*, 1984), and dopaminergic (Léna *et al.*, 2005) functioning. Thus, whether or not a defense mechanism is involved (Hartman *et al.*, 2001), the neurobiology would have to accommodate the maintenance of some mind-control processes. From a psy-choanalytic perspective, the often encountered rather brutal awakening could correspond to a rudimentary, "impulsive" (p 5, Laplanche and Pontalis, 2006) acting out like manifestation.

This particular slow wave sleep-associated mental activity has to be compared with that of the short sleep stage (a few seconds to a few minutes in humans) that occurs just prior to REM sleep, when returning from stage IV to ascending stage II. It shows slow wave patterns (spindles and particularly K complexes of stage II) interspersed with REM sleep patterns: EEG activation without eye movement (Goldsteinas *et al.*, 1966; Lairy, 1966; Lairy *et al.*, 1968; Salzarulo *et al.*, 1968). Even when the subject is behaviorally awakened from it, it is difficult to establish psychological contact. Finally, verbal reports do not reveal visual content but a "feeling of indefinable discomfort, anxious perplexity and harrowing worry" (p 279, Lairy *et al.*, 1968). Although Foulkes thinks that this result is debatable (personal communication, 1998), it is noteworthy that in this sleep period occurring just prior to REM sleep, mice (Glin *et al.*, 1991), rats (Benington *et al.*, 1994; Bjorvatn *et al.*, 1998; Depoortere and Loew, 1973; Gaillard *et al.*, 1977; Gottesmann, 1964, 1967, 1972; Kleinlogel, 1990; Neckelmann and Ursin, 1993; Weiss and Adey, 1965), and cats (Gottesmann *et al.*, 1984) also show an intermediate stage characterized by deep slow wave sleep patterns (high-amplitude cortical spindles) associated with REM sleep patterns (low-frequency hippocampal theta rhythm). This stage is characterized by the lowest level of thalamic transmission of all sleep-waking stages (Gandolfo *et al.*, 1980), an index of strong ascending deactivation. This sleep period is really a transitional stage, since the spindles are of significantly higher amplitude and duration than they are during slow wave sleep, the theta rhythm is of significantly lower frequency than during REM sleep, and the reticular arousal threshold is significantly higher than during slow wave sleep and significantly lower than during REM sleep (Piallat and Gottesmann, 1995). Finally, this stage is permanently induced by acute intercollicular transections performed on rats (Gottesmann *et al.*, 1980) and cats (Gottesmann *et al.*, 1984) [for review, see Gottesmann (1996)] and is massively extended at expense of REM sleep by low doses of barbiturates both in rats (Gottesmann, 1964, 1967) and in cats (Gottesmann *et al.*, 1984).

All results to date show that this short-lasting period corresponds to a transient functional disconnection of the forebrain from the brainstem (a physiological cerveau isolé-like preparation Bremer, 1935), with the cortex being nearly completely inactivated and the inhibitory control influences being at their lowest level (Demetrescu *et al.*, 1966). It is understandable that in this very particular and necessarily short-lasting stage (for teleological reasons, the subject being massively disconnected from the environment), the low mental content is unusual and anxiety ridden.

To conclude recent findings confirm that true dreams (mainly visual hallucinatory activity, see later) occur soon after falling asleep (Fosse and Domhoff, 2007; Nielsen, 2003; Nielsen *et al.*, 2005; Takeuchi *et al.*, 1999, 2001): "dreaming...does occur with fairly substantial frequency at sleep onset" (p 285, Foulkes *et al.*, 1966). The dreams are most often short-lasting and followed by transient awakening. Certainly, because of their proximity with the waking state, these dreaming periods are fragile and sensitive to interferences (noise, movement, irruption of ideas). Their short latency of appearance after sleep onset strongly suggests that they mainly occur during descending stage I which, according to Nielsen's (2003) assertion, comprises "covert" REM sleep criteria such as the well known low-voltage EEG activity and the "slow, pendular rhythmical swings of the eyeballs from one side of the orbit to the other" (Dement, 1964). "During descending stage I lasting from one-half to 5 minutes... there is a regressive content and there is partial or complete loss of reality contact. During descending stage II there is a paradoxical return to nonregressive content and complete loss of contact with external reality. There is a relatively destructuralized 'ego' (during descending stage I) and a relatively restructuralized ego (during descending stage II)" (p 241–242) (Vogel *et al.*, 1966), Kahn and Hobson (2005) prefer to speak about "self" instead of "ego." As also pointed out long ago by Foulkes and Vogel (1965) and by Vogel *et al.*(1966)—and contrary to the unpleasant and often threatening content of dreams occurring during true REM sleep (Revonsuo, 2003), a notion now somewhat questioned (Malcolm-Smith *et al.*, 2007) (see later)—in descending stage I there are roughly equivalent amounts of positive and negative affects (contrary to what occurs in narcoleptic patients Fosse *et al.*, 2002, in whom dreams are described as "terrific hypnagogic hallucinations" Auerbach, 2007). This phenomenon of positive as well as negative affects could be the consequence of the maintenance of certain previous neurobiological waking processes that totally disappear during REM sleep episodes occurring after established slow wave sleep. Other complementary criteria of REM sleep during this descending stage I include the deactivation of the posterior cingulate gyrus (Kaufman *et al.*, 2006) and the fact that, during sleep, the P_{300} event-related potential can only be recorded in both stages (Niiyama *et al.*, 1994).

B. The REM Dreaming Sleep Stage. Are There Neurobiological and Psychological Processes Supporting Psychoanalytic Theory?

1. *REM Sleep Neurobiology*

Although it is possible that "dreaming and REM sleep are controlled by different brain mechanisms" (Solms, 2000), as REM sleep can occur without dreaming after prefrontal leukotomy (Solms, 2000) and dreaming can take place outside of REM sleep (see above and lucid dreaming Voss *et al.*, 2009), dreaming essentially occurs during REM sleep. This was first incidentally described by Loomis *et al.* 1937), and later definitively established by Kleitman's group (Aserinsky and Kleitman, 1953; Dement and Kleitman, 1957a, 1957b), who showed that dreaming is associated with REM [it was therefore called REM sleep (Cohen and Dement, 1965)]. As irregular and superficial breathing and, most importantly, rapid low-voltage EEG (as during waking) are observed during this stage, it has also been called paradoxical sleep (Jouvet, 1965). This sleep stage was later shown, following a first observation by Klaue (1937), to be associated with muscular atonia (Berger, 1961; Dement, 1958; Jouvet *et al.*, 1959) (for details see Gottesmann, 2001). Several recent results show that dreaming seems to only occur in the physiological background of REM sleep (Nielsen, 2000; Takeuchi *et al.*, 1999, 2001), including during certain sleep stages when some criteria of REM sleep are present but covert (Nielsen, 2000). Indeed, certain REM sleep characteristics, like atonia, are known to occasionally occur during slow wave sleep, and this phenomenon is reinforced after REM sleep deprivation (Tinguely *et al.*, 2006; Werth *et al.*, 2002). Other REM sleep criteria, like erection, are able to occur during REM sleep deprivation (Karacan, 1965) as covert REM sleep criteria. As recalled previously (Gottesmann, 2001), Dement wrote long ago: REM sleep is like a symphony, and some instruments of the orchestra can be suppressed without hindering the occurrence of music.

Dreaming is generated by forebrain activation and particularly involves dopaminergic processes: neuroleptics decrease dreaming (Solms, 2000), A_{10} area dopaminergic neurons fire by bursts during REM sleep (Dahan *et al.*, 2007) and thereby release high levels of dopamine (Suaud-Chagny *et al.*, 1992), and increases in central dopamine promote vivid dreaming and nightmares (Thompson and Pierce, 1999). These activating processes are underpinned by pontine and mesopontine neuron firing that is nearly specific to REM sleep (Datta *et al.*, 2002; Gottesmann, 1969; McCarley and Hobson, 1971; Sakai, 1988; Sakai and Crochet, 2003; Steriade and McCarley, 1990; Vertes, 1977).

Aside from the low-voltage rapid EEG which is similar to that occurring in the waking state (Aserinsky and Kleitman, 1953; Dement, 1958; Dement and Kleitman, 1957a; Jouvet *et al.*, 1959), and the occurrence of gamma rhythm (Corsi-Cabrera *et al.*, 2003; Llinas and Ribary, 1993; Perez-Garci *et al.*, 2001), cortical activation is evidenced by different processes. First, there is neuron firing, which occurs at the same level as during active waking (Evarts, 1965). Second is surface negativity (index of neuron depolarization) (Kawamura and Sawyer, 1965; Tabushi *et al.*, 1966; Wurtz, 1965); Third is increased blood pressure, as first shown long ago in humans (Snyder *et al.*, 1963, 1964). This process is probably responsible for the observed increase in cortical temperature (Kawamura and Sawyer, 1965; Parmeggiani, 2007).

Current knowledge has confirmed these already classic results by a global, tomographically demonstrated activation of the cortex, particularly the medial

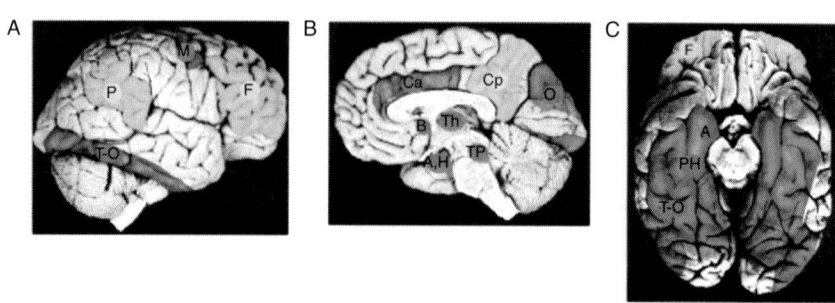

FIG. 6. Schematic neuroimaging representation of brain activity during REM sleep. Increased activity vs. waking is shown in red. Decreased activity is represented in blue. (A) lateral view. (B) medial view. (C) ventral view. A. H.: Amygdala and hypothalamus. B: basal forebrain. Ca: anterior cingulate gyrus. Cp: posterior cingulate gyrus. F: prefrontal cortex. M: motor cortex. P: supramarginal cortex. Ph: parahippocampic gyrus. O: Occipital-lateral cortex. Th: thalamus. T-O: temporo-occipital extrastriate cortex. TP: pontine tegmentum. Reprinted from Schwartz and Maquet (2002). *Trends in Cognitive Sciences*, with permission.

prefrontal cortex, extrastriate visual cortex, and limbic system (Braun *et al.*, 1997; Madsen *et al.*, 1991; Maquet and Franck, 1997; Maquet *et al.*, 1996, 2004; Nofzinger *et al.*, 1997); a good overview can be found in (Schwartz and Maquet, 2002) (Fig. 6). The specific activation of the cortical saccadic eye movement system (Hong *et al.*, 1995) is of interest because dreaming mainly occurs at the same time as the eye movements (Aserinsky and Kleitman, 1953), which are also generated by pontine spikes (Brooks and Bizzi, 1963; Datta *et al.*, 1998; Farber *et al.*, 1980; Gottesmann, 1969; Jouvet and Michel, 1959; McCarley *et al.*, 1978), also observed in humans by phasic waves (Fernandez-Mendoza *et al.*, 2009; McCarley and Ito, 1983; Peigneux *et al.*, 2001; Salzarulo *et al.*, 1975; Wehrle *et al.*, 2005), which were long ago correlated with electrophysiological phasic activation (Kiyono and Iwama, 1965; Satoh, 1971). These brainstem originating spikes could be responsible for reflex-like eye movements, such as those observed in congenitally blind persons (Amadeo and Gomez, 1966; Gross *et al.*, 1965), while the cortical area that controls eye movements could be responsible for dreaming-induced eye movements, which have been described as related to dream content (Doricchi *et al.*, 2007; Herman *et al.*, 1984; Hong *et al.*, 1997; Miyauchi *et al.*, 2009; Roffwarg *et al.*, 1962). Moreover, during REM sleep, the activated amygdala (Maquet and Franck, 1997), which sends abundant projections to the anterior cingulate cortex and extrastriate visual cortex, certainly modulates these structures by its involvement in emotional features (Dang-Vu *et al.*, 2007).

Finally, cortical glutamate is released during REM sleep at the same levels as during waking (Gottesmann, 2006a; Léna *et al.*, 2005) (Fig. 7). Acetylcholine is released at the level of only quiet waking (Marrosu *et al.*, 1995). This decrease could contribute to REM sleep-associated cognitive disturbances (Perry and

Piggott, 2003; Sarter and Bruno, 2000)—older results have shown deficits in intelligence quotient (IQ) scores after administration of anticholinergic compounds (Cartwright, 1966)—and the occurrence of hallucinatory activity (Cartwright, 1966), which has been recently confirmed (Collerton *et al.*, 2005).

The main difference with waking, however, lies in the inhibitory processes acting in the forebrain. As early as 1964, Evarts (1964) postulated that there are cortical inhibitory control processes that disappear during REM sleep. Evarts wrote that this kind of pattern "may result from a reduction in the effectiveness of some frequency-limiting mechanism which acts to stabilize discharge during waking" (p 170). Two years later, Demetrescu *et al.* (1966) also described cortical inhibition processes becoming minimal during REM sleep (Fig. 8).

More recently, it was shown that although the gamma rhythm occurs during REM sleep as it does during waking in humans (Llinas and Ribary, 1993; Ribary *et al.*, 1991) its synchronization is suppressed over cortical areas (Corsi-Cabrera *et al.*, 2003; Perez-Garci *et al.*, 2001) and between the hippocampus and cortex (Cantero *et al.*, 2004); further, the intra-hippocampal coherence is decreased during 95% of REM sleep (Montgomery *et al.*, 2008). The latter is an index of intracerebral disconnections which disturb coordinated brain activities.

Another electrophysiological criterion of cortical disinhibition was provided by prepulse inhibition experiments. Indeed, early results established that the recovery cycle of cortical evoked potentials during REM sleep in cats is significantly shorter than it is during waking (Allison, 1968; Rossi *et al.*, 1965). Recent studies in humans showed identical disinhibition (Kisley *et al.*, 2003), which indicate a loss of efficiency in forebrain, primarily cortical, functioning.

The following experimental demonstration of cortical disinhibition during REM sleep was based upon the functioning of monoaminergic neurons (Fig. 7). Indeed, noradrenergic (Aston-Jones and Bloom, 1981a; Hobson *et al.*, 1975) and serotonergic (McGinty and Harper, 1976; McGinty *et al.*, 1974; Rasmussen *et al.*, 1984) neurons become silent during REM sleep, and noradrenaline release in the prefrontal cortex (Léna *et al.*, 2005) and serotonin release in the frontal cortex (Portas *et al.*, 1998a) are minimal during REM sleep. In the nucleus accumbens (Léna *et al.*, 2005) and amygdala (Park, 2002), the release of noradrenaline is also strongly decreased during REM sleep. Dopamine release is also decreased in the prefrontal cortex when compared to waking, while it is maximal in the nucleus accumbens (Léna *et al.*, 2005). These two phenomena are related to the prefrontal deactivation (Brake *et al.*, 2000; Jackson *et al.*, 2001; Takahata and Moghaddam, 2000). In addition, in the mesolimbic tract, the burst mode of dopaminergic neuron firing releases more dopamine (Gonon, 1988; Suaud-Chagny *et al.*, 1992) (see also Schultz, 2007), and the neurons of at least the dopaminergic A_{10} area fire by bursts during REM sleep (Dahan *et al.*, 2007; Miller *et al.*, 1983) (see above).

The imagery approach has also shown that some cortical areas become deactivated during REM sleep (Fig. 6). This is the case with the dorsolateral

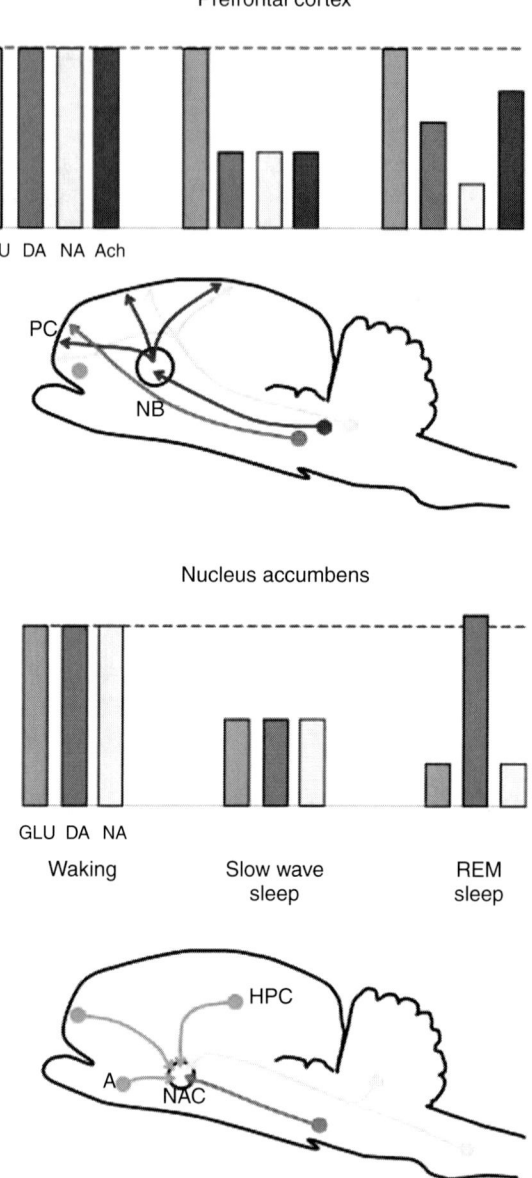

Fig. 7. *(Continued)*

prefrontal cortex (Braun *et al.*, 1997; Maquet *et al.*, 1996), part of the parietal cortex (Maquet *et al.*, 1996), and the primary visual cortex (Braun *et al.*, 1998) (Fig. 9). This last observation (although today questioned Hong *et al.*, 2009) is of interest because it indicates a disconnection from sensory inputs which is reinforced by a presynaptic inhibition of thalamic sensory afferents, particularly during the eye movement bursts of REM sleep (Dagnino *et al.*, 1969; Gandolfo *et al.*, 1980; Ghelarducci *et al.*, 1970; Iwama *et al.*, 1966; Steriade, 1970) (Fig. 9) when dreaming most often occurs. The thalamic blockade of sensory afferents during the eye movement bursts was recently confirmed in humans (Wehrle *et al.*, 2007). This strong filtering of sensory afferents could explain why there is no resetting of the gamma rhythm under sensory stimulation during REM sleep (Llinas and Ribary, 1993). The decrease of sensory afferents favors hallucinations (Behrendt and Young, 2005)

Finally, the posterior cingulate cortex, which is not part of the limbic system, is also deactivated during REM sleep (Maquet *et al.*, 1996, 2004) (Fig. 6). It is of interest that there is only one behavioral case in which the prefrontal dorsolateral cortex and the posterior cingulate cortex are deactivated together: when piano concertists are so involved in their playing that they "lose themselves" (Parsons *et al.*, 2005) and are thus somewhat disconnected from the environment.

Conclusion. The cerebral petrol (activation) is present during REM sleep. However, driver control is massively impaired because of the decrease—nearly the disappearance—of forebrain inhibitory influences. The disequilibrium between the antagonistic but complementary activation–inhibition pair must influence mind functioning.

2. *REM Sleep Consciousness*

Dreaming is a hallucinatory (also sometimes rather called illusionary Hawkins, 1966) activity which mainly involves visual activity; auditory elements

---

FIG. 7. Schematic drawing of neurotransmitter release during sleep-waking stages. In the prefrontal cortex (PC), glutamate (GLU in green) is principally released from local cells. Its concentration does not vary during sleep-waking stages (as in schizophrenia). In the nucleus accumbens (NAC), glutamate is released from the prefrontal cortex, hippocampus (HPC), and amygdala (A). Its release decreases during REM sleep (as in schizophrenia). Dopamine in the prefrontal cortex originates from the midbrain ventral tegmental (A_{10}) area. It is lower during REM sleep than during waking (as in schizophrenia). In the nucleus accumbens, its concentration is maximal (as in schizophrenia). Prefrontal cortex noradrenaline—originating from the pontine locus coeruleus (A_6 nucleus)—is decreased during REM sleep. It is also decreased in the nucleus accumbens, which is innervated by the medulla oblongata A_2 nucleus (there is a deficit of this neuromodulator in schizophrenia). During REM sleep, cortical acetylcholine released from the basalis (Meynert) nucleus (NB) only reaches the level seen during quiet waking (acetylcholine deficit favors schizophrenic symptoms). Reprinted from Gottesmann (2006a). *Neuroscience*, with permission.

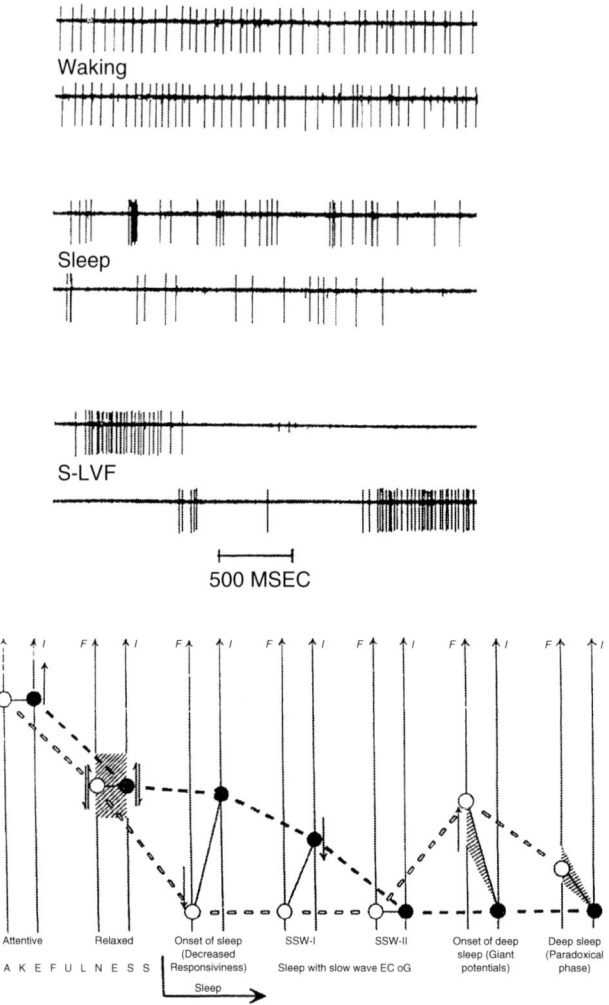

FIG. 8. Electrophysiological criteria of cortical regulation during sleep-waking stages. Top. Pyramidal cell recording in monkeys. During waking, there is regular high-frequency firing. During slow wave sleep (sleep), the firing frequency decreases and becomes irregular. During REM sleep (S-LVF), high frequency bursts of discharges alternate with long silences. Reprinted from Evarts (1964). *Journal of Neurophysiology*, with permission. Bottom. Thalamo-cortical responsiveness. Activating and inhibitory influences acting at the cortical level are high during waking. They both decrease during slow wave sleep, becoming minimal during deep slow wave sleep prior to REM sleep. During REM sleep, the cortex is again activated while being disinhibited. Reprinted from Demetrescu *et al.* (1966). *Electroencephalography and Clinical Neurophysiology*, with permission. These two old results are sufficient to explain the irrational mentation during REM sleep because of the loss of cortical control processes.

occur in about 60% of dreams, and motor and tactile sensations are present in 15%. Smell and taste content appears in fewer than 5% of dreams (recalled in Schwartz and Maquet, 2002).

The phenomenological study of dreaming was extensively undertaken during the first decades after the discovery of REM sleep. It showed that REM sleep periods with numerous eye movements are correlated with active dream reports, whereas REM sleep with fewer eye movements are correlated with passive dream reports (Berger and Oswald, 1962; Dement and Wolpert, 1958b). These eye movements are most numerous in the final nighttime REM sleep episodes, except with depression, in which case there is a tendency for early REM sleep occurrence (Cartwright *et al.*, 2003; Gottesmann and Gottesman, 2007; Gresham *et al.*, 1965; Hartmann *et al.*, 1966; Rush *et al.*, 1986). Among the pioneering studies, (Dement and Wolpert, 1958a) showed that "in 45 instances, long continuous dreams are more likely to be recalled in the absence of gross body movements while fragmented dreams are more likely to be recalled if the periods contain body activity" (p 545). Moreover, "memory of dream decays rapidly in time after cessation of the dream experience and dream recall becomes fragmented before undergoing complete extinction" (p 605–606, Wolpert and Trosman, 1958). In this first period of REM sleep studies, Goodenough *et al.* (1959) observed that, although both self-defined dreamers and non-dreamers reported dreams after provoked awakening from REM sleep (70% reported dreaming when awakened during the eye movement period, and 35% did so when awakened during eye quiescence), there was a significant difference in recall between the two groups. They noticed that, specifically in non-reporters, there was some alpha rhythm during the eye movements, which suggests some maintained waking neurobiological processes. In addition, even when subjects were awakened during nighttime REM sleep periods, recall was better during the last REM sleep periods (Shapiro *et al.*, 1963). Finally, the last study from this early research period showed that (1) when "non-reporters" that had been awakened during REM sleep reported mental content, they labeled it as thought; (2) non-reporters recalling dreams had a very high arousal threshold; and (3) subjects reporting nothing (non-reporters) did not have an elevated arousal threshold (Lewis *et al.*, 1966), which nevertheless seems in opposition with the above mentioned presence of some alpha rhythm during the eye movements.

Today, our knowledge of the phenomenology of dreaming has advanced significantly. For example, Schredl (2007) distinguishes six models of dream recall. First, historically, the repression hypothesis (Freud, 1900a), which will be developed below. Second, the lifestyle hypothesis (Schonbar, 1965): people who are "more field-independent, introverted and creative" retain their dreams better. Third, the interference hypothesis (Cohen and Wolfe, 1973): "the fewer interferences (external noises, internal distractions like thoughts about the upcoming day) that are active in the time span between awakening and recording

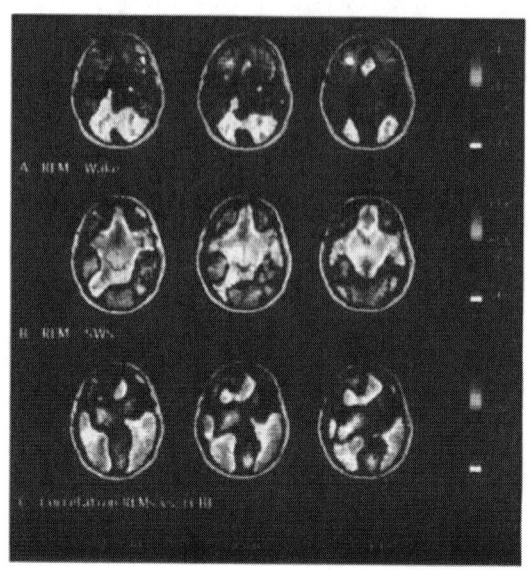

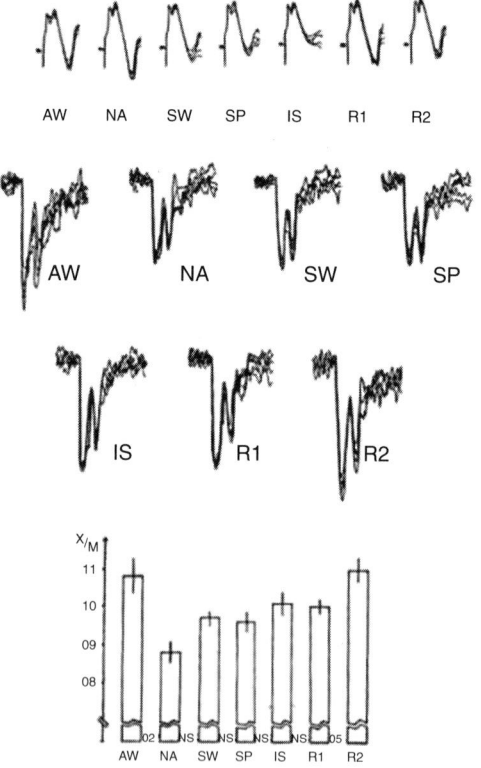

AW   NA   SW   SP   IS   R1   R2

AW   NA   SW   SP

IS   R1   R2

Fɪɢ. 9. *(Continued)*

and telling or mentally rehearsing the dream, the greater the chance of recalling the dream experience" (p 82). Thus, field-independent and introverted persons should be better recallers. Fourth, the salience hypothesis (Cohen and MacNeilage, 1974), which suggested that the more vivid and impressive dreams are, the better they will be recalled. Fifth, the arousal-retrieval model (Koulack and Goodenough, 1977), which postulated that at least brief arousal is necessary to transfer the dream from short-term memory into long-term memory. This means that at least a short arousal period is obligatory for dream recall. Therefore, dreams of the final REM sleep period are better recalled than the content of earlier REM sleep periods. Then, the process of retrieval is important. Once again, the more vivid a dream is, the better the recall will be. Finally, following the state-shift model, Koukkou and colleagues (Koukkou and Lehman, 1983) hypothesized that "forgetting of dreams is a function of the magnitude of the difference between states during storage and recall" (p 221). The more the brain is activated, as is the case during REM sleep when compared to waking, the better the recall will be.

From a neurobiological standpoint, several concomitant phenomena are possibly related to dream recall, and particularly to its most often immediate or rapid forgetting. First, cortisol is released in the early morning, when the last and longest REM sleep period occurs. Dreams to be recalled must first be memorized, and cortisol impairs memory processes (Kirschbaum et al., 1996; Payne and Nadel, 2004). Perhaps more importantly, noradrenaline release reappears in the few seconds preceding behavioral arousal (Aston-Jones and Bloom, 1981a), and this neuromodulator is involved in higher integrated processes (Arnsten and Dudley, 2005; Arnsten and Goldman-Rakic, 1987; Berridge et al., 2006; Goldman-Rakic and Brown, 1981; Milstein et al., 2007). Its reappearance should instantaneously restore processes such as the cortical (Corsi-Cabrera et al., 2003; Perez-Garci et al., 2001) and hippocampo-cortical (Cantero et al., 2004)

FIG. 9. Sensory deafferentation during REM sleep. Top. While the extrastriate visual cortex is activated during REM sleep, the primary visual cortex (V_1 area) is deactivated. (A) REM sleep vs. waking. (B) REM sleep vs. slow wave sleep. (C) rapid eye movements correlated with cerebral blood flow (negative correlation). Reprinted from Braun et al. (1998). Science, with permission. Bottom. Upper line. Thalamic postsynaptic responsiveness in the rat. The amplitude of the r_1 component is higher during non-active waking (NA), without theta rhythms, than during active waking (AW) with theta rhythms. During REM sleep, responsiveness is decreased during the rapid eye movement periods (R2) when compared to REM sleep without eye movement (R_1). Below. Presynaptic responsiveness. The amplitude of the first (t_1) component is enhanced during active waking and during the rapid eye movement periods of REM sleep. This is an index of inhibitory cortical influences on sensory afferents during active waking and of pontine-originating inhibition during REM sleep eye movements. See text. (SW: slow waves; SP: spindles; IS: intermediate stage. Reprinted from Gandolfo et al. (1980). Brain Research Bulletin with permission.

coupled gamma rhythm, thus re-establishing intracerebral functional relations. Further, variations in the recovery of dorsolateral prefrontal activation at REM sleep outcome could affect the corresponding recovery of efficiency and favor dream forgetting processes (Balkin *et al.*, 2002). Finally, endogenous increases in CaMKII (Cao *et al.*, 2008; Wang *et al.*, 2008) upon arousal, or still related transient β-noradrenergic receptor decrease of sensitivity (Kindt *et al.*, 2009), could explain the non-recording of dreams in the waking memory.

Several authors have underlined the positive effects of dream forgetting. The most important such argument is that it prevents memory overloading. Greenberg and Liederman (1966) stated that "dreaming serves the purpose of preparing the short-term memory perceptual areas for the next day's sensory input by transferring the daily input to a more permanent storage area. Dreaming may then be part of the process of transferring the day's residue to the long-term storage" (p 521). Nowadays, this memorization process is known to involve the hippocampus, since its waking-induced long-term potentiation in subsequent REM sleep periods upregulates the activity-dependent gene *zif-268* in this structure, as well as in other limbic structures (amygdala and entorhinal cortex) and in several areas of the neocortex. This gene activation, which is involved in synaptic plasticity, propagates from the hippocampus, since locally tetracaine suppresses the extra-hippocampic gene activation (Ribeiro *et al.*, 2002). While post-acquisition neuronal reverberation depends mainly on slow-wave sleep episodes (but also occurs during REM sleep Louie and Wilson, 2001; Maquet *et al.*, 2000), the transcriptional events capable of promoting long-lasting memory storage are only triggered during the ensuing REM sleep (Ribeiro and Nicolelis, 2004). This work is important since it specifically addresses REM sleep processes instead of global sleep (Basheer *et al.*, 1997; Cirelli and Tononi, 1998; O'Hara *et al.*, 1993; Pompeiano *et al.*, 1994, 1995). In some ways, the "clearing" of memory was also supported by Crick and Mitchison (1983), for whom the purpose of REM sleep "is to remove certain undesirable modes of interaction in networks of cells in the cerebral cortex. We postulate that this is done in REM sleep by a reverse learning mechanism, so that the trace in the brain of the unconscious dream is weakened, rather than strengthened by the dream" (p 111). "We dream in order to forget" (p 112). Accordingly, as postulated by Newman and Evans (1965) and by Hawkins (1966) as early as the 1960s, when REM sleep suppresses certain memory "traces" it is analogous to in computers "the clearing out of storage area, so that it would be empty during the next active waking part of the cycle for new data storage" (p 254, Gaardner, 1966) In contrast, other memory processes would be reinforced (Hennevin *et al.*, 1995, 2007; Pace-Schott, 2003; Smith, 1995), although this point has been strongly questioned by others (Morrison and Sanford, 2003; Siegel, 2001; Vertes, 2004; Vertes and Eastman, 2000; Vertes and Siegel, 2005). The analogy of brain functioning during REM sleep with computers was more recently developed by Jouvet (1992, 1998), who postulated the readjustment of genetic behavioral programs and also that mental retardation would involve deficits in this process due to decreased REM sleep (Clausen *et al.*, 1977; Grubar, 1983; Petre-Quadens and Jouvet, 1967). More recently, Dang-Vu *et al.* (2007) indirectly supported Jouvet's hypothesis: "Dreaming involves the genetic reprogramming of cortical networks that might promote the maintenance of psychological individuality despite potentially adverse influences from the waking experiences" (p 98). Finally, when the usual dream forgetting—the clearing up—process fails in predisposed subjects, it could open the door to schizophrenia, with the hallucinations entering waking consciousness and being taken for reality (Kelly, 1998). As previously recounted (Gottesmann, 2005a): when the French poet Gérard de Nerval (1808–1855) drifted into madness, he described "an over-welling of the dream into reality."

Another point is the fragmentation of dreams, "the discontinuity associated with contextual misbinding" (p 95, Dang-Vu *et al.*, 2007). Already in the 1960s (Gottesmann, 1967, 1970, 1971), we hypothesized that memory processes, stored as some kind of "traces," could enter the dreamer's consciousness according to the urgency provided by their load (by comparison with a condenser). As the dream development progresses, the load of another trace would, in turn, become more important and, by increased pressure, enter the dreamer's consciousness (p 114, Gottesmann, 1967); this would most often be behaviorally shown by gross body movements, indicating the end of a dream (Dement and Wolpert, 1958a; Wolpert and Trosman, 1958). Only traumatic events could give rise to traumatic neurosis (Freud, 1920) and induce memory traces that could rapidly recover their load and thus repetitively, over successive nights, enter the dreamer's consciousness (Gottesmann, 1967, 1970, 1971; Hartmann, 1967; Ross *et al.*, 1994). Current neurobiological data show that, in association with the hippocampus for recent episodic events, and with the medial prefrontal cortex for long-term memories, the influences of the activated amygdala (Maquet and Franck, 1997) on the extrastriate visual cortex (Braun *et al.*, 1998; Madsen *et al.*, 1991) seem to support the emotional components of such recurrent dreams (Nielsen and Stenstrom, 2005; Pace-Schott, 2007).

This question leads to the more general question of the manifest content of successive dreams occurring in a given night that was first studied by Dement and Wolpert (1958b): "No single dream was ever exactly duplicated by another dream, nor were the dreams of a sequence ever perfectly continuous, one taking up just where the preceding had left off. For the most part, each dream seemed to be a self-continuing drama relatively independent of the preceding or following dream. Nonetheless, the manifest content of nearly every dream exhibited some obvious relationship to one or more dreams occurring on the same night. Only contiguous dreams were obviously related. . . There was also considerable overlapping of relationships" (p 569).

In this pioneering period, Trosman *et al.* (1960) also observed that "a direct relationship among the manifest contents in the dreams of a night was rarely noted. However, the appearance of unique elements at similar points in a sequence, the lateral similarities of events from night to night and the confluence of similar dimensions suggest an organization of the manifest contents into regular patterns" (p 606). Finally, Rechtschaffen *et al.* (1963) observed that "manifest elements tended to be repeated during a night on occasions when preoccupations of the recent past remain so intense that they 'press' for discharge throughout the sleep period" (p 546). Such results can be interpreted as in posttraumatic events, although of lower intensity. All of these phenomena could be related to a given loss of prefrontal controls of mind functioning (deficit of working memory, attention and self-awareness) because of its dorsolateral deactivation.

A final point to be mentioned is the speed of mental functioning that takes place during REM sleep. "Acceleration of thinking belongs to dreaming as well as to insanity, and to moments of high emotion, of utmost turmoil" (p 162, Maury, 1861). In the medial prefrontal cortex of rats, during sleep there is an at least six-fold accelerated replay of waking-established learning processes (Euston *et al.*, 2007). This high-speed functioning could perhaps partly explain the well-known unexpected, instantaneous dream of Maury (1861); see translation in (Gottesmann, 1999, 2007a). From a neurobiological standpoint, the decrease in prefrontal cortex functioning, as shown by neuroimagery, and the decrease in cortical inhibitory control processes, as also evidenced by electrophysiological and neurochemical criteria, can explain the unrestrained and uncontrolled "hyperassociative" (Hobson and Pace-Schott, 2002) mental processes occurring in an otherwise activated forebrain. However, in spite of the fragmentation of dreams, partly consecutive to this hyperassociative ("hyperconnective" Hartmann, 2007) mental functioning, in many dreams there is continuity with waking mental activity, as already alluded to by Dement and Wolpert (1958b). Recent studies have revealed the persistence of dream content (e.g., characters, types of social interactions, themes Domhoff, 2007) not only during the same night (Rechtschaffen *et al.*, 1963; Trosman *et al.*, 1960), "but over months, years, and decades, which supports the idea that dreams are coherent and meaningful. Studies of dream journals also contribute to the evidence that dreams are usually not bizarre by showing that many aspects of dream content are continuous with waking conceptions and concerns, with 'concerns' defined in a general way that covers wishes, interests, worries, and fears" (Domhoff, 2007). However, it must be underlined that, most of the time, sooner or later in dreams there is falling apart, a lost of coherence and most often, some dream content is partially rebuilt upon arousal, see below (Blechner, 2006; Gottesmann, 1967).

To conclude, it is important to recall, in relation to the following discussion of the psychoanalytic approach to the signification of dreams, the similarities between dreaming and schizophrenia in terms of mentation and their common neurobiological background (Gottesmann, 1999, 2005a, 2006a; Gottesmann and Gottesman, 2007). As carefully analyzed by Hobson and his colleagues (Hobson *et al.*, 1998), dreaming is characterized by "sensory hallucinations, bizarre imagery, diminished reflective awareness, orientational instability, intensification of emotion and instinctual behaviors," all of which resemble schizophrenic symptoms. This analogy has long been noted by philosophers (Kant, Schopenhauer) neurophysiopsychiatrists (Jackson, Ey), and writers (Maury) (for review see Gottesmann, 2006a). Moreover, there are numerous strong neurobiological similarities between the two kinds of mental activity. (1) The alpha rhythm is nearly absent from the classical EEG (0.5–25 c/s) during REM sleep, exactly as in schizophrenia during waking (Stassen *et al.*, 1999). This near absence reflects a habituation deficit related to disinhibition processes, which has been well established during REM sleep as well as in this disease. (2) The gamma rhythm is no longer synchronized over cortical areas during REM sleep (Cantero *et al.*, 2004; Massimini *et al.*, 2005; Perez-Garci *et al.*, 2001); this is an index of intracerebral disconnections that is also characteristic of schizophrenia (Kubicki *et al.*, 2008; Meyer-Lindenberg *et al.*, 2001, 2005; Peled *et al.*, 2000; Young *et al.*, 1998). (3) Spontaneous firing of cortical neurons in animals (Evarts, 1964), as well as prepulse inhibition of event-related potentials in both animals (Allison, 1968; Rossi *et al.*, 1965) and humans (Kisley *et al.*, 2003), indicates a profound disinhibition process at work during REM sleep, as it does in schizophrenia (Kisley *et al.*, 2003). (4) On emergence from dreaming there is a lack of differentiation between self- and hetero-sensory stimulation (tickle), as in the schizophrenia waking state (Blagrove *et al.*, 2006). (5) There is a deactivation of the dorsolateral prefrontal cortex during both REM sleep (Braun *et al.*, 1997; Maquet *et al.*, 1996) and schizophrenia (Berman *et al.*, 1993; Buschbaum *et al.*, 1982; Fletcher *et al.*, 1998; Weinberger *et al.*, 1986), which can explain the decrease in or loss of self-conscious awareness in both states. (6) During REM sleep, the primary visual cortex is deactivated (Braun *et al.*, 1998) (however, see Hong *et al.*, 2009) and there is presynaptic inhibition of thalamic sensory afferents (Dagnino *et al.*, 1969; Gandolfo *et al.*, 1980; Ghelarducci *et al.*, 1970; Iwama *et al.*, 1966; Steriade, 1970). These deficits of sensory constraints favor the occurrence of schizophrenic hallucinations (Behrendt and Young, 2005). This sensory deafferentation also explains the increased threshold for pain that is often reported during psychotic episodes (Griffin and Tyrrell, 2003). (7) Noradrenergic and serotonergic neurons become silent during REM sleep (see above). Both of these neuromodulators mainly inhibit cortical neurons and are in deficit in schizophrenia (Friedman *et al.*, 1999; Linner *et al.*, 2002; Silver *et al.*, 2000; Van Hes *et al.*, 2003). This is another example of a disturbance of cortical inhibitory control processes in both states. (8) During REM sleep, the prefrontal dopamine concentration is decreased relative to waking, while glutamate is unchanged (Léna *et al.*, 2005); both of these observations are also true of schizophrenia (Abi-Dargham and Moore, 2003;

Lauriat *et al.*, 2005). (9) During REM sleep in rats the level of dopamine in the nucleus accumbens is maximal while the level of glutamate is minimal (Léna *et al.*, 2005); this is also true of schizophrenia (Grace, 2000; Heresco-Levy, 2000; MacKay *et al.*, 1982), Further, both of these neurochemical features induce not only psychotic symptoms but also vivid dreaming as well (Larsen and Tandberg, 2001; Perry and Piggott, 2003; Reeves *et al.*, 2001; Thompson and Pierce, 1999). (10) The acetylcholine concentration seen in the cortex during REM sleep in cats is lower than it is during active waking. Such a decrease is known to promote hallucinations (Collerton *et al.*, 2005) and cognitive deficits (Perry *et al.*, 1999; Raedler *et al.*, 2003; Sarter and Bruno, 2000), both of which are observed in schizophrenia.

3. *The contribution of psychoanalysis to the understanding of dreaming*

It must be particularly emphasized that the first powerful published psychoanalytic work concerned dreaming, and was more precisely devoted to "The *interpretation* of dreams" (Freud, 1900a). The difference between the phenomenology of dreaming and the original psychoanalytic approach concerns the search for meaning and possible underlying symbolism.

Probably since REM sleep first appeared phylogenetically, dreaming has fascinated human beings, who undoubtedly questioned themselves from the beginning about its signification. Indeed, analyses of detailed behavioral components of dreaming in animals, and descriptions of dreaming in humans, are already found in Lucrece (1900); for translation, see (Gottesmann, 2001, 2010a). A few centuries later, in the Talmud, it was written that "an uninterpreted dream is like an unread letter" (p 1, Fromm, 1953).

Another significant assertion of Freud relates to our main interest: "The forgetting of dreams remains inexplicable unless the power of the psychical censorship is taken into account" (p 517, Freud, 1900a)

The hypotheses of Freud concerning dreaming have raised some questions.

First, although the day's residue related to hippocampal-mediated memories (Nielsen and Stenstrom, 2005) is indeed often encountered in dreams, a significant source of dreaming was hypothesized by Freud to be a collection of memories from infancy, all occurring "before the latency period at nearly the fifth year" (p 74, Freud, 1938). It is understandable that "the therapeutic action of psychoanalysis ... (partly)... consists in work on dreams as pictographic and symbolic representations of implicit pre-symbolic and pre-verbal experiences" (p 945, Mancia, 2003). However, the dream sources are also supposed to be primal phantasies (Urphantasien): "I call such phantasies—of the observation of sexual intercourse between the parents, of seduction, of castration, and others—'primal phantasies' " (p 269, Freud, 1915 see also p 32–38 Freud, 1918). These primal scenes (Urszenes), such as castration—although probably often imaginary (however, think about castration performed by Arabs on their African slaves, about eunuch harem-keepers, and about castration performed to obtain castrati, the last one of whom, Moreschi, died in 1922), as today children dream about aggression—occurred occasionally in primeval times; in this regard, Laplanche and Pontalis (2006) underlined the analogy with Œdipe's complex (p 333). Freud wrote "that the earliest impressions, received at a time when a child was scarcely yet capable of speaking, produce at some time or another effects of a compulsive character without themselves being consciously remembered. We believe we have a right to make the same assumption about the earliest experiences of the whole humanity" (p 130, Freud, 1938). Today, taking a similar view, Revonsuo (2003) points out that some studies have observed up to 63% of threatening content in men's dreams (p 108), underlining that "the ancestral environment in which the human brain evolved included frequent dangerous events that constitute extreme threats to human reproductive success. They thus presented serious selection pressures to ancestral human populations and fully activated the threat-simulation mechanisms . . . Dream consciousness is specialized in the simulation of threatening events" (p 86). "Recurring, realistic threat simulations led to improved threat perception and avoidance skills and therefore increased the probability of successful reproduction of any given individual. Consequently, the threat-simulation system was selected for during our evolutionary history" (p 99, Revonsuo, 2003). "The dream-protection system can be seen as an ancestral defense mechanism comparable to other biological defense mechanisms whose function is to automatically elicit efficient protective responses when the appropriate cues are encountered" (p 104). In 1897, Manaceine (1897) found about 50% of unpleasant dreams (quoted by Kleitman, 1939). A recent study, however, only found a level of 20% of threatening events in dreams, even after recent chronic exposure to life-threatening events (Malcolm-Smith *et al.*, 2007). Is it possible that the emotional investment of hypothesized primal event memories be higher than recent life memories?

Whatever the percentage of cases, these new findings raise certain questions. First, the wish fulfilment of these dreams is not obvious. However, as underlined by Darcourt (2006), "during sleep, unconscious tensions take advantage of the decreased level of censorship to allow representations that did not cross the barrier during waking to enter the consciousness. This intrusion allows a decrease in tension (and is thus a pleasure in the Freudian sense) in the unconscious, but is a source of tension in the consciousness, which perceives it negatively and protects itself by transforming the representations into more acceptable ones" (p 211). Second, there seem nevertheless to be contradictions between the common dream threat content, the decreased affect intensity as postulated by Freud (see above), and the biological findings. Indeed, to date the only strong neurobiological similarities that have been found are between dreaming and schizophrenia (see above). The following points can be underlined. The nucleus accumbens, which is mainly involved in positive schizophrenic symptoms (hallucinations, delusions, and bizarre thought processes, which are also common in dreams) is regulated by equilibrated influences originating from the prefrontal cortex, ventral hippocampus, and amygdala. Grace (2000) has shown that, in schizophrenia, there is an abnormal decrease in ventral hippocampus glutamatergic output influences (Fig. 10). The decrease in hippocampal influences (also observed during REM sleep Hobson and Pace-Schott, 2002; Schwartz, 2003) prevents prefrontal cortex glutamatergic influences from acting on the nucleus accumbens. Consequently, these two decreases promote the invasion of the nucleus accumbens by amygdala glutamatergic influences, with their emotional load; tomographic results have confirmed that the amygdala has powerful influences during REM sleep (Maquet and Franck, 1997). Consequently, a decreased emotional load in dreaming as postulated by Freud is open to discussion.

In connection with sexuality and with respect to primal phantasies, as observed previously (Gottesmann, 2007a), although references to this behavior and to sexual symbols (see, for example p 356–357 Freud, 1900a) are nearly overrepresented in the Traumdeutung, and although one of his patients spoke about continuous "prickly feelings and overexcitement in my parts" (p 586, Freud, 1900a), there is no indication that erection systematically occurs upon emergence from morning awakening (all the more so, since last REM period is most often followed by the definitive arousal). This is most unexpected, because Freud's theory about sexuality encountered strong resistance from the intellectual community, and erection as a biological characteristic of the dreaming sleep stage would have been of great help to the author. What is most surprising is that Freud had an extensive knowledge of classical literature (see his beautiful Chapter 1 in the Traumdeutung), and he particularly mentioned Hervey de Saint Denys (1867) who, after Lucretius, described such activity during sleep. Although neurobiology has recently established that this activity is not related to libido, since "erections occur during REM sleep in all normal healthy males from infancy to old age" (Hirskowitz and Schmidt, 2005), in vegetative patients (Oksenberg et al., 2000), as well as in animals (Schmidt et al., 1994), was it in Freud a form of psychological repression by censorship?

In relation to dream-work, the two often combined processes of condensation and displacement of psychological events could by themselves explain the bizarreness of dreams. Yu (2001) established a parallel between the bizarreness and broken experiences of dreaming with a similar phenomenon observed by Penfield (1958) after temporal stimulation. However, there is a major difference between the two: In the ecmnesia phenomenon, there is a "subconscious recall of past experiences" (Penfield and Perot, 1963), whereas during dreaming there is nearly always an original hallucinatory creation. In fact, brain function is so profoundly disturbed during REM sleep that, rather than attributing to a potential

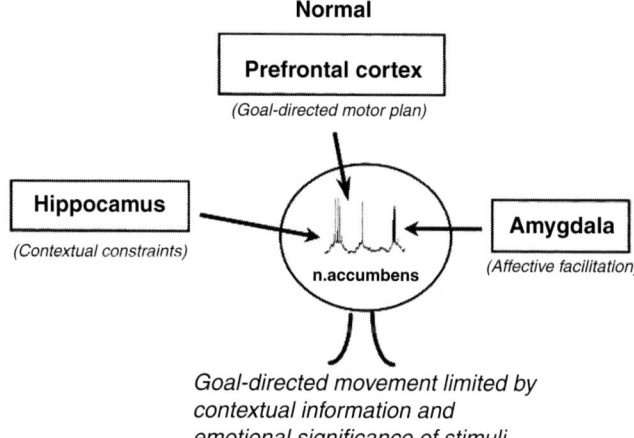

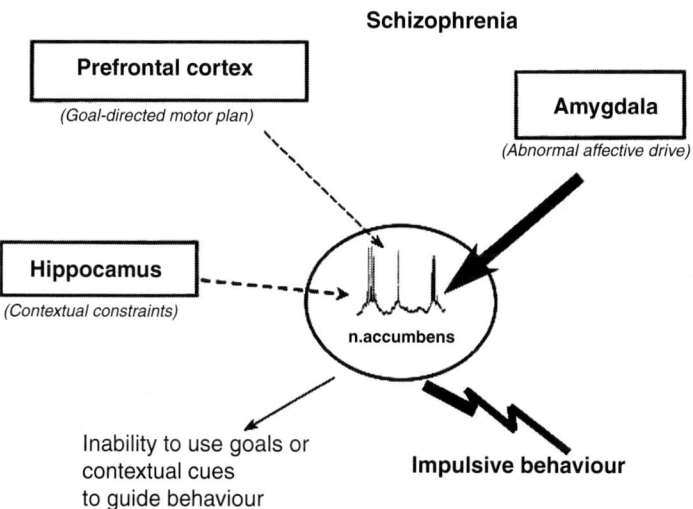

FIG. 10. Similar regulation of the nucleus accumbens during REM sleep and in schizophrenia. During waking in normal subjects, there is an equilibrium between the hippocampus, prefrontal cortex, and amygdala influences on the nucleus accumbens. During REM sleep and in schizophrenia, the hippocampal output is reduced, if not suppressed, the prefrontal cortex is deactivated, and the amygdala is activated. Moreover, in both states, there is an abnormally high accumbental level of dopaminergic function, and a decreased level of glutamate function (see text). Reprinted from Grace (2000). *Brain Research Reviews*, with permission.

metaphorical censor the ability to transform an unconscious wish, most often loaded in affects, into (at least to a first approximation) disorganized mind content which needs interpretation, it seems better to consider these neurobiological dysfunctionalities to be responsible for the psychotic-like mentation of REM sleep (Gottesmann, 2005a, 2006a). First, among the most recent forebrain structures, the dorsolateral prefrontal cortex is deactivated during REM sleep as in schizophrenia (see above). Thus, cognitive performance could only be impaired. Moreover, the different forebrain structures are disconnected from one another, as evidenced by the observed uncoupled intracortical gamma rhythm, particularly between the frontal cortex and the perceptive areas. Finally, the major neurochemical influences acting at the cortical level are decreased (dopamine and acetylcholine) or nearly absent (noradrenaline and serotonin) when compared to waking. In such a generally activated forebrain, it is understandable that there are psychological activities (once again, the "engine" is running because of metabolic petrol), but rather difficult to imagine a censorship directed by a kind of metaphoric superego ("which arises from an identification with the father taken as a model" (p 54, Freud, 1923) that would be able to control brain functioning like, possibly, during waking, and even able to disguise unconscious wishes within symbolic representations (Boarg, 2006). However, it is noteworthy that Yu (2006) thinks that "these high order functions can largely be accomplished by the ventral visual system in the temporo-occipital pathway, which is also implicated in dream-work and distortion" (p 56, Yu, 2006) since, as shown by split-brain experiments, "many parts of the brain can work independently and 'intelligently' without the ego's awareness" (p 56). However, in such clinical cases all homolateral connections are preserved.

As emphasized by Freud, "the chief feature of dreams and insanity lies in their eccentric trains of thought and their weakness of judgment" (Radestock quoted p 91 by Freud, 1900a). In addition, not only philosophers and writers, but neuropsychiatrists like Henri Ey (1967), have underlined the relationship between dreaming and madness. Indeed, Ey said, "it is obvious, it cannot be but obvious that dream and madness spurt out from the same source" (p 575), which is an interesting commentary since it is based on both psychological and neurobiological perspectives. Moreover, Freud not only recalled Hughlings Jackson's (a famous neurophysiopathologist) assertion "Find out all about dreams and you will find out all about insanity" (p 569, Freud, 1900a), but himself saw an "indisputable analogy between dreaming and insanity" (p 92). Thus, within the context of deep mind dysfunction, high integrated control processes (censorship) which would, despite everything, lead to meaningful mental content, again is rather improbable. In fact, it seems more accurate to say that there is "more censorship in the waking analysis of dreams than during the dream itself" (p 17, Blechner, 2006).

However, it must be recognized that there is, during REM sleep, maintenance of some criteria of cognitive processes, which could explain that "in the case of dream distortion or disguise, there is (sometimes) coexistence in the dreamer of the original belief and an *additional* belief that what he or she is *really* doing is something *other* than what appears in the content of the belief' (p 51, Petocz, 2006). For example, the P_{300} evoked potential component appears when, in a series of stimuli, one is unpredictably changed (in intensity, duration etc.) and the subject must detect it. The amplitude of this component is thought to reflect "synchronous activation of multiple distant sites and the updating of consciousness" (Dehaene *et al.*, 2001). "P_{300} may be visible in REM sleep (but with a different topographic distribution), presumably reflecting the altered extent of cognitive processing relative to wakefulness" (Colrain and Campell, 2007).

Another problem is the generally rapid evanescence, or even non-recall, of dreams. Freud (1925) compared the forgetting processes with the children's game called "mystic writing-pad." Current neurobiological knowledge, once again, provides little support for the influence of a psychological censorship that would refuse dream recording in the waking memory and proceed by a repressive mechanism. As documented above in the paragraph related to the phenomenology of dreaming, the high level of cortisol, the recovery of waking noradrenaline function prior to behavioral arousal, the possible endogenous increase of CaM-KII and/or β-noradrenergic receptor desensization, and the less rapid recovery of crucial prefrontal functions and corresponding persistence of hippocampo-prefrontal relation deficiencies, seem sufficient to at least partly explain this difficulty in memorizing. We have instead brought up the idea of a "physiological censorship" (Gottesmann, 2006b; Gottesmann and Gottesman, 2007), among other purposes intended to prevent memory overloading, particularly with respect to unusual bizarre content; once again, dreaming is a psychotic-like mental activity. Schizophrenia is treated by pharmacotherapy. There seems to be a natural protection against this abnormal brain functioning. During REM sleep, the global cortical disorganization, and particularly the deactivation of the dorsolateral prefrontal cortex and its late recovery of function at arousal, could be a self-protection mechanism. From a teleological perspective, this forgetting mechanism was probably generated by "the necessity" (Monod, 1970), dictated by evolution, of avoiding daytime irruptions of mental disturbances that could induce disorganized behavior (Gottesmann, 2007b; Kelly, 1998)

It is perhaps noteworthy that, although there was new interest in the phenomenology of dreaming as soon as it became possible in 1953 to detect its occurrence by electrophysiological criteria, as early as the 1960s two researchers tried to explain psychological and psychoanalytic findings related to dreaming using then-available neurophysiological results. Indeed, Hernandez-Peon (1967; Hernandez-Peon and Sterman, 1966) wrote several very interesting papers linking REM sleep, dreaming, hallucinations, and psychoanalysis. He noted that the waking neurophysiological system inhibits the dreaming system, which is able to appear after REM sleep-induced forebrain disinhibition. Based on his own

electrophysiological (evoked potentials) and pharmacological (picrotoxin) findings, he dissociated activating from inhibitory subcortical influences acting in the
forebrain during waking. "Release of inhibition acting upon the memory systems
account for the manifest content of dreams, and release of inhibition acting upon
the emotional and motivational systems accounts for the latent content of
dreams" (p 636, Hernandez-Peon, 1966). In the same way, Gottesmann (1967,
1970, 1971), taking into account Demetrescu *et al.*'s (1966) demonstration of
activating and inhibitory influences acting at the cortical level, wrote that the
high levels of both kinds of influence during waking could explain the logical
mentation, and the decreases in both during slow wave sleep could support "only
slight mental activity governed by the same processes as that of wakefulness."
Further, just prior to REM sleep, during the "intermediate phase" (or intermediate stage in the rat), "a particular distribution of the ascending facilitatory and
inhibitory influences...(near absence of both) could be at the origin of the
anxious consciousness" (p 125, Gottesmann, 1967). During REM sleep, the
strongly activated but disinhibited cortex could lead to "mental activity that is
'illogical' or based on a logic that is different from that of wakefulness" (p 124),
and "certain 'mnemonic traces' could be activated during REM sleep and not
during waking" (p 124). "Certain 'traces' would maintain an important load
leading to repetitive dreams" (p 124). Finally, upon emergence from REM
sleep, after recovery of the waking activating and inhibitory neurophysiological
influences, "the subject's perception field, in the presence of apparently illogical
elements, would tend: either to refuse to acknowledge (forgetting), or to modify
them so as to render them 'logical'. Only the oneiric contents that are sufficiently
intense, or pregnant, would be re-recorded without any alteration in the 'logical'
mnesic system of waking and would therefore seem illogical" (p 124–125). In the
same topic these papers were followed 10 years later by those of Hobson's team
(Hobson and McCarley, 1977; McCarley and Hobson, 1977).

   One of the main problems in discussing the psychoanalysis-neurobiology
relationship is that "there seems to be anxiety among some people that if we do
not accept all of Freud, we must accept nothing" (p. 19, Blechner, 2006). This is
for example the case of Hobson: "Freud returns? Like a bad dream" (Hobson,
2004; Hobson and Pace-Schott, 1999) and Aron (2007), who refers to "Folk
psychology" when speaking about repression processes. Once again, it is improbable in view of today's neurobiological knowledge that, during REM sleep, there
is a powerful, organized, and lucid yet unconscious censorship that would
"accomplish partial release of disruptive psychological tensions" (p 588, Jones,
1968). Forebrain function is too disorganized. Indeed, although Johnson (2006)
observed that "frontal function is not absent during dreaming, but is reduced,
allowing for emergence of threatening wishes into disguised awareness" (p 37)
(see above), he overlooks the other major (electrophysiological and neurochemical) forebrain disturbances. In relation to dream content, Antrobus (2003) speaks

about "noisy activity," and Ardito (2003) about "nonsense elements." However, there is a spontaneous or elaborated attempt to explain our own dreams, since "upon awakening, the interpretative process accelerates as the dormant verbal and meaning modules of the left temporal and prefrontal cortices become active" (p 116, Antrobus, 2003). Due to the increased excitability of some memory traces, obviously in some way supported by affects, episodic events can enter REM sleep consciousness. It remains to be demonstrated, however, that dreaming leads to phantasmatic wish-fulfilment and that it "accomplishes partial release of disruptive psychological tensions" (p 588, Jones, 1968), since repetitive dreams show the failure of such "safety valve" function.

IV. Conclusion

During evolution, it can be hypothesized that, building on vigilance processes already available in the animal kingdom, consciousness appeared, primarily based on perception categorization, prior to the emergence of unconscious processes. With recent discoveries of the neurosciences, knowledge about the neurobiological basis of consciousness has made remarkable progress. It has already opened up a large vista, not on a somewhat hypothetical unconscious, but rather on the brain mechanisms underlying unconscious processes that are possibly specific to the human species and that were so well fleshed out and partially cleared up by Sigmund Freud. However, in spite of the extraordinary intellectual construction of the Traumdeutung, as well as modern neurobiological results pointing to the active psychological repression of unwanted memories, as of today this process appears to occur during waking rather than during REM sleep dreaming.

Acknowledgments

I thank Dr. Peter Follette for his improvement of the English.

References

Abi-Dargham, A., and Moore, H. (2003). Prefrontal DA transmission at D1 receptors and the pathology of schizophrenia. *Neuroscientist* **9**, 404–416.

Adey, W. R., Segundo, J. P., and Livingston, R. B. (1957). Corticofugal influences on intrinsic brain stem conduction in cat and monkey. *J. Neurophysiol.* **20**, 1–16.

Allison, T. (1968). Recovery cycles of primary evoked potentials in cats sensorimotor cortex. *Experentia* **24**, 240–241.

Amadeo, M., and Gomez, A. (1966). Eye movements, attention and dreaming in the congenetically blind. *Canad. Psychiat. Ass.* **11**, 501–507.

Anderson, M. C., and Green, C. (2001). Suppressing unwanted memories by executive control. *Nature* **410**, 366–369.

Anderson, M. C., Ochsner, K. N., Kuhl, B., Cooper, J., Robertson, E., Gabrieli, S. W., Glover, G. H., and Gabrieli, J. D. E. (2004). Neural systems underlying the suppression of unwanted memories. *Science* **303**, 232–235.

Antrobus, J. S. (2003). How does the dreaming brain explain the dreaming mind. In: Sleep and Dreaming. Scientific Advances and Reconsiderations (E. Pace-Schott, M. Solms, M. Blagrove, and S. Harnad, eds.), Cambridge University Press, Cambridge, pp. 115–118.

Apud, J. A., Mattay, V., Chen, J., Kolachana, B. S., Callicott, J. H., Rasetti, R., Alce, G., Iudicello, J. E., Akbar, N., Egan, M. F., Goldberg, T. E., and Weinberger, D. R. (2007). Tolcapone improves cognition and cortical information processing in normal human subjects. *Neuropsychopharmacology* **32**, 1011–1020.

Apud, J. A., and Weinberger, D. R. (2007). Treatment of cognitive deficits associated with schizophrenia: potential role of catechol-o-methyltransferase inhibitors. *CNS Drugs* **21**, 535–557.

Araneda, R., and Andrade, R. (1991). 5-hydroxytryptamine2 and 5-hydroxytryptamine 1A receptors mediate opposing responses on membrane excitability in the association cortex. *Neuroscience* **40**, 399–412.

Ardito, R. B. (2003). Dreaming as an active construction of meaning. In: Sleep and Dreaming (E. Pace-Schott, M. Solms, M. Blagrove, and S. Harnad, eds.), Cambridge University Press, Cambridge, pp. 118–119.

Arnsten, A. F., and Dudley, A. G. (2005). Methylphenidate improves prefrontal cortical cognitive function through alpha2 adrenoceptor and dopamine D1 receptor actions: relevance to therapeutic effects in attention deficit hyperactivity disorder. *Behav. Brain Funct.* **1**, 2.

Arnsten, A. F., and Goldman-Rakic, P. S. (1987). Noradrenergic mechanisms in age-related cognitive deficit. *J. Neur. Transm.* (Suppl. 241), 317–324.

Aron, A. R. (2007). The neural basis of inhibition in cognitive control. *Neuroscientist* **13**, 1–15.

Aserinsky, E., and Kleitman, N. (1953). Regularly occurring periods of eye motility, and concomitant phenomena during sleep. *Science* **118**, 273–274.

Aston-Jones, G., and Bloom, F. E. (1981a). Activity of norepinephrine-containing neurons in behaving rats anticipates fluctuations in the sleep-waking cycle. *J. Neurosci.* **1**, 876–886.

Aston-Jones, G., and Bloom, F. E. (1981b). Norepinephrine-containing locus coeruleus neurons in behaving rat exhibit pronounced responses to non-noxious environmental stimuli. *J. Neurosci.* **1**, 887–900.

Aston-Jones, G., and Gold, J. I. (2009). How we say no: norepinephrine, inferior frontal gyrus, and response inhibition. *Biol. Psychiat.* **65**, 548–549.

Auerbach, S. (2007). Dreams and dreaming in disorders of sleep. In: The New Science of Dreaming (D. Barrett and P. McNamara, eds.), Vol. 1, Praeger, Wesport, pp. 221–243.

Bagetta, G., De Sarro, G. B., Priolo, E., and Nistisco, G. (1988). Ventral tegmental area: site through which dopamine D2-receptor agonists evoke behavioral and electrocortical sleep. *Br. J. Pharmacol.* **85**, 860–866.

Balkin, T. J., Braun, A. R., Wesensten, N. J., Jeffries, K., Varga, M., Baldwin, P., Belenky, G., and Herscovitch, P. (2002). The process of awakening: a PET study of regional brain activity patterns mediating the re-establishment of alertness and consciousness. *Brain* **125**, 2308–2319.

Basheer, R., Sherin, J. E., Saper, C. B., Morgan, J. E., McCarley, R., and Shiromani, P. (1997). Effects of sleep on wake-induced c-fos expression. *J. Neurosci.* **17**, 9746–9750.

Batini, C., Moruzzi, G., Palestini, M., and Rossi, G. F. (1958). Persistent patterns of wakefulness in the pretrigeminal midpontine preparation. *Science* **128**, 30–32.

Batini, C., Moruzzi, G., Palestini, M., and Rossi, G. F. (1959). Effects of complete pontine transections on the sleep wakefulness rhythm: the midpontine pretrigeminal preparation. *Arch. Ital. Biol.* **97**, 1–12.

Batsel, H. L. (1960). Electroencephalographic synchronization and desynchronization in the chronic "cerveau isolé" of the dog. *Electroencephalogr. Clin. Neurophysiol.* **12**, 421–430.

Batsel, H. L. (1964). Spontaneous desynchronization in the chronic cat "cerveau isolé." *Arch. Ital. Biol.* **102**, 547–566.

Behrendt, R. P., and Young, C. (2005). Hallucinations in schizophrenia, sensori impairment and brain disease: an unified model. *Behav. Brain Sci.* **27**, 771–787.

Belardetti, F., Borgia, R., and Mancia, M. (1977). Proencephalic mechanisms of ECoG desynchronization in cerveau isolé cats. *Electroencephalogr. Clin. Neurophysiol.* **42**, 213–225.

Benington, J. H., Kodali, S. K., and Heller, H. C. (1994). Scoring transitions to REM sleep in rats based on the phenomena of pre-REM sleep: an improved analysis of sleep structure. *Sleep* **17**, 28–36.

Berger, H. (1929). Ueber das electroenkephalogram des Menschen. *Arch. Psychiat. Nervenkr.* **87**, 527–570.

Berger, R. (1961). Tonus of extrinsic laryngeal muscles during sleep and dreaming. *Science* **134**, 840.

Berger, R. J., and Oswald, I. (1962). Eye movements during active and passive dreams. *Science* **137**, 601.

Berman, K. F., Doran, A. R., Pickar, D., and Weinberger, D. R. (1993). Is the mechanism of prefrontal hypofunction in depression the same as in schizophrenia? Regional cerebral blood flow during cognitive activation. *Br. J. Psychiat.* **162**, 183–192.

Berridge, C. W., Devilbiss, D. M., Andrzejewski, M. E., Arnsten, A. F., Kelley, A. E., Schmeichel, B., Hamilton, C., and Spencer, R. C. (2006). Methylphenidate preferentially increases catecholamine neurotransmission within the prefrontal cortex at low doses that enhance cognitive function. *Biol. Psychiat.* **60**, 1111–1120.

Bjorvatn, B., Fagerland, S., and Ursin, R. (1998). EEG power densities (0.5–20 Hz) in different sleep-wake stages in the rat. *Physiol. Behav.* **63**, 413–417.

Blagrove, M., Blakemore, S. J., and Thayer, B. R. J. (2006). The ability to self-tickle following rapid eye movement sleep dreaming. *Conscious Cogn.* **15**, 285–294.

Blake, H., and Gerard, R. W. (1937). Brain potentials during sleep. *Am. J. Physiol.* **119**, 692–703.

Blechner, M. J. (2006). A post-Freudian psychoanalytic model of dreaming. *Neuro-Psychoanalysis* **8**, 17–20.

Boarg, S. (2006). Freudian dream theory, dream bizareness, and the disguise-censor controversy. *Neuro-Psychoanalysis* **8**, 5–17.

Bonvallet, M., and Allen, M. B. (1963). Prolonged spontaneous and evoked reticular activation following discrete bulbar lesions. *Electroencephalogr. Clin. Neurophysiol.* **15**, 969–988.

Bosinelli, M. (1995). Mind and consciousness during sleep. *Brain Res.* **69**, 195–201.

Bosinelli, M., and Cigogna, P. C. (2003). REM and NREM mentation: Nielsen's model once again supports the supremacy of REM. In: Sleep and Dreaming Scientific Advances and Reconsiderations (E. Pace-Schott, M. Solms, M. Blagrove, and S. Harnard, eds.), Cambridge University Press, Cambridge, pp. 124–125.

Brake, W. G., Flores, G., Francis, D., Meaney, M. J., Srivastava, L. K., and Gratton, A. (2000). Enhanced nucleus accumbens dopamine and plasma corticosterone stress responses in adult rats with neonatal excitotoxic lesions to the medial prefrontal cortex. *Neuroscience* **96**, 687–695.

Braun, A. R., Balkin, T. J., Wesensten, N. J., Carson, R. E., Varga, M., Baldwin, P., Selbie, S., Belenky, G., and Herscovitch, P. (1997). Regional cerebral blood flow throughout the sleep-wake cycle: an [15]0 PET study. *Brain* **120**, 1173–1197.

Braun, A. R., Balkin, T. J., Wesensten, N. J., Gwardry, f, Carson, R. E., Varga, M., Baldwin, P., Belenky, G., and Herscovitch, P. (1998). Dissociated pattern of activity in visual cortices and their projections during human rapid eye movement sleep. *Science* **279**, 91–95.

Bremer, F. (1935). Cerveau "isolé" et physiologie du sommeil. *C. R. Soc. Biol.* **118**, 1235–1241.

Bremer, F. (1936). Nouvelles recherches sur le mécanisme du sommeil. *C. R. Soc. Biol.* **122**, 460–464.

Bremer, F., and Chatonnet, J. (1949). Acetylcholine et cortex cérébral. *Arch. Int. Physiol.* **57**.

Brooks, D. C., and Bizzi, E. (1963). Brain stem electrical activity during deep sleep. *Arch. Ital. Biol.* **101**, 648–665.

Bubnoff, N., and Heidenhain, R. (1881). Ueber Erregungs-Hemmungsvorgänge innerhalb der motorischen Hirncentren. In: Arch. Gesam. Physiol. (E. F. W. Pflüger, ed), pp. 137–202.

Buschbaum, M. S., Ingvar, D. H., Kessler, R., Waters, R. N., Cappelletti, J., Van Kammen, D. P., King, A. C., Johnson, J. L., Manning, R. G., Flynn, R. W., Bunney, W. E. J., and Sokoloff, L. (1982). Cerebral glucography with positron tomography, use in normal subjects and in patients with schizophrenia. *Arch. Gen. Psychiat.* **39**, 251–259.

Buser, P. (2007). Neuroscience et psychanalyse. *Rev. Fr. Psychanal.* **71**, 359–367.

Buzsaki, G., Bickford, R. G., Ponomareff, G., Thal, L. J., Mandel, R., and Cage, F. H. (1988). Nucleus basalis and thalamic control of neocortical activity in the freely moving rat. *J. Neurosci.* **8**, 4007–4026.

Buzsaki, G., and Gage, F. H. (1989). The nucleus basalis: a key structure in neocortical arousal. In: Central Cholinergic Synaptic Transmission (M. Froescher and U. Misgeld, eds.), Kirhäuser Verlag, Basel, pp. 159–171.

Cantero, J. L., Atienza, M., Madsen, J. R., and Stickgold, R. (2004). Gamma EEG dynamics in neocortex and hippocampus during human wakefulness and sleep. *NeuroImage* **22**, 1271–1280.

Cao, X., Wang, H., Mei, B., An, S., Yin, L., Wang, L. P., and Tsien, J. (2008). Inducible and selective erasure of memories in the mouse brain via chemical-genetic manipulation. *Neuron* **60**, 353–366.

Cartwright, R. D. (1966). Dream and drug-induced fantasy behavior. *Arch. Gen. Psychiat.* **15**, 7–15.

Cartwright, R., Baehr, E., Kirkby, J., Pandi-Perumal, S. R., and Kabat, J. (2003). REM sleep reduction, mood regulation and remission in untreated depression. *Psychiat. Res.* **121**, 159–167.

Castellucci, V. F., and Kandel, E. R. (1974). A quantal analysis of the synaptic depression underlying habituation of the gill-withdrawal reflex in aplysia. *Proc. Nat. Acad. Sci. USA* **71**, 5004–5008.

Ceechi, M., Giorgetti, M., Bacciottini, L., Giovannini, M. G., and Blandina, P. (1998). Increase of acetylcholine release from cortex of freely moving rats by administration of histamine into the nucleus basalis magnocellularis. *Inflamm. Res.* **47**(Suppl. 1), S32–S33.

Celesia, G. G., and Jasper, H. H. (1966). Acetylcholine released from cerebral cortex in relation to sate of activation. *Neurology* **16**, 1053–1063.

Cirelli, C., and Tononi, G. (1998). Differences in gene expression between sleep and waking as revealed by mRNA differential display. *Mol. Brain Res.* **56**, 293–305.

Clausen, J., Sersen, E. A., and Lidsky, A. (1977). Sleep patterns in mental retardation: Down's syndrome. *Elecroenceph. Clin. Neurophysiol.* 183–191.

Cohen, H. B., and Dement, W. C. (1965). Sleep: changes in threshold to electroconvulsive shocks in rats after deprivation of "paradoxical" phase. *Science* **150**, 1318–1319.

Cohen, D. B., and MacNeilage, P. F. (1974). A test of the salience hypothesis of dream recall. *J. Consult. Clin. Psychol.* **42**, 699–703 (quoted by Schredl 2007).

Cohen, D. B., and Wolfe, G. (1973). Dream recall and repression: evidence for an alternative hypothesis. *J. Consult. Clin. Psychol.* **41**, 349–355 (quoted by Schredl 2007).

Collerton, D., Perry, E., and McKeith, I. (2005). A novel perception and attention deficit model for recurrent visual hallucinations. *Brain Behav. Sci.* **28**, 737–757.

Colrain, I. M., and Campell, K. B. (2007). The use of evoked potentials in sleep research. *Sleep Med. Rev.* **11**, 277–293.

Cordeau, J. P., Moreau, A., Beaulnes, A., and Laurin, C. (1963). EEG and behavioral changes following microinjections of acetylcholine and adrenaline in the brain stem of cats. *Arch. Ital. Biol.* **101**, 30–47.

Corsi-Cabrera, M., Miro, E., del Rio Portilla, Y., Perez-Garci, E., Villanueva, Y., and Guevera, M. (2003). Rapid eye movement sleep dreaming is characterized by uncoupled EEG activity netween frontal and perceptual cortical regions. *Brain Cogn.* **51**, 337–345.

Creutzfeldt, O., Baumgartner, G., and Schoen, L. (1956). Reaktionen einzelner Neurons des senso-motorischen cortex nach elektrischen Reizen. I Hemmung und Erregung nach direkten und contralateralen Einzelreizen. *Arch. Psychiat. Nervenkr.* **194**, 597–619.

Crick, F., and Koch, C. (1995). Are we aware of neural activity in primary visual cortex? *Nature* **375**, 121–123.

Crick, F., and Michison, G. (1983). The function of dream sleep. *Nature* **304**, 111–114.

Crisp, A. H. (1996). The sleepwalking/night terrors syndrome in adults. *Postgrad. Med. J.* **72**, 599–604.

Cuculic, Z., and Himwich, H. E. (1968). An examination of a possible cortical cholinergic link in the EEG arousal reaction. *Prog. Brain Res.* **28**, 27–39.

Dagnino, N., Favale, E., Loeb, C., Manfredi, M., and Seitun, A. (1969). Presynaptic and postsynaptic changes in specific thalamic nuclei during deep sleep. *Arch. Ital. Biol.* **107**, 668–684.

Dahan, L., Astier, B., Vautrelle, N., Urbain, N., Kocsis, B., and Chouvet, G. (2007). Prominent burst firing of dopaminergic neurons in the ventral tegmental area during paradoxical sleep. *Neuropsychopharmacology* **32**, 1232–1241.

Dang-Vu, T., Schabus, M., Desseilles, M., Schwartz, S., and Maquet, P. (2007). Neuroimaging of REM and dreaming. In: The New Science of Dreaming (D. Barrett and P. McNamara, eds.), Vol. 1, Praeger, Wesport, pp. 95–113.

Darcourt, G. (2006). La psychanalyse peut-elle encore être utile à la psychiatrie. Odile Jacob, Paris.

Datta, S., Siwek, D. F., Patterson, E. H., and Cipolloni, P. B. (1998). Localization of pontine PGO wave generation sites and their anatomical projections in the rat. *Synapse* **30**, 409–423.

Datta, K., Spoley, E. E., Mavanji, V., and Patterson, E. H. (2002). A novel role of pedunculopontine tegmental kainate receptors: a mechanism of rapid eye movement sleep generation in the rat. *Neuroscience* **114**, 157–164.

Dehaene, S., Naccache, L., Cohen, L., Le Bihan, D., Mangin, J. F., Poline, J. B., and Rivière, D. (2001). Cerebral mechanisms of word masking and unconscious repetition priming. *Nat. Neurosci.* **4**, 752–758.

Dement, W. C. (1958). The occurrence of low voltage fast electroencephalogram patterns during behavioral sleep in the cat. *Electroencephalogr. Clin. Neurophysiol.* **10**, 291–296.

Dement, W. C. (1964). Eye movements during sleep. In: The Oculomotor System (M. Bender, ed.), Harper and Row, New York, pp. 366–415.

Dement, W., and Kleitman, N. (1957a). Cyclic variations in EEG during sleep and their relation to eye movements, body motility, and dreaming. *Electroencephalogr. Clin. Neurophysiol.* **9**, 673–690.

Dement, W., and Kleitman, N. (1957b). The relation of eye movements during sleep to dream activity: an objective method for the study of dreaming. *J. Exp. Psychol.* **53**, 339–346.

Dement, W., and Wolpert, E. A. (1958a). The relation of eye movements, body motility, and external stimuli to dream content. *J. Exp. Psychol.* **55**, 543–553.

Dement, W., and Wolpert, E. A. (1958b). Relationship in the manifest content of dreams occurring on the same night. *J. Neural Transm.* **126**, 568–578.

Demetrescu, M., Demetrescu, M., and Iosif, G. (1966). Diffuse regulation of visual thalamo-cortical responsiveness during sleep and wakefulness. *Electroencephalogr. Clin. Neurophysiol.* **20**, 450–469.

Depoortere, H., and Loew, D. M. (1973). Motor phenomena during sleep in the rat: a mean of improving the analysis of the sleep-wakefulness cycle. In: The Nature of Sleep (U. Jovanovic, ed.), Gustav Fisher, Stuttgart, pp. 101–104.

Depue, B. E., Banich, M. T., and Curran, T. (2006). Suppression of emotional and nonemotional content in memory. *Psychol. Sci.* **17**, 441–447.

Depue, B. E., Curran, T., and Banich, M. T. (2007). Prefrontal regions orchestrate suppression of emotional memories via a two-phase process. *Science* **317**, 215–219.

Descarries, L., Beaudet, A., and Watkins, K. C. (1975). Serotonin nerve terminals in the adult rat neocortex. *Brain Res.* **100**, 563–588.

Descarries, L., Gisinger, V., and Steriade, M. (1997). Diffuse transmission by acetylcholine in the CNS. *Prog. Neurobiol.* **53**, 603–625.

Di Chiara, G. P., Vargiu, M. L., Argiolas, A., and Gessa, G. L. (1976). Evidence for dopamine receptors in the mouse brain mediating sedation. *Nature* **264**, 564–567.

Doricchi, F., Iaria, G., Silvetti, M., Figliozzi, F., and Siegler, I. (2007). The "ways" we look at dreams: evidence from unilateral spatial neglect (with an evolutionary account of dream bizarreness. *Exp. Brain Res.* **178**, 450–461.

Domhoff, G. W. (2007). Realistic stimulation and bizarreness in dream content. The new science of dreaming. (D. Barret and P. McNamara, eds.), praeger, westport, pp. 1–27.

Espana, R. A., Reiser, K. M., Valentino, R. J., and Berridge, C. W. (2005). Organisation of hypocretin/orexin efferents to locus coeruleus and basal forebrain arousal-related structures. *J. Comp. Neurol.* **481**, 165–183.

Euston, D. R., Tatsuno, M., and McNaughton, B. L. (2007). Fast-forward playback of recent memory sequences in prefrontal cortex during sleep. *Science* **318**, 1147–1150.

Evarts, E. V. (1960). Spontaneous and evoked activity of single units in visual cortex during sleep and waking. *Fed. Proc.* **19**, 290.

Evarts, E. V. (1964). Temporal patterns of discharge of pyramidal tract neurons during sleep and waking in the monkey. *J. Neurophysiol.* **27**, 152–171.

Evarts, E. V. (1965). Neuronal activity in visual and motor cortex during sleep and waking. In: Aspects anatomo-fonctionnels de la physiologie du sommeil (CNRS, ed.), CNRS, Paris, pp. 189–212.

Ey, H. (1967). La dissolution du champ de la conscience dans le phénomène sommeil-veille et ses rapports avec la psychopathologie. *Pres. Med.* **75**, 575–578.

Farber, J., Marks, G. A., and Roffwarg, H. P. (1980). REM sleep PGO-type waves are present in the dorsal pons of the albino rat. *Science* **209**, 615–617.

Fernandez-Mendoza, J., Lozano, B., Seijo, F., Santamarta-Liébana, E., Ramos-Platon, M. J., Vela-Bueno, A., and Fernandez, G. F. (2009). Evidence of subthalamic PGO-like waves during REM sleep in humans: a deep brain polysomnographic study. *Sleep* **32**, 1117–1126.

Fletcher, P. C., McKenna, P. J., Frith, C. D., Grasby, P. M., Friston, K. J., and Dolan, R. J. (1998). Brain activation in schizophrenia during a graded memory task studied with functional neuroimaging. *Arch. Gen. Psychiat.* **55**, 1001–1008.

Foote, S. L., Freedman, R., and Oliver, A. P. (1975). Effects of putative neurotransmitters on neuronal activity on monkey auditory cortex. *Brain Res.* **86**, 229–242.

Fosse, R., and Domhoff, G. W. (2007). Dreaming as non-executive orienting: a conceptual framework for consciousness during sleep. In: New Science of Dreaming (D. Barrett and P. McNamara, eds.), Vol. 2, Praeger, Westport, pp. 49–78.

Fosse, R., Stickgold, R., and Hobson, J. A. (2001). Brain-mind states: reciprocal variation in thoughts and hallucinations. *Psychol. Sci.* **12**, 30–36.

Fosse, R., Stickgold, R., and Hobson, J. A. (2002). Emotional experience during rapid-eye-movement sleep in nacolepsy. *Sleep* **25**, 724–732.

Foulkes, W. D. (1962). Dream reports from different stages of sleep. *J. Abnorm. Soc. Psychol.* **65**, 14–25.

Foulkes, D., Spear, P. S., and Symonds, J. D. (1966). Individual differences in mental activity at sleep onset. *J. Abnorm. Psychol.* **71**, 280–286.

Foulkes, W. D., and Vogel, G. (1965). Mental activity at sleep onset. *J. Abnorm. Psychol.* **70**(231), 243.

Frederickson, R. C. A., Jordan, L. M., and Phillis, J. W. (1971). The action of noradrenaline on cortical neurons: effects of pH. *Brain Res.* **35**, 556–560.

Freud, S. (1895). Project for a Scientific Psychology (S. Edition, ed.), Vol. 1, The Hogard Press, London.

Freud, S. (1897). Letter to Wilhem Fliess (22 December). In: La naissance de la Psychanalyse (M. Bonaparte, ed.), Presses Universitaires de France, Paris, pp. 211–213.

Freud, S. (1900a). The Interpretation of Dreams. The Hogart Press, London.

Freud, S. (1900b). L'interprétation des rêves. Presses Universitaires de France, Paris.

Freud, S. (1901). The Psychopathology of Everyday Life. (S. Edition, ed.), Vol. 6, The Hogart Press, London.

Freud, S. (1911). Formulations on the Two Principles of Mental Functioning (S. Edition, ed.), Vol. XII, The Hogarth Press, London.

Freud, S. (1914). Zur Einführung des Narzissismus (G. Werke, ed.), Vol. X, Fischer Verlag, Frankfurt am Main, pp. 137–170.

Freud, S. (1915). A case of paranoia running counter to the psychoanalytic theory of the disease. (S. Edition, ed.), Vol. XIV, The Hogard Press, London.

Freud, S. (1918). The dream and the primal scene. In: Freud S. From the History of an Infantile Neurosis (S. Edition, ed.), Vol. XVII, The Hogart Press, London, pp. 29–47.

Freud, S. (1920). Beyong the Pleasure Principle. The Hogart Press, London.

Freud, S. (1923). The Ego and the Id (S. Edition, ed.), The Hogart Press, London.

Freud, S. (1925). A Note upon the "Mystic Writing-Pad," Standart ed., Vol. 19, The Hogart Press, London, pp. 227–232.

Freud, S. (1938). Moses and Monotheism (S. Edition, ed.), Vol. VII, The Hogart Press, London.

Friedman, J. I., Adler, D. N., and Davis, K. L. (1999). The role of norepinephrine in the physiopathology of cognitive disorders: potential applications to the treatment of cognitive dysfunction in schizophrenia and Alzheimer's disease. Biol. Psychiat. 46, 1243–1252.

Fromm, E. (1953). Le langage oublié. Payot, Paris.

Fuxe, K., Rambert, F. A., Ferraro, L., O'Connor, W., Laurent, P., Agnati, L. F., and Tangagnelli, S. (1996). Preclinical studies with modafinil. Evidence for vigilance enhancement and neuroprotection. Drugs Today 32, 313–326.

Gaardner, K. (1966). A conceptual model of sleep. Arch. Gen. Psychiat. 14, 253–260.

Gaillard, R., Del Cul, A., Naccache, L., Vinckier, F., Cohen, L., and Dehaene, S. (2006). Nonconscious semantic processing of emotional words modulates conscious access. Proc. Nat. Acad. Sci. USA 103, 7524–7529.

Gaillard, J. M., Martinoli, R., and Kafi, S. (1977). Automatic analysis of states of vigilance in the rat. Sleep 1976. Karger, Basel, pp.455–457.

Gais, S., Albouy, G., Darsaud, A., Rauch, G., Schabus, M., Sterpenich, V., Peigneux, P., and Maquet, P. (2008). Hippocampal-neocortical interactions in long-term consolidation depend on sleep. J. Sleep Res. 17(Suppl. 1), 11.

Gallopin, T., Fort, P., Eggermann, E., Caul, B., Luppi, P. H., Roissier, J., Audimat, E., Mühlethaler, M., and Seraphin, M. (2000). Identification of sleep- promoting neurons in vitro. Nature 404, 922–925.

Gandolfo, G., Arnaud, C., and Gottesmann, C. (1980). Transmission in the ventrobasal complex of rat during the sleep-waking cycle. Brain Res. Bull. 5, 921–927.

Gastaut, H., and Broughton, R. (1964). A clinical and polygraphic study of episodic phenomena during sleep. Recent Advan. Biol. Psychiat. 7, 197–221.

Gessa, G. L., Proceddu, M. L., Collu, M., Mereu, G., Serra, M., Ongini, G. L., and Bioggio, G. (1985). Sedation and sleep by high doses of apomorphine after blockade of D-1 receptors by SCH 23390. Eur. J. Pharmacol. 109, 269–274.

Ghelarducci, B., Pisa, M., and Pompeiano, M. (1970). Transformation of somatic afferent volleys across the prethalamic and thalamic components of the lemniscal system during the rapid eye movements of sleep. Electroencephalogr. Clin. Neurophysiol. 29, 348–357.

Giulledge, A. T., and Stuart, G. J. (2005). Cholinergic inhibition of neocortical pyramidal neurons. J. Neurosci. 25, 10308–10320.

Glin, L., Arnaud, C., Berracochéa, D., Galey, D., Jaffard, R., and Gottesmann, C. (1991). The intermediate stage of sleep in mice. *Physiol. Behav.* **50**, 951–953.

Goldman-Rakic, P. S., and Brown, R. M. (1981). Regional changes of monoamines in cerebral cortex and subcortical structures of aged rhesus monkeys. *Neuroscience* **6**, 177–187.

Goldsteinas, L., Guennoc, A., and Vidal, J. C. (1966). Nouvelles données cliniques sur le vécu des "phases intermédiaires" du sommeil. *Rev. Neurol.* **115**, 507–511.

Gonon, F. G. (1988). Nonlinear relationship between impulse flow and dopamine released by rat midbrain dopaminergic neurons as studied by *in vivo* electrochemistry. *Neuroscience* **24**, 19–28.

Goodenough, D. R., Shapiro, A., Holden, M., and Steinschriber, L. (1959). A comparison of "dreamers" and "nondreamers": eye movements, electroencephalograms, and the recall of dreams. *J. Abnorm. Soc. Psychol.* **59**, 295–302.

Gottesmann, C. (1964). Données sur l'activité corticale au cours du sommeil profond chez le Rat. *C. R. Soc. Biol.* **158**, 1829–1834.

Gottesmann, C. (1967). Recherche sur la psychophysiologie du sommeil chez le Rat. Presses du Palais Royal, Paris (still available, discussion and summary in English).

Gottesmann, C. (1969). Etude sur les activités électrophysiologiques phasiques chez le Rat. *Physiol. Behav.* **4**, 495–504.

Gottesmann, C. (1970). La psychophysiologie du sommeil. *Bull. Psychol.* **24**, 520–528.

Gottesmann, C. (1971). Psychophysiologie du sommeil. *Ann. Psychol.* **71**, 451–488.

Gottesmann, C. (1972). Le stade intermédiaire du sommeil chez le Rat. *Rev. EEG Neurophysiol.* **3**, 65–68.

Gottesmann, C. (1996). The transition from slow wave sleep to paradoxical sleep: evolving facts and concepts of the neurophysiological processes underlying the intermediate stage of sleep. *Neurosci. Biobehav. Rev.* **20**, 367–387.

Gottesmann, C. (1999). Neurophysiological support of consciousness during waking and sleep. *Prog. Neurobiol.* **59**, 469–508.

Gottesmann, C. (2001). The golden age of rapid eye movement sleep discoveries. I. Lucretius-1964. *Prog. Neurobiol.* **65**, 211–287.

Gottesmann, C. (2005a). Dreaming and schizophrenia. A common neurobiological background. *Sleep Biol. Rhythms* **3**, 64–74.

Gottesmann, C. (2005b). The Golden Age of Rapid Eye Movement Sleep Discoveries. 1965–1966. Nova Science Publishers, Inc., New York.

Gottesmann, C. (2006a). The dreaming sleep stage: a new neurobiological model of schizophrenia? *Neuroscience* **140**, 1105–1115.

Gottesmann, C. (2006b). Neurobiological disturbances during the dreaming-sleep stage can explain dream bizarreness without recourse to censorship. *Neuro-Psychoanalysis* **8**, 27–32.

Gottesmann, C. (2007a). A neurobiological history of dreaming. In: The Science of Dreaming (P. McNamara and D. Barrett, eds.), Vol. 1, Greenwood Publishers, Westport, pp. 1–51.

Gottesmann, C. (2007b). Schizophrenia: a conjectured daytime psychobiological invasion by rapid eye movement (REM) sleep. Schizophrenia Research Forum.

Gottesmann, C. (2009). Neurobiology of conscious and unconscious processes during waking and sleep. *Psyche* **15**, 92–108.

Gottesmann, C. (2010a). Historical perspective: the development of the science of dreaming. *Inter. Rev. Neurosci.*

Gottesmann, C. (2010b). Neurobiology of unconscious processes during waking and dreaming. In: Psychiatry Research Trends: Dreams and Geriatric Psychiatry (D. M. Montez, ed.), Vol. 1, In press, Nova Science Publishers, New York.

Gottesmann, C., Gandolfo, G., and Zernicki, B. (1984). Intermediate stage of sleep in the cat. *J. Physiol. Paris* **79**, 359–372.

Gottesmann, C., and Gottesman, I. I. (2007). The neurobiological characteristics of the rapid eye movement (REM) dreaming sleep stage are candidate endophenotypes of depression, schizophrenia, mental retardation and dementia. *Prog. Neurobiol.* **81**, 237–250.

Gottesmann, C., User, P., and Gioanni, H. (1980). Sleep: a physiological cerveau isolé stage? *Waking Sleeping* **4**, 111–117.

Grace, A. A. (2000). Gating information flow within the limbic system and the pathophysiology of schizophrenia. *Brain Res. Rev.* **31**, 330–341.

Greenberg, R., and Leiderman, P. H. (1966). Perceptions, the dream process and memory. An up-to-date version of notes on a mystic writing pad. *Compreh. Psychiat.* **7**, 517–523.

Gresham, S. C., Agnew, H. W. J., and Williams, R. L. (1965). The sleep of depressed patients. An EEG and eye movement study. *Arch. Gen. Psychiat.* **13**, 503–507.

Griffin, J., and Tyrrell, I. (2003). Human givens: A new approach to emotional health and clear thinking HG Publishing, Paris.

Gross, J., Byrne, J., and Fisher, C. (1965). Eye movements during emergent stage 1 EEG in subjects with lifelong blindness. *J. Neural. Transm.* **141**, 365–370.

Gross, D. W., and Gotman, J. (1999). Correlation of high-frequency oscillations with the sleep-wake cycle and cognitive activity in humans. *Neuroscience* **94**, 1005–1018.

Grubar, J. C. (1983). Sleep in mental deficiency. *Rev. EEG Neurophysiol.* **13**, 107–114.

Hartman, D., Crisp, A. H., Sedwick, P., and Borrow, S. (2001). Is there a dissociative process in sleepwaking and night terrors. *Postgrad. Med. J.* **77**, 244–249.

Hartmann, E. (1967). The Biology of Dreaming. Ch. C. Thomas, Springfield.

Hartmann, E. (2007). The nature and functions of dreaming. In the new science of dreaming. (D. Barret and P. McNamera, eds.), praeger, westport, pp. 171–192.

Hartmann, E., Verdone, P., and Snyder, F. (1966). Longitudinal studies and dreaming patterns in psychiatric patients. *J. Neural. Transm.* **142**, 117–126.

Hawkins, D. R. (1966). A review of psychoanalytic dream theory in the light of recent psychophysiological studies of sleep and dreaming. *Br. J. Med. Psychol.* **39**, 85–104.

Helmholtz, Hv. (1860). Handbuch der physiologischen Optik ('Unconscious Conclusions', p 601, quoted by Pavlov, I.: Oeuvres Choisies p 225). Leipzig, Voss.

Hennevin, E., Hars, B., Maho, C., and Bloch, V. (1995). Processing of learned information in paradoxical sleep: relevance for memory. *Behav. Brain Res.* **69**, 125–135.

Hennevin, E., Huetz, C., and Edeline, J. M. (2007). Neural representations during sleep: from sensory processing to memory traces. *Neurobiol. Learn. Mem.* **87**, 416–440.

Heresco-Levy, U. (2000). N-methyl-D-aspartate (NMDA) receptor-based treatment approaches in schizophrenia: the first decade. *Int. J. Neuropharmacol.* **3**, 243–258.

Herman, J. H., Erman, M., Boys, R., Peiser, L., Taylor, M. E., and Roffwarg, H. (1984). Evidence for a directional correspondence between eye movements and dream imagery in REM sleep. *Sleep* **7**, 52–63.

Hernandez-Peon, R. (1966). A neurophysiological model of dreams and hallucinations. *J. Neural. Transm.* **141**, 623–650.

Hernandez-Peon, R. (1967). Neurophysiology, phylogeny, and functional significance of dreaming. In: Physiological Correlates of Dreaming (E. Neurol., ed.), Vol. Suppl. 4, Academic Press, New York, pp. 106–125.

Hernandez-Peon, R., and Sterman, M. B. (1966). Brain functions. *Ann. Rev. Psychol.* **17**, 363–394.

Hervey de Saint Denys, M. J. L. (1867). Les rêves et les moyens de les diriger. Amyot, Paris (republished (1964) Tchou, Paris).

Hirskowitz, M., and Schmidt, M. H. (2005). Sleep-related erections: clinical perspectives and neural mechanisms. *Sleep Med. Rev.* **9**, 311–329.

Hobson, J. A. (2004). Freud returns? Like a bad dream. *Sci. Am.* **290**, 89.

Hobson, J. A., and McCarley, R. W. (1977). The brain as a dream state generator: an activation-synthesis hypothesis of the dream process. *Am. J. Psychiat.* **134**, 1335–1348.

Hobson, J. A., McCarley, R. W., and Wyzinski, P. W. (1975). Sleep cycle oscillation: reciprocal discharge by two brainstem neuronal groups. *Science* **189**, 55–58.

Hobson, J. A., and Pace-Schott, E. (1999). Response to commentaries. *Neuro-Pasychoanalysis* **1**, 206–224.

Hobson, J. A., and Pace-Schott, E. F. (2002). The cognitive neuroscience of sleep: neuronal systems, consciousness and learning. *Nat. Rev. Neurosci.* **3**, 679–693.

Hobson, J. A., Stickgold, R., and Pace-Schott, E. F. (1998). The neuropsychology of REM sleep dreaming. *NeuroReport* **9**, R1–R14.

Hofle, N., Paus, T., Reutens, D., Fiset, P., Gotman, J., Evans, A. C., and Jones, B. E. (1997). Regional cerebral blood flow changes as a function of delta and spindle activity during slow wave sleep. *J. Neurosci.* **17**, 4800–4808.

Hong, C. C. H., Gillin, J. C., Dow, B. C., Wu, J., and Buschbaum, M. S. (1995). Localized and lateralized cerebral glucose metabolism associated with eye movements during REM sleep and wakefulness: a positron emission tomography (PET) study. *Sleep* **18**, 570–580.

Hong, C. C. H., Harris, J. C., Pearlson, G. D., Kim, J.-S., Calhoun, V. C., Fallon, J. H., Golay, X., Gillen, J. S., Simmonds, D. J., van Zijl, P. C. M., Zee, D. S., and Pekar, J. J. (2009). fMRI evidence for multisensory recruitment associated with rapid eye movements during sleep. *Hum. Brain Map.* **30**, 1705–1722.

Hong, C. C. H., Potkin, S. G., Antrobus, J. S., Dow, B. C., Callaghan, R., and Gillin, J. C. (1997). REM sleep eye movements counts correlate with visual imagery in dreaming: a pilot study. *Psychophysiology* **34**, 377–381.

Hugelin, A., and Bonvallet, M. (1957). Etude expérimentale des interrelations réticulo-corticales: proposition d'une théorie de l'asservissement réticulaire à un contrôle diffus cortical. *J. Physiol. Paris* **49**, 1201–1223.

Iwama, K., Kawamoto, T., Sakkakura, H., and Kasamatsu, T. (1966). Responsiveness of cat lateral geniculate at pre- and postsynaptic levels during natural sleep. *Physiol. Behav.* **1**, 45–53.

Jackson, M. E., Frost, A. S., and Moghaddam, B. (2001). Stimulation of prefrontal cortex at physiologically relevant frequencies inhibits dopamine release in nucleus accumbens. *J. Neurochem.* **78**, 920–923.

Ji, D., and Wilson, M. (2007). Coordinated memory replay in the visual cortex and hippocampus during sleep. *Nat. Neurosci.* **10**, 100–107.

Johnson, B. (2006). Commentary on "Freudian dream theory, dream bizarreness, and the disguise-censor controversy." *Neuro-Psychoanalysis* **8**, 33–40.

Jones, E. (1953/1958). La vie et l'oeuvre de Sigmund Freud. La jeunesse de Freud (1856–1900). Presses Universitaires de France, Paris.

Jones, R. M. (1968). The psychoanalytic theory of dreaming-1968. *J. Neural. Transm.* **147**, 587–604.

Jones, B. (2003). Arousal systems. *Front. Biosc.* **8**, s438–s451.

Jones, B. E. (2004). Activity, modulation and role of basal forebrain cholinergic neurons innervating the cerebral cortex. *Prog. Brain Res.* **145**, 145–157.

Jouvet, M. (1965). Paradoxical sleep: a study of its nature and mechanisms. In: Sleep Mechanisms (K. B. Akert, J. P. Schadé, eds.), Vol. 18, pp. 20–62, Progress in Brain Research. Elsevier, Amsterdam.

Jouvet, M. (1992). Le sommeil et le rêve. Jacob, Paris, p. 220.

Jouvet, M. (1998). Paradoxical sleep as a programming system. *J. Sleep Res.* **7** (Suppl. 1), 1–5.

Jouvet, M., and Michel, F. (1959). Corrélations électromyographiques du sommeil chez le Chat décortiqué et mésencéphalique chronique. *C. R. Soc. Biol.* **153**, 422–425.

Jouvet, M., Michel, F., and Courjon, J. (1959). Sur un stade d'activité électrique rapide au cours du sommeil physiologique. *C. R. Soc. Biol.* **153**, 1024–1028.

Kaada, R. R. (1951). Somato-motor autonomic and electrocorticographic responses to electrical stimulation of "rhinencephalic"and other structures in primates, cat and dog. *Acta Physiol. Scand.* **83**(Suppl. 24), 1–285.

Kahn, D., and Hobson, J. A. (2005). State-dependent thinking: a comparison of waking and dreaming thoughts. *Conscious. Cogn.* 429–438.

Kales, A., and Jacobson, A. (1967). Mental activity during sleep: recall studies, somnambulism, and effects of rapid eye movement deprivation and drugs. *Exp. Neurol.* (Suppl. 4), 81–91.

Kandel, E. R. (1999). Biology and the future of psychoanalysis: a new intellectual framework for psychiatry revisited. *Am. J. Psychiat.* **156**, 505–524.

Karacan, I. (1965). Effect of Exciting Presleep Events on Dream Reporting and Penile Erections during Sleep. Brooklyn, New York.

Kaufman, J., Wehrle, R., Wetter, T. C., Holsboer, F., Auer, D. P., Pollmacher, T., and Czisch, M. (2006). Brain activation and hypothalamic functional connectivity during non-rapid eye movement sleep: an EEG/fMRI study. *Brain* **129**, 655–667.

Kawamura, H., and Sawyer, C. H. (1965). Elevation in brain temperature during paradoxical sleep. *Science* **150**, 912–913.

Kelly, P. H. (1998). Defective inhibition of dream event memory formation: a hypothesized mechanism in the onset and progression of symptoms of schizophrenia. *Brain Res. Bull.* **46**, 189–197.

Khateb, A., fort, P., Pegna, A., Jones, B., and Mühlethaler, M. (1995). Cholinergic nucleus basalis neurons are excited by histamine *in vitro. Neuroscience* **69**, 495–506.

Kindt, M., Soeter, M., and Verliet, B. (2009). Beyond extinction: erasing human fear responses and preventing the return of fear. *Nat. Neurosci.* **12**, 256–258.

Kirschbaum, C., Wolf, O. T., May, M., Wippich, W., and Hellhammer, D. H. (1996). Stress- and treatment-induced elevations of cortisol levels associated with impaired declarative memory in healthy adults. *Life Sci.* **58**, 1475–1483.

Kisley, M. A., Olincy, A., Robbins, E., Polk, S. D., Adler, L. E., Waldo, M. C., and Freedman, R. (2003). Sensory gating impairement associated with schizophrenia persists into REM sleep. *Psychophysiology* **40**, 29–38.

Kiyono, S., and Iwama, K. (1965). Phasic activity of cat's cerebral cortex during paradoxical sleep. *Med. J. Osa. Univers.* **16**, 149–159.

Klaue, R. (1937). Die bioelektrische Tätigkeit der Grosshirnrinde im normalen Schlaf und in der Narkose durch Schlafmittel. *J. Psychol. Neurol.* **47**, 510–531.

Kleinlogel, H. (1990). Analysis of the vigilance stages in the rat by fast Fourier transformation. *Neuropsychobiology* **23**, 197–204.

Kleitman, N. (1939). Sleep and Wakefulness. As Alternating Phases in the Cycle of Existence. The University Of Chicago Press, Chicago.

Kosofsky, B. E., and Molliver, M. E. (1987). The serotonergic innervation of cerebral cortex: different classes of axon terminals arise from dorsal and median raphe nuclei. *Synapse* **1**, 153–168.

Koukkou, D., and Lehman, D. (1983). Dreaming. The functional state-shift hypothesis. A neuropsychophysiological model. *Br. J. Psychiat.* **142**, 221–231.

Koulack, D., and Goodenough, D. (1977). Model for the recall of dreams upon waking, proposed to account for impairement of memory of dreams. *Ann. Med. Psychol.* **1**, 35–42.

Krnjevic, K., and Phillis, J. W. (1963). Actions of certain amines on cerebral cortex neurons. *Br. J. Pharmacol.* **20**, 471–490.

Krnjevic, K., Randic, M., and Straughan, D. W. (1966). Pharmacology of cortical inhibition. *J. Physiol. (Lond.)* **184**, 78–105.

Kroft, W., and Kuschinsky, K. (1991). Electroencephalographic correlates of the sedative effects of dopamine agonists presumably acting on autoreceptors. *Neuropharmacology* **30**, 953–960.

Kubicki, M., Styner, M., Gerig, G., Markant, D., Smith, K., MacCarley, R. W., and Shenton, M. E. (2008). Reduced interhemispheric connectivity in schizophrenic-tractography based segmentation of the corpus callosum. *Schizophr. Res.* **106**, 125–131.

Kurosawa, M., Sato, A., and Sato, Y. (1989). Stimulation of the nucleus basalis of Meynert increases acetylcholine release in the cerebral cortex in rats. *Neurosci. Lett.* **98**, 45–50.

Lairy, G. C. (1966). Données récentes sur la physiologie et la physiopathologie de l'activité onirique. Excerpta Medica International Congress Series. Madrid, pp. 8–9.

Lairy, G. C., Barros-ferreira, M., and Goldsteinas, L. (1968). Données récentes sur la physiologie et la physiopathologie de l'activité onirique. In: The Abnormalities of Sleep in Man (E. L. H. Gastaut, G. Berti Ceroni, G. Coccagna, ed.), Aulo Gaggi, Bologna, pp. 275–283.

Laplanche, J., and Pontalis, J. B. (2006). The Language of Psychoanalysis. Karnac Books, London.

Larsen, J. P., and Tandberg, E. (2001). Sleep disorders with Parkinson's disease: epidemiology and management. *CNS Drugs* **15**, 267–275.

Lauriat, T. L., Dracheva, S., Chin, B., Schmeidler, J., McInnes, L. A., and Haroutunian, V. (2005). Quantitative analysis of glutamate transporter mRNA expression in prefrontal and primary visual cortex in normal and schizophrenic brain. *Neuroscience* **137**, 843–851.

Lavin, A., and Grace, A. A. (2001). Stimulation of D_1-type dopamine receptors enhances excitability in prefrontal cortical pyramidal neurons in a state-dependent manner. *Neuroscience* **104**, 335–346.

Lehmann, C., Mueller, T., Federspiel, A., Hubl, D., Schroth, G., Huber, W., Strik, W., and Dierks, T. (2004). Dissociation between overt unconscious face processing in fusiform face area. *NeuroImage* **21**, 75–83.

Léna, I., Parrot, S., Deschaux, O., Muffat, S., Sauvinet, V., Renaud, B., Suaud-Chagny, M. F., and Gottesmann, C. (2005). Variations in the extracellular levels of dopamine, noradrenaline, glutamate and aspartate across the sleep-wake cycle in the medial prefrontal cortex and nucleus accumbens of freely moving rats. *J. Neurosci. Res.* **81**, 891–899.

Levy, B. J., and Anderson, M. C. (2008). Individual differences in the suppression of unwanted memories: the executive deficit hypothesis. *Acta Psychol.* **127**, 623–635.

Levy, R. B., Reves, A. D., and Aoki, C. (2006). Nicotinic and muscarinic reduction of unitary excitatory post synaptic potentials in sensory cortex: dual intracellular recording *in vitro*. *J. Neurophysiol.* **95**, 2155–2166.

Lewis, H. B., Goodenough, D. R., Shapiro, A., and Sleser, I. (1966). Individual differences in dream recall. *J. Abnorm. Psychol.* **71**, 52–59.

Libet, B., and Gleason, G. A. (1994). The human locus coeruleus and anxiogenesis. *Brain Res.* **634**, 178–180.

Lin, J. S., Roussel, B., Akaoka, H., Fort, P., Debilly, G., and Jouvet, M. (1992). Role of catecholamines in the modafinil and amphetamine induced wakefulness, a comparative pharmacological study in the cat. *Brain Res.* **591**, 319–326.

Lin, J. S., Sakai, K., and Jouvet, M. (1994). Hypothalamo-preoptic histaminergic projections in sleep-wake control in the cat. *Eur. J. Pharmacol.* **6**, 618–625.

Linner, L., Wiker, C., Wadenberg, M. L., Schalling, M., and Svensson, T. H. (2002). Noradrenaline reuptake inhibition enhances the antipsychotic-like effect of raclopride and potentiates D2-blockade-induced dopamine release in the medial prefrontal cortex of the rat. *Neuropsychopharmacology* **27**, 691–698.

Llinas, R., and Paré, D. (1991). Of dreaming and wakefulness. *Neuroscience* **44**, 521–535.

Llinas, R., and Ribary, U. (1993). Coherent 40 Hz oscillation characterizes dream state in humans. *Proc. Nat. Acad. Sci. USA* **90**, 2078–2081.

Loomis, A. L., Harvey, E. N., and Hobart, G. A. I. (1937). Cerebral states during sleep, as studied by human brain potentials. *J. Exp. Psychol.* **21**, 127–144.

Louie, K., and Wilson, M. A. (2001). Temporally structured replay of awake hippocampal ensemble activity during rapid eye movement sleep. *Neuron* **29**, 145–156.

Luciana, M., Collins, P. F., and Depue, R. A. (1998). Opposing role for dopamine and serotonin in the modulation of human spatial working memory functions. *Cereb. Cortex* **8**, 218–226.

Lucrece, T. C. (1900). De rerum natura OXONII. E. Translation Baily, C. Topographeo Clarendoniano, Cambridge University Press, Cambridge, 178 pp.

MacKay, A. V., Iversen, L. L., Rossor, M., Spokes, E., Bird, E., Arregui, A., and Snyder, S. (1982). Increased brain dopamine and dopamine receptors in schizophrenia. *Arch. Gen. Psychiat.* **39**, 991–997.

Madsen, P. L., Holm, S., Vorstrup, S., Friberg, L., Lassen, N. A., and Wildschiodtz, G. (1991). Human regional cerebral blood flow during rapid-eye-movement sleep. *J. Cerbr. Blood Flow Met.* **11**, 502–507.

Magnes, J., Moruzzi, G., and Pompeiano, O. (1961). Synchronization of the EEG produced by low frequency electrical stimulation of the region of the solitary tract. *Arch. Ital. Biol.* **99**, 33–67.

Maire, F. W., and Patton, H. D. (1954). Hyperactivity and pulmonary edema from rostral hypothalamic lesions in rats. *Am. J. Physiol.* **178**, 315–320.

Malcolm-Smith, S., Solms, M., Turnbull, O., and Tredoux, C. (2007). Threat in dreams: an adaptation. *Conscious. Cogn.* **17**, 1281–1291.

Manacéine, Md. (1897). Sleep: Its Physiology, Pathology, Hygiene, and Psychology. Walter Scott, London.

Manceau, A., and Jorda, M. (1948). Les centres mésodiencéphaliques du sommeil. *Sem. Hop.* **24**, 3193–3202.

Mancia, M. (2003). Dream actors in the theatre of memory: their role in the psychoanalytic process. *Int. J. Psychanal.* **84**, 945–952.

Manunta, Y., and Edeline, J. M. (1999). Effects of noradrenaline on frequency tuning of auditory cortex neurons during wakefulness and slow wave sleep. *Eur. J. Neurosci.* **11**, 2134–2150.

Maquet, P., Delguedre, C., Delfiore, G., Aerts, J., Peters, J. M., Luxen, A., and Franck, G. (1997). Functional neuroanatomy of human slow wave sleep. *J. Neurosci.* **17**, 2807–2812.

Maquet, P., and Franck, G. (1997). REM sleep and amygdala. *Mol. Psychiat.* **2**, 195–196.

Maquet, P., Laureys, S., Peigneux, P., Fuchs, S., Petiau, C., Phillips, C., Aerts, J., Del Fiore, G., Delguedre, C., Meulemans, T., Luxen, A., Franck, G., Van Der Linden, M., Smith, C., and Cleeremans, A. (2000). Experience-dependent changes in cerebral activation during human REM sleep. *Nat. Neurosci.* **3**, 831–836.

Maquet, P., Peters, J. M., Aerts, J., Delfiore, G., Degueldre, C., Luxen, A., and Franck, G., (1996). Functional neuroanatomy of human rapid-eye-movement sleep and dreaming. *Nature* **383**, 163–166.

Maquet, P., Ruby, P., Schw artz, S., Laurey, S., Albouy, T., Dang-Vu, T., Desseilles, M., Boly, M., Melchior, G., and Peigneux, P. (2004). Regional organisation of brain activity during paradoxical sleep. *Arch. Ital. Biol.* **142**, 413–419.

Marrosu, F., Portas, C., Mascia, M. F., Casu, M. A., Fa, M., Giagheddu, M., Imperato, A., and Gessa, G. L. (1995). Microdialysis measurement of cortical and hippocampal acetylcholine release during sleep-wake cycle in freely moving cats. *Brain Res.* **671**, 329–332.

Massimini, M., Ferrarelli, F., Huber, R., Esser, S. K., Singh, H., and Tononi, G. (2005). Breakdown of cortical effective connectivity during sleep. *Science* **309**, 2228–2232.

Maury, F. (1861). Le sommeil et les rêves. Didier ed., p. 156.

McCarley, R. W., and Hobson, J. A. (1971). Single neuron activity in giganto cellular tegmental field: selectivity of discharge in desynchronized sleep. *Science* **174**, 1250–1252.

McCarley, R. W., and Hobson, J. A. (1977). The neurobiological origin of psychoanalytic dream theory. *Am. J. Psychiat.* **134**, 1211–1221.

McCarley, R. W., and Ito, K. (1983). Intracellular evidence linking medial pontine reticular formation neurons to PGO wave generation. *Brain Res.* **280**, 343–348.

McCarley, R., Nelson, J. P., and Hobson, J. A. (1978). Ponto-geniculo-occipital(PGO) burst neurons: correlative evidence for neuronal generator waves. *Science* **201**, 269–272.

McCormick, D. A. (1992). Neurotransmitter actions in the thalamus and cerebral cortex and their role in neuromodulation of thalamocortical activity. *Prog. Neurobiol.* **39**, 337–388.

McGinty, D. J., and Harper, R. M. (1976). Dorsal raphe neurons: depression of firing during sleep in cats. *Brain Res.* **101**, 569–575.

McGinty, D. J., Harper, R. M., and Fairbanks, M. K. (1974). Neuronal unit activity and the control of sleep states. In: Advances in Sleep Research (E. D. Weitzman, ed.), Vol. 1, Spectrum, New York, pp. 173–216.

McGinty, D., and Sterman, M. (1968). Sleep suppression after basal forebrain lesions in the cat. *Science* **160**, 1253–1255.

Meulders, M., Massion, J., Colle, J., and Albe-Fessard, D. (1963). Effets d'ablations télencéphaliques sur l'amplitude des potentiels évoqués dans le centre médian par stimulation somatique. *Electroencephalogr. Clin. Neurophysiol.* **15**, 29–38.

Meyer-Lindenberg, A., Olsen, R. K., Kohn, P. D., Brown, T., Egan, M. F., Weinberber, D. R., and Berman, K. F. (2005). Regionally specific disturbance of dorsolateral prefrontal-hippocampal function connectivity in schizophrenia. *Arch. Gen. Psychiat.* **62**, 379–386.

Meyer-Lindenberg, A., Poline, J. B., Kohn, P. D., Holt, J. L., Egan, M. F., Weinberger, D. R., and Berman, K. F. (2001). Evidence for abnormal cortical functional connectivity during working memory in schizophrenia. *Am. J. Psychiat.* **158**, 1809–1817.

Miller, J. D., Farber, J., Gatz, P., Roffwarg, H., and German, D. (1983). Activity of mesencephalic dopamine and non-dopamine neurons across stages of sleep and waking in the rat. *Brain Res.* **273**, 133–141.

Milstein, J. A., Lehmann, O., Theobald, D. E., Dalley, J. W., and Robbins, E. (2007). Selective depletion of cortical noradrenaline by ant-domaine beta-hydroxylase-saporin impairs attentional function and enhances the effect of guafacine in the rat. *Psychophamacology* **190**, 51–63.

Miyauchi, S., Misaki, M., Kan, S., Fukunaga, T., and Koike, T. (2009). Human brain activity time-locked to rapid eye movements during REM sleep. *Exp. Brain Res.* **192**, 657–667.

Monod, J. (1970). Le hasard et la nécessité. Le Seuil, Paris.

Monroe, L. J., Rechtschaffen, A., Foulkes, D., and Jensen, J. (1965). Discriminability of REM and NREM reports. *J. Person. Soc. Psychol.* **2**, 456–460.

Montgomery, S. M., Sirota, A., and Buzsaki, G. (2008). Theta and gamma coordination of hippocampal networks during waking and rapid eye movement sleep. *J. Neurosci.* **28**, 6731–6741.

Monti, J. M., Jantos, H., and Fernandez, M. (1989). Effects of the selective D-2 receptor agonist, quipirole on sleep and wakefulness in the rat. *Eur. J. Pharmacol.* **169**, 61–66.

Morrison, A. R., and Sanford, L. D. (2003). Critical brain characteristics to consider in developing dream and memory theories. In: Sleep and Dreaming. Scientific Advances and Reconsiderations (E. Pace-Schott, M. Solms, M. Blagrove, and S. Harnad, eds.), Cambridge University Press, Cambridge, pp. 189–190.

Moruzzi, G., and Magoun, H. W. (1949). Brain stem reticular formation and activation of the EEG. *Electroencephalogr. Clin. Neurophysiol.* **1**, 455–473.

Moxon, K. A., Devilbiss, D. M., Chapin, J. K., and Waterhouse, B. D. (2007). Influence of norepinephrine on somatosensory neuronal responses in the rat thalamus: a combined modeling and *in vivo* multi-channel, multi-neuron recording study. *Brain Res.* **1147**, 105–123.

Muzur, A., Pace-Schott, E., and Hobson, J. A. (2002). The prefrontal cortex in sleep. *Trends Cogn. Sci.* **6**, 475–481.

Naccache, L., Gaillard, R., Adam, C., Hasboun, D., Clémenceau, S., Baulac, M., Dehaene, S., and Cohen, L. (2005). A direct record of emotions evoked by subliminal words. *Proc. Nat. Acad. Sci. USA* **102**, 7713–7717.

Nauta, W. J. H. (1946). Hypothalamic regulation of sleep in rats: an experimental study. *J. Neurophysiol.* **9**, 285–316.

Neckelmann, D., and Ursin, R. (1993). Sleep stages and EEG power spectrum in relation to acoustical stimulus threshold in the rat. *Sleep* **16**, 467–477.

Nelson, C. N., Hoffer, B. J., Chu, N. S., and Bloom, F. E. (1973). Cytochemical and pharmacological studies on polysensory neurons in the primate frontal cortex. *Brain Res.* **62**, 115–133.

Newman, E. A., and Evans, C. R. (1965). Human dream processes as analogous to computer programme clearance. *Nature* **206**, 534.

Nielsen, T. (2000). Cognition in REM and NREM sleep: a review and possible reconciliation of two models of sleep mentation. *Behav. Brain Sci.* **23**, 851–866.

Nielsen, J. B. (2003). A review of mentation in REM and NREM sleep: "covert" REM sleep as a possible reconciliation of two opposing models. In: Sleep and Dreaming. Scientific Advances and Reconsiderations (E. Pace-Schott, M. Solms, M. Blagrove, and S. Harnad, eds.) Cambridge University Press, Cambridge, pp. 59–74.

Nielsen, T. A., and Stenstrom, P. (2005). What are the memory sources of dreaming. *Nature* **437**, 1286–1289.

Nielsen, T., Stenstrom, P., Takeuchi, T., Saucier, S., Lara-Carrasco, B. A., solomonova, E., and Martel, E. (2005). Partial REM-sleep deprivation increases the dream-like quality of mentation from REM sleep and sleep onset. *Sleep* **28**, 1083–1089.

Niiyama, Y., Fujiwara, R., Satoh, N., and Hishikawa, Y. (1994). Endogenous components of event-related potential during NREM stage 1 and REM sleep in man. *Int. J. Psychophysiol.* **17**, 165–174.

Nofzinger, E. A., Mintun, M. A., Wiseman, M. B., and Kupfer, J. (1997). Forebrain activation during REM sleep: an FDG PET study. *Brain Res.* **770**, 192–201.

O'Hara, B. F., Young, K. A., Watson, F. L., Heller, H. C., and Kilduff, T. S. (1993). Immediate early gene expression in brain during sleep deprivation: preliminary observations. *Sleep* **16**, 1–7.

Oksenberg, A., Arons, E., Sazbon, L., Mizrahi, A., and Radwan, H. (2000). Sleep-related erections in vegetative state patients. *Sleep* **23**, 953–957.

Pace-Schott, E. (2003). Recent findings on the neurobiology of sleep and dreaming. In: Sleep and Dreaming. Scientific Advances and Reconsiderations (E. Pace-Schott, M. Solms, M. Blagrove, and S. Harnad, eds.), Cambridge University Press, Cambridge, pp. 335–350.

Pace-Schott, E. (2007). The frontal lobes and dreaming. In: The New Science of Dreaming (D. Barrett and P. McNamara, eds.) Vol. 1, Praeger, Westport, pp. 115–154.

Park, S. P. (2002). In vovo microdialysis measures of extracellular norepinephrine in the rat amygdala during sleep-wakefulness. *J. Korean Med. Sci.* **17**, 395–399.

Parmeggiani, P. L. (2007). REM sleep related increase in brain temperature: a physiologic problem. *Arch. Ital. Biol.* **145**, 13–21.

Parsons, L. M., Sergent, J., Hodges, D. A., and Fox, P. T. (2005). The brain basis of piano performance. *Neuropsychologia* **43**, 199–215.

Payne, J. L., and Nadel, L. (2004). Sleep, dreaming and memory consolidation: the role of the stress hormone cortisol. *Learn. Mem.* **11**, 671–678.

Peigneux, P., Laureys, S., Fuchs, S., Delbeuck, X., Degueldre, C., Aerts, J., Delfiore, G., Luxen, A., and Maquet, P. (2001). Generation of rapid eye movements during paradoxical sleep in humans. *NeuroImage* **14**, 701–708.

Peled, A., Geva, A. B., Kremen, W. S., Blankfeld, H. M., Esfandiarfard, R., and Nordahl, T. E. (2000). Functional connectivity and working memory in schizophrenia: an EEG study. *Int. J. Neurosci.* **106**, 47–61.

Penfield, W. (1958). The Excitable Cortex in Conscious Man. University Press, (quoted by Yu 2001) Liverpool.

Penfield, W., and Perot, P. (1963). The brain's record of auditory and visual experience. *Brain* **86**, 597–697. Quoted by Yu (2001).

Pepeu, G., and Bartholini, A. (1968). Effect of psychoactive drugs on the output of acetylcholine from the cerebral cortex of the cat. *Eur. J. Pharmacol.* **4**, 254–263.

Perez-Garci, E., del Rio-Portilla, Y., Guevara, M. A., Arce, C., and Corsi-cabrera, M. (2001). Paradoxical sleep is characterized by uncoupled gamma activity between frontal and perceptual cortical regions. *Sleep* **24**, 118–126.

Perry, E. K., and Piggott, M. A. (2003). Neurotransmitter mechanisms of dreaming: implication of modulatory systems based on dream intensity. In: Sleep and Dreaming. Scientific Advances and Reconsiderations (E. Pace-Schott, M. Solms, M. Blagrove, and S. Harnad, eds.), Cambridge University Press, Cambridge, pp. 202–204.

Perry, E., Walker, M., Grace, J., and Perry, R. (1999). Acetylcholine in mind: a neurotransmitter of consciousness? *Trends Neurosci.* **22**, 273–280.

Petocz, A. (2006). Short-changed on the Freudian metaphors. *Neuro-Psychoanalysis* **8**, 48–53.

Petre-Quadens, O., and Jouvet, M. (1967). Sleep in the mentally retarded. *J. Neurol. Sci.* **4**, 354–357.

Phillis, J. W., and Chong, G. C. (1965). Acetylcholine release from the cerebral and cerebellar cortices: its role in cortical arousal. *Nature* **207**, 1253–1255.

Phillis, J. W., Lake, N., and Yarbrough, G. (1973). Calcium mediation of the inhibitory effects of biogenic amines on cerebral critical neurons. *Brain Res.* **53**, 465–469.

Piallat, B., and Gottesmann, C. (1995). The reticular arousal threshold during the transition from slow wave sleep to paradoxical sleep in the rat. *Physiol. Behav.* **58**, 199–202.

Pompeiano, M., Cirelli, C., Arrighi, P., and Tononi, G. (1995). C-fos expression during wakefulness and sleep. *Neurophysiol. Clin.* **25**, 329–541.

Pompeiano, M., Cirelli, C., and Tononi, G. (1994). Immediate-early genes in spontaneous wakefulness and sleep: expression of c-fos and NGFI-A m RNA and protein. *J. Sleep Res.* **3**, 80–96.

Portas, C., Bjorvatn, B., Fagerland, S., Gronli, J., Mundal, V., Sorensen, E., and Ursin, R. (1998a). On-line detection of extracellular levels of serotonin in dorsal raphe nucleus and frontal cortex over the sleep/wake cycle in the freely moving rat. *Neuroscience* **83**, 807–814.

Portas, C. M., Rees, G., Howseman, A. M., Josephs, O., Turner, R., and Frith, C. D. (1998b). A specific role for the thalamus in mediating the interaction of attention and arousal in humans. *J. Neurosci.* **18**, 8979–8989.

Raedler, T., Knable, M., Jones, D., Urbina, R., Gorey, J., Lee, K., Egan, M., Coppola, R., and Weinberger, D. (2003). *In vivo* determination of muscarinic acetylcholine receptor availability in schizophrenia. *Am. J. Psychiatry* **160**, 118–127.

Rasmussen, K., Heym, J., and Jacobs, B. L. (1984). Activity of serotonin-containing neurons in nucleus centralis superior of freely moving cats. *Exp. Neurol.* **83**, 302–317.

Rauchs, G., Feyers, D., Maquet, P., and Collette, F. (2008). Does sleep favour forgetting of irrelevant information. *J. Sleep Res.* **17**(Suppl. 1), 10.

Reader, T. A., Ferron, A., Descarries, L., and Jasper, H. H. (1979). Modulatory role for biogenic amines in the cerebral cortex. Microiontopheric studies. *Brain Res.* **160**, 219–229.

Rechtschaffen, A., Vogel, G., and Shaikun, G. (1963). Interrelatedness of mental activity during sleep. *Arch. Gen. Psychiat.* **9**, 536–547.

Rees, G., Wojciulik, E., Clarke, K., Husain, M., Frith, C. D., and Driver, J. (2000). Unconscious activation of visual cortex in the damaged right hemisphere of a parietal patient with extinction. *Brain* **123**, 1624–1633.

Reeves, M., Lindholm, D. E., Myles, P. S., Fletcher, H., and Hunt, J. O. (2001). Adding ketamine to morphine for patient-controlled analgesia after major abdominal surgery: a double-blind, randomized trial. *Anesth. Analg.* **93**, 116–120.

Revonsuo, A. (2003). The reinterpretation of dreams: an evolutionary hypothesis of the function of dreaming. In: Sleep and Dreaming. Scientific Advances and Reconsiderations (E. Pace-Schott, M. Solms, M. Blagrove, and S. Harnad, eds.), Cambridge University Press, Cambridge, pp. 85–109.

Ribary, U., Ionnides, A. A., Singh, K. D., Hasson, R., Bolton, J., Lado, F., Mogilner, A., and Llinas, R. (1991). Magnetic field tomography of coherent thalamocortical 40 Hz oscillations in humans. *Proc. Nat. Acad. Sci. USA* **88**, 11037–11041.

Ribeiro, S., Mello, C. V., Vehlo, T., Gardner, T. J., Jarvis, E. D., and Pavlides, C. (2002). Induction of hippocampal long-term potentiation during waking leads to increased extrahippocampal zif-268 expression during ensuing rapid-eye-movement sleep. *J. Neurosci.* **22**, 10914–10923.

Ribeiro, S., and Nicolelis, M. A. L. (2004). Reverberation, storage and postsynaptic propagation of memory during sleep. *Learn. Mem.* **11**, 687–696.

Roffwarg, H. P., Dement, W. C., Muzio, J. N., and Fisher, C. (1962). Dream imagery: relationship to rapid eye movements of sleep. *Arch. Gen. Psychiat.* **7**, 235–258.

Rose, J., Schiffer, A. M., Dittrich, L., and Güntükün, O. (2010). The role of dopamine in maintenance and distractibility of attention in the "prefrontal cortex" of pigeons. *Neuroscience* **167**, 232–237.

Ross, R. J., Ball, W. A., Dinges, D. F., Kribbs, N. B., Morrison, A. R., Silver, S. M., and Mulvaney, F. D. (1994). Rapid eye movement sleep disturbance in posttraumatic stress disorder. *Biol. Psychiat.* **35**, 195–202.

Rossi, G. F., Palestini, M., Pisano, M., and Rosadini, G. (1965). An experimental study of the cortical reactivity during sleep and wakefulness. In: Aspects anatomo-fonctionnels de la physiologie du sommeil (CNRS, ed.), CNRS, Paris, pp. 509–532.

Rush, A. J., Erman, M. K., Giles, D. E., Schlesser, M. A., Carpenter, G., Vasavada, N., and Roffwarg, H. P. (1986). Polysomnographic findings in recently drug-free and clinical remitted depressed patients. *Arch. Gen. psychiat.* **43**, 878–884.

Sakai, K. (1988). Executive mechanisms of paradoxical sleep. *Arch. Ital. Biol.* **126**, 239–257.

Sakai, K., and Crochet, S. (2003). A neural mechanism of sleep and wakefulness. *Sleep Biol. Rhythms* **1**, 29–42.

Salzarulo, P., Lairy, G. C., Bancaud, J., and Munari, C. (1975). Direct depth recording of the striate cortex during REM sleep in man: are there PGO potentials? *Electroencephalogr. Clin. Neurophysiol.* **38**, 199–202.

Salzarulo, P., Vidal, J. C., and de Barros-Ferreira, M. (1968). Etude polygraphique des phases intermédiaires du sommeil des malades mentaux comparées aux phases de mouvements oculaires. *Rev. Sperm. Freniatria* **92**, 476–495.

Sarter, M., and Bruno, J. P. (2000). Cortical cholinergic inputs mediating arousal, attentional processing and dreaming: differential afferent regulation of the basal forebrain by telencephalic and brainstem afferents. *Neuroscience* **95**, 933–952.

Satoh, T. (1971). Direct cortical response and PGO spike during paradoxical sleep of the cat. *Brain Res.* **28**, 576–578.

Schmidt, M. H., Valatx, J. L., Schmidt, H. S., Wauquier, A., and Jouvet, M. (1994). Experimental evidence of penile erections during paradoxical sleep in the rat. *NeuroReport* **5**, 561–564.

Schonbar, R. A. (1965). Differential dream recall frequency as a component of "life style." *J. Consult. Psychol.* **29**, 468–474 (quothed by Schredl 2007).

Schredl, M. (2007). Dream recall: models and empirical data. In: New Science of Dreaming (D. Barrett and P. McNamara, eds.), Vol. 2, Praeger, Westport, pp. 79–114.

Schultz, W. (2007). Multiple dopamine functions at different time course. *Ann. Rev. Neurosci.* **20**, 259–288.

Schwartz, S. (2003). Are life episodes replayed during dreaming. *Trends Cogn. Sci.* **7**, 325–327.

Schwartz, S., and Maquet, P. (2002). Sleep imaging and the neuro-psychological assessment of dreams. *Trends Cogn. Sci.* **6**, 23–30.

Selden, N. R. W., Robbins, T. W., and Everitt, B. J. (1990). Enhanced behavioral conditioning to context and impaired behavioral and neuroendocrine responses to conditioned stimuli following ceruleocortical noradrenergic lesions: support for an attentional hypothesis of central noradrenergic function. *J. Neurosci.* **10**, 531–539.

Shapiro, A., Goodenough, D., Biederman, I., and Sleser, I. (1963). Dream recall as a function ofg method of awakening. *Psychosom. Med.* **25**, 174–180.

Siegel, J. M. (2001). The REM sleep-memory consolidation hypothesis. *Science* **294**, 1058–1063.

Silver, H., Barash, I., Aharon, N., Kaplan, A., and Poyurovsky, M. (2000). Fluvoxamine augmentation of antipsychotics improves negative symptoms in psychotic chronic schizophrenic patients: a placebo-controlled study. *Int. Clin. Psychopharmacol.* **15**, 257–261.

Sloan, N., and Jasper, H. H. (1950). Studies of the regulatory functions of the limbic cortex. *Electroencephalogr. Clin. Neurophysiol.* **2**, 317–327.

Smiley, J. F., and Goldman-Rakic, P. S. (1993). Heterogeneous targets of dopamine synapses in monkey prefrontal cortex demonstrated by serial section electron microscopy: a laminar analysis using the silver-enhanced diaminobenzidine sulfite (SEDS) immunolabeling technique. *Cereb. Cortex* **3**, 223–238.

Smith, C. (1995). Sleep states and memory processes. *Behav. Brain Res.* **69**, 137–145.

Snyder, F., Hobson, J. A., and Goldfrank, F. (1963). Blood pressure changes during human sleep. *Science* **142**, 1313–1314.

Snyder, F., Hobson, J. A., Morrison, D. F., and Goldfrank, F. (1964). Changes in respiration, heart rate, and systolic blood pressure in human sleep. *J. Appl. Physiol.* **19**, 417–422.

Solms, M. (2000). Dreaming and REM sleep are controlled by different brain mechanisms. *Behav. Brain Sci.* **23**, 843–850.

Stassen, H. H., Coppola, R., Gottesman, I. I., Torrey, E. F., Kuny, S., Rickler, K. C., and Hell, D. (1999). EEG differences in monozygotic twins discordant and concordant for schizophrenia. *Psychophysiology* **36**, 109–117.

Steinbusch, H. W. M., and Mulder, A. H. (1984). Immunohistochemical localization of histamine in neurons and mast cells in the rat brain. In: Classical Transmitters and Transmitter Receptors in the CNS (T. Bjorklund, T. Hökfelt, and M. J. Kuhar, eds.), Handbook of Chemical Neuroanatomy, Elsevier, Amsterdam, pp. 126–140.

Steriade, M. (1970). Ascending control of thalamic and cortical responsiveness. *Int. Rev. Neurobiol.* **12**, 87–144.

Steriade, M. (1996). Arousal: revisiting the reticular system. *Science* **272**, 225–226.

Steriade, M., Contreras, D., Amzica, F., and Timofeev, I. (1996). Synchronization of fast (30–40 Hz) spontaneous oscillations in the intrathalamus and thalamocortical networks. *J. Neurosci.* **16**, 2788–2808.

Steriade, M., and McCarley, R. W. (1990). Brainstem Control of Wakefulness and Sleep. Plenum Press, New York, p. 267.

Suaud-Chagny, M. F., Chergui, K., Chouvet, G., and Gonon, F. (1992). Relationship between dopamine release in the rat nucleus accumbens and the discharge activity of dopaminergic neurons during local in vivo application of amino acids in the ventral tegmental area. *Neuroscience* **49**, 63–72.

Szerb, J. C. (1967). Cortical acetylcholine release and electroencephalographic arousal. *J. Physiol. (Lond.)* **192**(2), 329–343.

Szymusiak, R., McGinty, D., Shepard, D., Shouse, M. N., and Sterman, M. (1990). Effects of systemic atropine sulfate administration on the frequency content of the cat sensorimotor EEG during sleep and waking. *Behav. Neurosci.* **104**, 217–225.

Tabushi, K., Hishikawa, Y., Ueyama, M., and Kaneko, Z. (1966). Cortical D.C. potential changes associated with spontaneous sleep in the cat. *Arch. Ital. Biol.* **104**, 152–162.

Takahata, R., and Moghaddam, B. (2000). Target-specific glutamatergic regulation of dopamine neurons in the ventral tegmental area. *J. Neurochem.* **75**, 1775–1778.

Takeuchi, T., Miyasita, A., Inugami, M., and Yamamoto, Y. (2001). Intrinsic dreams are not produced without REM sleep mechanims: evidence through elicitation of sleep onset periods. *J. Sleep Res.* **10**, 43–52.

Takeuchi, T., Ogilvie, R. D., Ferrelli, A. V., Murphy, T., Yamamoto, Y., and Inugami, M. (1999). Dreams are not produced without REM sleep mechanisms. *Sleep Res. Online* **2**(Suppl. 1), 279.

Taylor, M. J., Freemantle, N., Geddes, J. R., and Bhagwagar, Z. (2006). Early onset of selective serotonin reuptake inhibitor antidepressant action: systematic review and meta-analysis. *Arch. Gen. Psychiat.* **63**, 1217–1223.

Thompson, D. F., and Pierce, D. R. (1999). Drug-indiced nightmares. *Ann. Pharmacother.* **33**, 93–98.

Tierney, P. I., Thierry, A. M., Glowinski, J., Deniau, J. M., and Gioanni, Y. (2008). Dopamine modulates temporal dynamics of feedback inhibition in rat prefrontal cortex *in vivo*. Cereb. Cortex.

Tinguely, G., Huber, R., Borbely, A. A., and Achermann, P. (2006). Non-rapid eye movement sleep with low muscle tone as a marker of rapid eye movement sleep. *BMC Neurosci.* **7**, 2.

Tracy, R. L., and Tracy, L. N. (1974). Reports of mental activity from sleep stages 2 and 4. *Perc. Mot. Skills* **38**, 647–648.

Tranel, D., and Damasio, A. R. (1985). Knowledge without awareness: an autonomic index of facial recognition by prosopagnosics. *Science* **228**, 1453–1454.

Trosman, H., Rechtschaffen, A., Offenkrantz, W., and Wolpert, E. (1960). Studies in Psychophysiology. IV. Relations among dreams in sequence. *Arch Gen Psychiatry.* **3**, 602–607.

Trulson, M. E., Jacobs, B. L., and Morrison, A. R. (1981). Raphe unit activity during REM sleep in normal cats and in pontine lesioned displaying REM sleep without atonia. *Brain Res.* **226**, 75–91.

Van Hes, R., Smid, P., Stroomer, C. N., Tipker, K., Tulp, M. T., Van der Heyden, J. A., McCreary, A. C., Hesselink, M. B., and Kruse, C. G. (2003). SL V310, a novel, potential antipsychotic, combining potent dopamine d2 receptor antagonism with serotonin reuptake inhibition. *Bioorg. Med. Chem. Lett.* **13**, 405–408.

Vanderwolf, C. H. (1988). Cerebral activity and behavior control by central cholinergic and serotonergic systems. *Int. Rev. Neurobiol.* **30**, 225–340.

Vanni-Mercier, G., Gigout, S., Debilly, G., and Lin, J. S. (2003). Waking selective neurons in the posterior hypothalamus and their response to histamine H_3-receptor ligands: an electrophysiological study in freely moving cats. *Behav. Brain Res.* **144**, 227–241.

Vanni-Mercier, G., Sakai, K., and Jouvet, M. (1984). Neurones spécifiques de l'éveil dans l'hypothalamus postérieur. *C. R. Acad. Sci.* **298**, 195–200.

Vertes, R. P. (1977). Selective firing of rat pontine gigantocellular neurons during movement and REM sleep. *Brain Res.* **128**, 146–152.

Vertes, R. P. (2004). Memory consolidation in sleep; dream or reality. *Neuron* **44**, 135–148.

Vertes, R. P., and Eastman, K. E. (2000). The case against memory consolidation in REM sleep. *Behav. Brain Sci.* **23**, 867–876.

Vertes, R., and Siegel, J. (2005). Time for the sleep community to take a critical look at the purported role of sleep in memory processing. *Sleep* **28**, 1228–1229.

Villablanca, J. (1965). The electrocorticogram in the chronic cerveau isolé cat. *Electroencephalogr. Clin. Neurophysiol.* **19**, 576–586.

Villablanca, J. (1966). Behavioral and polygraphic study of "sleep" and "wakefulness" in chronic decerebrate cats. *Electroencephalogr. Clin. Neurophysiol.* **21**, 562–577.

Vogel, G., Foulkes, D., and Trosman, H. (1966). Ego functions and dreaming during sleep onset. *Arch. Gen. Psychiat.* **14**, 238–248.

von Economo, C. (1928). Theorie du sommeil. *J. Neurol. Psychiat* **7**, 437–464.

Voss, U., Holzmann, R., Tuin, I., and Hobson, J. A. (2009). Lucid dreaming: a state of consciousness with features of both waking and non-lucid dreaming. *Sleep* **32**, 1191–1200.

Wall, P. D., Glees, P., and Fulton, J. F. (1951). Corticofugal connexions of posterior orbital surface in rhesus monkey. *Brain* **74**, 66–71.

Wang, H., Feng, R., Wang, L. P., Li, F., Cao, X., and Tsien, J. (2008). CAMKII activation state underlies synaptic labile phase of LTP and short-term memory formation. *Curr. Biol.* **18**, 1546–1554.

Wang, Z., and McCormick, D. A. (1993). Control of firing mode of corticotectal and corticopontine layer V burst-generating neurons by norepinephrine, acetylcholine and 1S, 3R-ACPD. *J. Neurosci.* **13**, 2199–2216.

Warren, R. A., and Dykes, R. W. (1996). Transient and long-lasting effects of iontophoretically administered norepinephrine on somatosensory cortical neurons in halothane-anesthetized cats. *Can. J. Physiol. Pharmacol.* **74**, 38–57.

Watanabe, T., Taguchi, Y., Shiosaka, S., Tanaka, J., Kubota, H., Terano, Y., Tohyama, M., and Wada, H. (1984). Distribution of the histaminergic neuron system in the central nervous system of rats: a fluorescent immunohistochemical analysis with histidine decarboxylase as a marker. *Brain Res.* **295**, 13–25.

Waterhouse, B. D., Azizi, S. A., Burne, R. A., and Woodward, D. (1990). Modulation of rat cortical area 17 neuronal responses to moving visual stimuli during norepinephrine and serotonin microiontophoresis. *Brain Res.* **514**, 276–292.

Wehrle, R., Czisch, M., Kaufmann, C., Wetter, T. C., Holsboer, F., Auer, D. P., and Pollmächer, T. (2005). Rapid eye movement-related brain activation in human sleep: a functional magnetic resonance imaging study. *NeuroReport* **16**, 853–857.

Wehrle, R., Kaufmann, C., Wetter, T. C., Holsboer, F., Auer, D. P., Pollmacher, T., and Czisch, M. (2007). Functional microstates within human REM sleep: evidence from fMRI of a thalamocortical network specific for phasic periods. *Eur. J. Neurosci.* **25**, 863–871.

Weinberger, D. R., Berman, K. F., and Zec, R. F. (1986). Physiological dysfunction of dorsolateral prefrontal cortex in schizophrenia. 1. Regional cerebral blood flow evidence. *Arch. Gen. Psychiat.* **43**, 114–124.

Weiss, T., and Adey, W. R. (1965). Excitability changes during paradoxical sleep in the rat. *Experentia* **21**, 1–4.

Werth, E., Achermann, P., and Borbely, A. A. (2002). Selective REM sleep deprivation during daytime: II. Muscle atonia in non-REM sleep. *Am. J. Physiol. Regul. Integr. comp. Physiol.* **283**, 527–532.

Wigren, H. K., Schepens, M., Matto, V., Stenberg, D., and Porkka-Heiskanen, T. (2007). Glutamatergic stimulation of the basal forebrain elevates extracellular adenosine and increases the subsequent sleep. *Neuroscience* **147**, 811–823.

Wikler, A. (1952). Pharmacological dissociation of behavior and EEG "sleep patterns" in dogs: morphine, N-allylmorphine and atropine. *Proc. Soc. exp. Biol. Med.* **79**, 261–265.

Wolpert, E. A., and Trosman, H. (1958). Studies in Psychophysiology of dreams. *A.M.A. Arch. Neurol. Psychiat.* **79**, 603–606.

Wurtz, R. H. (1965). Steady potential shifts during wakefulness and sleep. *Electroencephalogr. Clin. Neurophysiol.* **18**, 528.

Yamatodani, A., Inagaki, N., Panula, P., Itowi, N., Watanabe, V., and Wada, H. (1991). Structure and functions of the histaminergic neuron system. In: Handbook of Experimental Pharmacology. Histamine and Histamine Antagonists (B. Uvnäs, ed.), Vol. 97, Springer, New York, pp. 243–283.

Young, C. E., Beach, T. G., Falkai, P., and Honer, W. G. (1998). SNAP-25 deficit and hippocampal connectivity in schizophrenia. *Cereb. Cortex* **8**, 261–268.

Yu, C. K. C. (2001). Neuroanatomical correlates of dreaming: the supramarginal gyrus (dream work). *Neuro-Psychoanalysis* **2**, 47–59.

Yu, C. K. C. (2006). The dream censor: illusion or homunculus. *Neuro-Psychoanalysis* **8**, 53–59.

Zajonc, R. B. (1962). Response suppression in perceptual defense. *J. Exp. Psychol.* **64**, 206–214.

THE USE OF DREAMS IN MODERN PSYCHOTHERAPY

Clara E. Hill* and Sarah Knox†

*Department of Psychology, University of Maryland, College Park, MD, USA
†College of Education, Marquette University, Milwaukee, WI, USA

We review theories of dream work. We also reviewthe empirical research about how dreams are used in psychotherapy, as well as the process and outcome of different models of dream work. Finally, we review how dream content can be used to understand client, the role of culture in dream work, client and therapist dreams about each other, and training therapists to do dream work.

Given that clients seek help for puzzling, terrifying, creative, and recurrent dreams, therapists need to feel competent working with dreams in psychotherapy. Unfortunately, therapists often feel unprepared for this task because dreams are typically not addressed in clinical training. In this chapter, we hope to provide therapists with information about the existing knowledge regarding working with dreams in psychotherapy, so that they can feel more confident working with dreams. We first describe the various theories of dream work, and then we examine the empirical evidence about dream work in psychotherapy.

First, let us clarify important terms we will use in this chapter. Although the more commonly used term in the literature is "dream interpretation," we use the term "dream work." Dream interpretation implies that therapists are the active agents in interpreting the client's dream, whereas dream work simply implies that dreams are a focus of attention during psychotherapy sessions, with both therapist and client actively engaged in exploring the dream. Dream work can refer either to events within therapy in which the focus is on dreams, as is typical in psycho-dynamically oriented psychotherapy (and of course outside of therapy), or to an entire approach to therapy (e.g., Jungian therapy or imagery rehearsal therapy [IRT]). In addition, we use the term "therapist" to refer to the person providing help (although she or he might be referred to as an analyst or a counselor in the cited literature), and we use the term "client" (rather than "patient") to refer to the person presenting his or her dream in psychotherapy.

I. Theories of Dream Work

A number of models have been developed over the last 100 or more years for working with dreams in psychotherapy. We first describe models developed for individual psychotherapy (focusing on psychoanalytic/psychodynamic, cognitive, and other models), and then describe models for group treatment.

A. INDIVIDUAL PSYCHOANALYTIC/PSYCHODYNAMIC THERAPY

The early psychoanalysts recognized the power of dreams, strongly calling for therapists to work with clients' dreams in therapy to illuminate both conscious and unconscious conflicts. Perhaps most notably, in his *The Interpretation of Dreams* (1900/ 1966), Sigmund Freud suggested that the primary purpose of dreams is to satisfy primitive, infantile wishes. Unacceptable to our conscious minds, he proposed that such wishes are repressed during waking life. According to Freud, however, we cannot censor our thoughts during sleep, and thus these wishes emerge in our dreams, often in distorted form (e.g., rather than dreaming of a boss directly, one might dream of a dangerous tiger). According to Freud, then, dreams provide ideal therapeutic fodder, serving as the "royal road" for examining the unconscious. His most powerful approach for working with dreams was free association, in which the dreamer says whatever comes to mind, with as much honesty as possible. Through these associations to dream images, the origins of the dreamer's intrapsychic conflicts are revealed. In his work with patients, Freud listened to the dream and then to the patient's associations to specific images, and offered an interpretation using his knowledge of the dreamer and of dreams' symbolic meanings.

Presenting an alternate view, Carl Jung (1964, 1974) believed dreams to be a normal and creative expression of one's unconscious mind. Asserting that dreams serve a compensatory function, Jung stated that dreams reflect issues that are unexpressed during waking life. He thus believed that dreams can provide a vital means of uniting the conscious and unconscious by making dreamers aware of hidden feelings. Dream interpretation remains one of the central components in Jungian therapy, although Jung did not define specific procedures for dream work. Rather, he supported therapists' working with dreams in whatever way was most useful for the dreamer. Jung himself frequently used associations, portrayal of dreams through artistic expressions, and interpretation of dreams via archetypes and myths.

A third notable early dream theorist was Alfred Adler (1936, 1938, 1958). Believing personality to be a unitary construct, Adler asserted that the conscious and unconscious minds are the same, and thus the individual's waking personality is reflected in dreams. According to Adler, dreams are an expression of the conscious mind and provide the person with reassurance, security, and protection against damage to self-worth (e.g., a dream in which the person is able to fend off an attacker leaves the person feeling a sense of agency). Of primary importance, as well, is the emotion stimulated by the dream, which Adler believed allowed the dreamer to find resolutions to problems (e.g., a dream in which the dreamer resolved a difficult situation would provide confidence that s/he could similarly resolve situations in waking life, even if s/he could not remember the dream). Thus, dreams are a way of preparing for future activities or events and fulfill a problem-solving role. Unfortunately, Adler provided no clear guidelines for working with dreams in therapy.

Several new psychoanalytic models for understanding dreams have been proposed in the last 30 years (Fosshage, 1983, 1987; Garma, 1987; Glucksman, 1988; Glucksman and Warner, 1987; Lippman, 2000; Natterson, 1980, 1993; Schwartz, 1990). Diverging from the earlier Freudian tradition and reflecting more recent research, these theorists now propose that the manifest content of dreams reflects the dreamer's waking life rather than distortions from the unconscious. Modern Jungian authors (Beebe, 1993; Bonime, 1987; Bosnak, 1988; Johnson, 1986) have maintained much of Jung's original theory, but provide more explicit guidelines for how to work with dreams in therapy. Contemporary Adlerians (Bird, 2005; Lombardi and Elcock, 1997) have likewise provided more explicit detail for applying Adler's theory, including the replacement of fixed symbolism with an individualized understanding of dream metaphors, an emphasis on providing encouragement and positive interpretations, and a redefinition of the interpreter's role as a collaborator rather than an expert. In this revised model, the therapist nurtures the dreamer's understanding of her/his dream, as well as nurturing the ensuing ability to use this new knowledge to gain insight about events in life.

B. INDIVIDUAL COGNITIVE THERAPY

Emerging in the second half of the 20th century was Aaron Beck's theory of cognitive patterns in dreams (1971). Stating not only that dreams parallel an individual's waking thoughts, Beck also posited that waking cognitions influence dreams. Although Beck acknowledged that dreams have many functions and that the dreamer does not gain insight from every dream, he nevertheless believed that some dreams in particular clarify an individual's problem and may reflect dysfunctional attitudes. According to Beck, dreams bring automatic, unrealistic thoughts to the dreamer's awareness, and so can be used to help clients recognize their distorted thinking.

More recently, other cognitive therapists have developed models for using dreams in therapy. As an example, Arthur Freeman and Beverly White (2004) described a method for using dreams as a standard homework task in cognitive-behavioral therapy (CBT). In this approach, the dream represents an idiosyncratic dramatization of the dreamer's view of both self and the world. Freeman and White also provide 15 guidelines for conducting CBT dream work. They assert, for instance, that dreams should be understood thematically rather than symbolically; thus, the ideas or images present in clients' dreams should be taken at face value and not as symbolic representations of something or someone else. In addition, they posit that clients' affective responses to their dreams parallel affective responses to waking life events. Freeman and White also state that dreams may be particularly useful when clients are "stuck" in therapy, and that clients should be encouraged to establish a system and routine for collecting and logging their dream content. Furthermore, in seeking to understand their dreams, clients should try to discern a "moral" or primary theme from the dreams.

C. OTHER INDIVIDUAL APPROACHES

A number of other dream approaches, representing various theoretical perspectives, have been developed. Phenomenologists hold that dreams reflect conscious experiences and can be examined just as experiences in waking life (Boss, 1958, 1963; Craig and Walsh, 1993). Gestalt therapists such as Fritz Perls (1969) and Erving and Miriam Polster (1973) attend to the here and now and ask dreamers to imagine that each part or image of the dream is a part of themselves and to have a dialogue amongst the parts, believing that these disparate parts must be integrated for the person to become whole. Eugene Gendlin (1986) and Alvin Mahrer (1990) described experiential approaches for helping dreamers re-experience the feelings in their dreams and thus begin to accept and integrate the feelings. Gayle Delaney (1991, 1993), Ann Faraday (1972, 1974), and Lillie Weiss (1986, 1999) developed models incorporating elements of Gestalt and Jungian theories and connecting dreams closely to waking life problems.

Finally, Clara Hill (1996, 2004) integrated many of the previous theories into her cognitive-experiential dream model. Her model rests on the assumptions that (1) dreams are a continuation of waking thought without immediate input from the external world; (2) dreams' meaning is personal, and thus standard symbols or dream dictionaries are likely not useful; (3) working with dreams requires therapist and client collaboration; (4) dreams are useful for helping people understand themselves more deeply; (5) dreams consist of cognitive, emotional, and behavioral components; and (6) therapists must have sound basic helping skills before they can effectively apply the dream model. Integrating experiential, psychoanalytic, Gestalt, and behavioral approaches to dream work, Hill's model rests on three stages (exploration, insight, and action). In the exploration stage, the therapist helps the client deeply and sequentially explores a few dream images by progressing through four steps (description, re-experiencing, association, and waking life triggers). Once several images have been thoroughly explored, the therapist helps the client construct the dream's meaning in terms of the phenomenological experience of the dream, the dream's connection to waking life, or the inner dynamics (i.e., parts of self, conflicts from childhood, spiritual-existential concerns). Once the therapist and client have co-created some meaning for the dream, the therapist helps the client talk about how she or he would like to change the dream. The therapist then bridges from the changes in the dream to changes in waking life (i.e., helps the client apply possible changes in the dream to possible changes in waking life), and then helps the client determine how to go about actually making such changes.

D. DREAM GROUPS

In addition to the theories focusing on dream work with individual clients, there has been a growing interest in groups formed for the purpose of sharing and understanding dreams (Hillman, 1990). The major model of group dream work was developed by Montague Ullman (1987), whose approach emphasizes safety and discovery in group dream work. Importantly, the dreamer must feel safe enough with the group to disclose what may be quite intimate material. To foster such safety, all members acknowledge that the dreamer has absolute control of the dream work process at every stage. Discovery arises from the group members all adopting the dream as their own, a process that consists of four stages: (1) the dreamer describes a dream and the group asks questions to obtain a clear sense of the dream; (2) group members project their own material and their own associations onto the dream and its images; (3) the dreamer then responds to the group's input; and (4) during a later meeting, the dreamer shares any further thoughts s/he had with the group.

Building on Ullman's method, Donald Wolk (1996) created an integrative technique that uses psychodrama as a means to help participants connect their dreams to present life circumstances. After the group selects a member's dream on which to focus, the dreamer retells the dream in the first person, present tense. Next, group members ask questions to clarify the content of and feelings related to the dream. Group members then share their feelings about the dream as if it were their own, thus becoming integral contributors to the process. The focus then shifts to group members working on the dream images as if they were their own, and as if they were metaphorical expressions of something about their lives. Next, the dreamer responds to the group's feelings and offered metaphors, knowing that s/he is the ultimate authority on the many possible meanings of the dream, as well as on what s/he is willing to examine further with the group. Finally, the group leader assists the dreamer in selecting a part of the dream s/he wishes to address, then helps her/him set the scene and select dream characters and objects from among the other group participants. After the enactment, the dreamer is requested to write a comprehensive account of her/his experience of the group dream process.

In his similar approach, Jeremy Taylor (1992, 1998) asserted that anonymity must be maintained whenever dreams are discussed beyond the group. Furthermore, he posited that only the individual dreamer may definitively determine the meaning of her/his dream, that dreams may have more than one meaning, and that group members should always begin with the phrase, "If it were my dream ..." when referring to another person's dream.

A cognitive approach to group dream work is Barry Krakow's IRT (Krakow, 2004; Krakow and Zadra, 2006) for distressing dreams and posttraumatic nightmares. The three or four, approximately 2-hour group sessions that comprise this approach consist of two primary components. The first involves education and cognitive restructuring to help clients reconceptualize their disturbing dreams as a learned sleep disorder. Once they begin to see that these nightmares may have initially had an important function but have become habitual, clients begin to see that they can alter the behavior. In the second component, clients are taught imagery rehearsal. They choose a nightmare, determine how they would change it into a new dream, and then rehearse this new dream during the therapy session and as homework. Krakow asserted that this technique accelerates the client's once-dormant imagery system which in itself is healing, such that not only the targeted, but also other disturbing, dreams are also positively affected. Importantly, this model is an educational approach and does not encourage a re-experiencing of the disturbing dream. In fact, clients are specifically advised to avoid rehearsing old nightmares, given that exposure is contraindicated. In addition, clients for whom the trauma is too recent or who insist on working with extremely negative nightmares tend not to do well in this approach.

Finally, another option for group dream work arises from an adaptation of Hill's cognitive-experiential model (Wonnell, 2004). This approach maintains the three-stage structure, and group members offer input in all the stages, using the Ullman phrase, "If it were my dream . . ." to reinforce the dreamer's control over her/his dream. Sharing some features of the Ullman, Wolk, and Taylor methods, the Hill model provides more detailed guidelines for the dreamwork process, especially in the exploration stage, which may prove helpful for newly formed groups or new members of established groups.

E. Summary

Clearly, then, dream theories have arisen from many theoretical perspectives, and for both individual and group therapy, thereby attesting to the value of working with dreams in therapy. The diversity of these models demonstrates that theoreticians agree on no single, "correct" way to work with dreams. Although the plethora of approaches is a sign that the field is expanding and is vital, empirical validation of these theories is crucial. We thus turn now to the empirical research on dream work in psychotherapy.

II. Empirical Research on the Demographics of Dream Work in Psychotherapy

In this section, we review research about what might be considered the demographics of dream work. Specifically, we cover what we know about the extent of dream work in psychotherapy, client factors in dream work, therapist activities used in dream work, and who volunteers for dream work.

A. How Much Dream Work Occurs in Psychotherapy?

According to several surveys (Crook and Hill, 2003; Fox, 2002; Huermann et al., 2009; Keller et al., 1995; Schredl et al., 2000), most therapists reported that they attend to dreams at least occasionally, although dreams were rarely a major focus of therapy. For example, cognitively oriented therapists in the Crook and Hill (2003) study reported that about 15% of clients had talked about dreams in the past year and that they had spent about 5% of therapy time working on these dreams. A comparison of the mostly cognitive-behavioral therapists in Crook and Hill (2003) with a psychoanalytic sample (Hill *et al.*, 2008) revealed that the latter group worked with dreams considerably more than did the former group: The psychoanalytic sample talked about dreams with about half of their clients and such work occupied about half of the time in therapy, suggesting that therapists whose theoretical orientation values dream work are more likely to use it.

B. CLIENT FACTORS IN DREAM WORK

Therapists were most likely to focus on dreams with clients who had troubling recurrent dreams or nightmares, were psychologically minded, were interested in learning about their dreams, had posttraumatic stress syndrome (PTSD), or were seeking growth (Crook and Hill, 2003). Relatedly, clients who indicated having discussed dreams in therapy had higher dream recall, more positive attitudes toward dreams, and more encouragement from therapists to talk about their dreams than clients who did not discuss dreams in therapy (Crook-Lyon and Hill, 2004). Clients who reported that they had not talked about dreams in their therapy sessions either indicated that other issues were more pressing or that bringing dreams into therapy had never occurred to them (Crook Lyon and Hill, 2004).

C. HOW DO THERAPISTS WORK WITH DREAMS?

In terms of how they actually work with dreams, cognitively oriented therapists reported that they most often listened if clients brought in dreams, explored connections between dream images and waking life, asked for a description of the images, and collaborated with clients to construct interpretations of dreams (Crook and Hill, 2003). Likewise, psychoanalytically oriented therapists also frequently engaged in these four activities, but in addition often encouraged clients to associate to dream images, worked with conflicts represented in dreams, interpreted dreams in terms of waking life and past experiences, invited clients to tell dreams, encouraged clients to re-experience feelings in dreams, used dream images as metaphors later in therapy, and mentioned to clients that they were willing to work with dreams (Hill *et al.*, 2008). Similarly, clients who discussed dreams indicated that therapists most often helped them interpret their dreams, relate their dreams to waking life, and associate to dream images (Crook Lyon and Hill, 2004). Hence, although both cognitively and psychoanalytically oriented therapists used many activities to work with dreams, they most often focused on exploring and understanding the dreams; they rarely addressed how clients might change their dreams or make changes in waking life based on their understanding of dreams.

One interesting finding in the previous paragraph is that psychoanalytically oriented therapists invited clients to tell dreams and also mentioned that they liked to work with dreams. Two other studies also provided preliminary evidence that clients are more likely to talk about dreams if therapists explicitly encourage them to bring dreams into therapy (Crook-Lyon and Halliday, 1992; Hill, 2004).

Although these reports of how dreams are used in therapy are informative, most of the studies involved surveys of therapists and clients retrospectively recalling events. Thus, the data might represent attitudes more than the actual occurrence of dream work. To more directly answer the question of how dream work actually occurs in therapy, then, we are currently conducting a study within ongoing psychotherapy where therapists indicate after every therapy session whether a dream was mentioned and what activities were used to work with the dream. This study should provide preliminary information about how often dreams are presented in therapy and what methods therapists use to work with these dreams.

D. Who Volunteers for Dream Work?

Two studies provide evidence that not everyone wants to do dream work. In Hill *et al.* (1997), 336 undergraduates obtained extra credit for participating in a study in which they completed a wide range of self-report psychological measures and kept dream journals for 2 weeks. After completing the study, students were asked whether they would like to volunteer for no credit to work on a dream with a therapist in training. Of the 336 participants in the larger study, 109 (32%) indicated a willingness to participate and then 65 (19%) actually did participate. The students who were most likely to volunteer to participate were women, had high estimated dream recall, positive attitudes toward dreams, and high levels of absorption (i.e., capacity to restructure one's phenomenal field), and were open to new experiences. In a similar type of study in Taiwan, Tien *et al.* (2006) obtained a slightly higher participant rate of 177 of 574 (31%) students agreeing to participate in a dream session. Those students who volunteered had more positive attitudes toward dreams than those who did not volunteer. These findings are consistent with those reported above that clients were more likely to bring dreams into therapy if they had positive attitudes toward dreams, and thus emphasize the importance of attitudes toward dreams in deciding whether or not to ask a client to work with dreams in therapy.

III. Empirical Research on Models of Dream Work

Many case studies, both anecdotal and empirical, indicate the appropriateness and effectiveness of working with dreams with a wide range of clients (e.g., clients with trauma, homelessness, sexual problems, depression, masochism, obsession) in both individual and group therapy (see review in Hill and Spangler, 2007). Eudell-Simmons and Hilsenroth (2005) also reviewed a number of case studies indicating that dreams themselves change as a function of successful psychotherapy. For example, Caroppo *et al.* (1997) reported that the last 18 dreams of one client were more adaptive and integrated than were the client's first 18 dreams in therapy. In Dimaggio *et al.* (1997), pleasant emotions in dreams increased as the client improved. Thus, at least according to case studies (which have inherent bias in terms of selection factors), dream work appears to produce salutary results.

Fortunately, we also now have a solid body of research on larger, randomly selected samples indicating the effectiveness of dream work. This empirical work has primarily been conducted on two models—Hill's cognitive-experiential approach and Krakow's IRT—and so we turn now to a review of this research.

A. RESEARCH ON HILL'S COGNITIVE-EXPERIENTIAL DREAM MODEL

One caveat we acknowledge is that studies on the Hill model have mostly involved single sessions of dream work or brief therapy involving dream work, all with recruited clients presenting dreams, rather than dream work within naturalistic ongoing psychotherapy with non-recruited clients. Studying recruited clients in single sessions or brief therapy allowed Hill and colleagues to control extraneous variables and isolate variables of interest, and thus provide evidence about the effectiveness of dream work. Generalizing to ongoing psychotherapy, however, is premature.

1. *Outcomes of Dream Work*

The outcomes of dream work using Hill's model have been assessed in several ways, including (1) session quality, (2) the goals of dream work (e.g., insight, action ideas, target problems, and attitudes toward dreams), and (3) broader outcomes for general psychotherapy (e.g., symptom change, changes in interpersonal functioning, decreases in depression, well-being, communication).

i. Session Quality

The quality of sessions involving dream work has been assessed by client and therapist ratings of depth, working alliance, and satisfaction, typically using measures completed immediately after sessions. In 12 studies, clients consistently rated the quality of dream sessions (using the Hill model of dream work) significantly higher than regular therapy sessions (see review by Hill and Goates, 2004). It would seem that clients felt better about the quality of the sessions when they focused on dreams than when they focused on other topics.

ii. Goals of Dream Work

With regard to the specific goals of dream work, gains in insight have been assessed through several methods (open-ended questions of clients, standard measures of insight and understanding, and ratings of insight reflected in interpretations given by clients of their dreams). From studies using these various approaches to investigating the Hill model comes convincing evidence (see review in Hill and Goates, 2004) that clients gained insight into their dreams. Interestingly, in Hill *et al.* (2006), clients had a moderate level of insight into their dreams prior to sessions and gained insight after both the exploration and insight stages of dream work, and also reported gaining additional insight at a 2-week follow-up. These findings reflect that clients might be stuck prior to sessions in terms of understanding their dreams, but rapidly become unstuck in their ability to keep thinking about their dreams

Hill and colleagues have also assessed changes in the quality of clients' action ideas following dream sessions (again see review in Hill and Goates, 2004). They found that clients became more clear and focused about what they could do differently in their waking lives based on what they learned about themselves in the dream sessions. Interestingly, the quality of action ideas was lower than insight both before and after sessions, suggesting that action does lag behind insight.

Another dream-related variable relates to changes in the target problem reflected in the dream. Clients are asked after sessions (because they often do not know before sessions) to describe the target problem reflected in the dream and then rate their functioning on the target problem both for the current time and also retrospectively with regard to their functioning on this problem before the session. In Hill *et al.* (2006), clients reported increases in functioning on their target problems after a dream session, suggesting that clients felt that working with their dreams directly helped them resolve problems in waking life.

Researchers have also used a more standardized measure (impact of specific events) to assess changes in specific target complaints. Here, clients reported improvements in relation to divorce in Falk and Hill (1995) and loss in Hill *et al.* (2000).

Yet another dream-related outcome is change in attitudes toward dreams. Tien *et al.* (2006) applied the Hill model in Taiwan and found that volunteer clients presenting dreams reported better attitudes toward dreams after two to three dream sessions than did controls who did not receive a dream session.

iii. Broader Outcomes

In terms of broader outcomes for therapy as a whole, some research has found decreases in general symptoms (Diemer *et al.*, 1996; Hill *et al.*, 2000; Wonnell and Hill, 2005) and in depression (Falk and Hill, 1995), as well as increases in existential well-being when spiritual insight was the focus of the dream work (Davis and Hill, 2005). Mixed results have been reported for changes in interpersonal functioning (Diemer *et al.*, 1996; Hill *et al.*, 2000). In their investigation of group dream work with separated and divorced women, Falk and Hill (1995) found that those in dream groups scored higher in self-esteem and insight than did those in the wait-list control at the final assessment. Kolchakian and Hill (2002) found increases in other dyadic perspective taking but no changes in dyadic adjustment, primary communications, and self-dyadic perspective with couples' dream work.

In sum, consistent and positive changes have been reported in session quality and on outcomes that are specifically focused on dream work (e.g., insight, action ideas, target problems, and attitudes toward dreams). Less clear evidence has been reported on outcomes not specifically targeted in dream work (e.g., depression, anxiety, and self-esteem). Given that dreams may not necessarily reflect these broader outcomes, it is not surprising that fewer changes have been found in broader outcomes than in outcomes specific to dream work.

2. *The Process of Dream Work*

Now that we have established positive outcomes for Hill's model for dreamwork, we present evidence regarding the process of dream work. Specifically, we focus first on components of the model, and then review more general process components (client involvement, therapist input, other therapist characteristics, and the development of insight).

i. Components of the Model

A number of experimental studies have been conducted examining components of the exploration, insight, and action stages. In a study involving description of dream images only, association to dream images only, or description and association in the exploration stage, Hill *et al.* (1998) found slightly more benefit in terms of outcome for the association-only condition, but in general found that both description and association were helpful. In terms of the insight stage, no differences were found in outcomes for waking life versus parts-of-self interpretations (Hill et al., 2001), nor were differences found in nonspiritual outcomes for waking life versus spiritual interpretations, although spiritual interpretations led to more spiritual insight (Davis and Hill, 2005). In terms of the action stage, Wonnell and Hill (2000) found that clients who completed all three stages (exploration, insight, action) had better action ideas and rated sessions higher on problem solving than did clients who only completed the exploration and insight stages. Furthermore, Wonnell and Hill (2005) found that intention to carry out action plans was predicted by the client's perception of how much the therapist used action skills, the level of client involvement, and the level of difficulty of the action plan. Implementation of action was predicted by the level of difficulty of the action plan and the intent to act.

Another way of examining components of the model has been through qualitative investigations that involved asking open-ended questions of participants who experienced dream work. In four studies (Hill et al., 1997, 2000, 2003; Tien et al., 2006), clients mentioned that gaining insight, making links to waking life, hearing a new or "objective" perspective, experiencing feelings/catharsis, and hearing ideas for changes were helpful components of working with dreams. Interestingly, few clients mentioned hindering aspects; when they did, there was no consistency in what they did not like, suggesting that variables unique to the session, client, or therapist rather than the model itself were problematic.

ii. Client Involvement

Four studies (Diemer et al., 1996; Hill et al., 2006; Wonnell and Hill, 2000, 2005) found evidence that client involvement (i.e., active engagement in the session, actively exploring, coming up with insights, and generating action ideas) is related to the outcome of individual dream work, although one study (Falk and Hill, 1995) did not find that client involvement was related to outcome of group dream work.

iii. Therapist Input

Therapist's input was mentioned in three aforementioned qualitative studies (Hill et al., 1997, 2000, 2003) as a helpful component of the dreamwork process. In addition, two studies (Heaton et al., 1998; Hill et al., 2003) found that volunteer clients gained more from working with a therapist than they did from using the same approach in a self-help format. We note, however, that a small subgroup of clients in the latter study preferred working by themselves. Liking the therapist was mentioned in two qualitative studies (Hill et al., 2000, 2003) as a helpful component of the process. One study (Hill et al., 2006) found evidence that therapist adherence to the model and competence using the model were related to session outcome. In contrast, Hill et al. (2003) did not find evidence for the effects of therapist input (interpretations in the insight stage and action ideas in the action stage) when they compared empathy alone and empathy plus input. Furthermore, Hill *et al.* (2007) found no differences between an empathy condition and an empathy and input condition for clients of East Asian descent, although clients who were more anxiously attached and lower on Asian values had better outcomes in the empathy-only condition, whereas clients who were less anxiously attached and higher on Asian values had better outcomes in the empathy and input condition. It is likely that clients in the earlier sets of studies enjoyed working with a therapist, but the empathy might have been the crucial factor. Hence, although it appears that the therapists' empathic presence is beneficial for most clients, the exact helpful components of therapist interventions are less clear.

iv. The Development of Insight

Additional evidence for the effects of specific process components was presented in a series of three case studies (two of whom gained a lot of insight, and one gained very little insight) examining how insight develops in dream sessions (Hill *et al.*, 2007; Knox *et al.*, 2008). The two insight-gained clients were very motivated and involved in the sessions, nonresistant, trusting of others, and affectively present but not overwhelmed by affect. In addition, their therapists were able to skillfully use probes for insight and manage countertransference reactions toward the clients. In contrast, the client who did not gain insight was resistant, untrusting, and emotionally overwhelmed in the session, and the therapist was not skillful in conducting the session and was not able to manage her negative countertransference. In another examination, Baumann and Hill (2008) found that therapists' interpretations, self-disclosures, and probes for insight were associated with high levels of client insight in the next speaking turn in the insight stage of dream sessions, suggesting that these are helpful interventions for facilitating insight. Across studies, therapist probes for insight appear to be particularly helpful.

v. Summary of Process Evidence

All components of the Hill model (exploration, insight, and action) appear to be helpful. Furthermore, it is helpful for clients to gain insight, make links to waking life, hear a new or "objective" perspective, experience feelings/catharsis, and hear ideas for changes. It also appears that client involvement and motivation are key components of dream work using the Hill model. Finally, if clients are to gain insight, they need to not be overwhelmed by affect in the session and be open to and trusting of the therapist. Furthermore, therapist presence and perhaps empathy are important, along with the ability to use probes for insight.

3. *Predicting Who Benefits from Dream Work*

We have some knowledge regarding what types of clients achieve the greatest benefit from dream work. First, clients with positive attitudes seem to have positive outcomes (Hill *et al.*, 2001, 2006; Zack and Hill, 1998). Taken together with the finding that the people who volunteered for dreams sessions had more positive attitudes toward dreams than those who did not volunteer (Hill *et al.*, 1997), valuing dreams may be an important precondition for dream work. A second important variable is the salience of dreams, in that clients who profited most from dream work presented dreams that seemed potent or powerful to them (Hill *et al.*, 2006). Third, self-efficacy for working with dreams seems important (Hill *et al.*, 2008), in that clients needed to feel that working with dreams would help them accomplish their goals.

In addition, in Hill *et al.*'s (2006, 2008) studies, clients who profited most from dream sessions had poor initial functioning on the problem reflected in the dream, low initial insight into the dream, and poor initial action ideas related to the dream. Hence, clients who had more to gain in terms of their functioning related to the specific dream gained the most from the sessions.

The valence of the dream has garnered less consistent results. Zack and Hill (1998) found the best session outcomes when dreams were moderately unpleasant or extremely pleasant, and the worst outcomes when dreams were moderately pleasant or extremely unpleasant. Hill and colleagues (2001), in contrast, found that session outcomes were best when dreams were pleasant. No relationship between dream valence and session outcome emerged in Hill *et al.* (2003). Perhaps, as Hill *et al.* (2007) suggested, dreams should be categorized into several types (positive interpersonal, negative interpersonal, interpersonal agency, interpersonal nightmares, non-interpersonal dreams, all others) rather than by valence. Furthermore, Hill *et al.* found more positive process and outcome for clients with positive, agency, and non-interpersonal dreams than for clients with negative dreams and nightmares.

Minimal evidence exists for the importance of other client characteristics (e.g., sex/gender, race/ethnicity, psychological mindedness) and other dream-related characteristics (e.g., recency, vividness, arousal, distortion) in terms of outcome of dream sessions (see also review in Hill and Goates, 2004).

In conjunction with the findings presented in the section on the demographics of dream work in naturally occurring therapy, these results suggest that it is best to do dream work with clients who have positive attitudes toward dreams, high self-efficacy or confidence in their ability to work with their dreams, who have salient dreams that are puzzling or dreams that reflect underlying concerns, who have low insight and action ideas related to the dreams, and who are willing to discuss dreams in therapy.

B. Empirical Research on Imagery Rehearsal Therapy (IRT)

Barry Krakow and colleagues have conducted a number of studies demonstrating the effectiveness of IRT in reducing nightmare frequency/intensity and increasing sleep quality in survivors of sexual assault (Krakow *et al.*, 2000, 2001), adolescent girls in a residential facility (Krakow *et al.*, 2001), crime victims with PTSD (Krakow *et al.*, 2001), and nightmare patients (Germain and Nielson, 2003). These studies have shown not only positive outcomes but also the maintenance of changes over ∼3 months. Interestingly, these same studies also found that symptoms of anxiety, depression, and PTSD decreased after successful nightmare treatment. Furthermore, Germain *et al.* (2004) demonstrated that the new dreams created by clients contained fewer negative elements and more positive elements and mastery than did the nightmares.

In their summary of this body of literature, Krakow and Zadra (2006) noted that about 70% of clients reported clinically meaningful improvements in nightmare frequency, with the percentage increasing to 90% when clients regularly used the techniques for 2–4 weeks. Krakow (2004) noted that the results are best for those clients who do not have major psychiatric distress or disorders. For example, in Krakow *et al.* (2001), one-third of sexual assault survivors dropped out of IRT before initiating treatment or very early in treatment, suggesting that IRT did not resonate well for them. No work, however, has yet been done to dismantle this approach and thereby determine the relative effectiveness of its various components (e.g., education about nightmares as a learned behavior, imagery rehearsal).

C. EMPIRICAL RESEARCH ON OTHER METHODS OF DREAM WORK

In a comparison of their four-step group method and Ullman's group method, Shuttleworth-Jordan and Saayman (1989) found that therapists and clients were more involved and experienced less tension or loss of control in the former than the latter method. Furthermore, three studies have shown the effectiveness of systematic desensitization in reducing nightmare frequency and intensity (Celucci and Lawrence, 1978; Kellner *et al.*, 1992; Miller and DiPlato, 1983), although one could question whether systematic desensitization is actually dream work.

IV. Empirical Research in Other Areas Related to Dreams and Psychotherapy

There are a number of other ways that dreams can be used in psychotherapy. We focus here on just a few of these applications.

A. Therapist Use of Dream Content to Understand Clients

Eudell-Simmons and Hilsenroth (2005) suggested that therapists examine the content of dreams to better understand their clients. Given that dreams provide information about the person, and clients are often invested in their dreams, examining the content of dreams can be a nonintrusive way of assessing personality problems. Relatedly, a substantial amount of evidence exists showing that dream content differs for different diagnostic groups (see reviews in Hill, 1996; Van de Castle, 1994), allowing therapists to assess whether their clients' dreams are similar to those of clients with depression, hysteria, schizophrenia, chronic brain syndrome, or a history of sex offenses.

Research regarding the prevalence of interpersonal themes in dreams may also prove beneficial for therapists. The typical dream, for instance, involves other people and feelings about these people (Hall and Van de Castle, 1966). Interestingly, the response of others in dream narratives was typically to reject and oppose the dreamer, whereas the responses of self were typically to feel anxious, ashamed, and helpless (Popp Diguer et al., 1998; Popp et al., 1998).

Dreams can also be used by therapists to understand aspects of the therapeutic process. From a psychodynamic perspective, Bradlow and Bender (1997) suggested that the first dream presented in analysis reflects crucial themes. Furthermore, Gillman (1993) described three types of undisguised transference dreams (a response to a break in the analytic barrier, a defense against an emerging transference neurosis, and reflection of a specific character defense). In addition, Sirois (1994) suggested that client dreams often signal sensitive moments in therapy, especially occurring when the client perceives the therapist's interventions as traumatic. Finally, clients sometimes present dreams about termination (Oremland, 1973). Intriguing as these observations are, empirical research is needed to increase our understanding of the role of dreams in psychotherapy (see also later section on client dreams about therapists).

B. Culture, Dreams, and Psychotherapy

1. Dream Work with Men

Men and women have different dream experiences. Men have lower dream recall than women (Cowen and Levin, 1995; Schredl, 2000), and men's dreams contain more aggression, anxiety, achievement, and work-related themes than do women's dreams (Schredl and Piel, 2005; Van de Castle, 1994).

Aaron Rochlen (2004) modified Hill's cognitive-experiential model for men. He included strategies to overcome men's resistance, such as providing more explanations about why each of the stages of dream work is necessary, encouraging men to move beyond concrete thinking in their work with dreams, providing models for men who are emotionally constricted, and recognizing when clients are too focused on action. Rochlen and Hill (2005) tested this model among men with different levels of gender role conflict: Men with high gender role conflict discussed conflicts between work and family, restrictive emotionality, and preoccupation with achievement and competition in sessions more often than did men with low gender role conflict. The outcome of sessions, however, was not different for men who had high versus low gender role conflict. These results suggest that once men agree to dream work, they find it helpful regardless of their level of gender role conflict. Of course, as reviewed earlier, it is difficult to get men to volunteer to work on their dreams.

2. Dream Work with East Asian Clients

Hill *et al.* (2007) successfully used dream work with East Asian clients. They found, however, no support for the oft-cited premise that East Asian clients should benefit more from a directive than nondirective approach. In fact, there were no overall outcome differences between a nondirective approach (i.e., therapists provided only empathic responses such as probes and reflections) compared with a directive approach (i.e., therapists provided input in addition to empathy, such that they gave probes, reflections, interpretations, and suggestions for action). Client variables, however, did moderate the results: Clients who were more anxiously attached and lower on Asian values did better in the empathy-only (nondirective) condition, whereas clients who were less anxiously attached and higher on Asian values had better outcomes in the empathy + input (directive) condition.

Sim *et al.* (2010) did an additional analysis of the data of those East Asian women in the Hill *et al.* sample who were first- and second-generation students. They found that interpersonal issues and academic/postgraduation/career issues were typical for both subgroups, but that first-generation Asian women more often disclosed issues related to immigration/cultural/adjustment and physical/health than did second-generation women. In terms of action ideas, both subgroups typically talked about making interpersonal behavioral changes, but first-generation Asian women talked more about changing thoughts and feelings than did second-generation Asian women. Hence, not only might race/ethnicity play a role, but also immigration status may play a role in what clients talk about in dream sessions.

3. Spirituality and Dream Work

Dreams have long been regarded as reflections of spirituality (Davis, 2004; Jung, 1964; Van de Castle, 1994), but not much is known about the relationship between spiritually centered dream work and therapeutic outcome. In one study, Davis and

Hill (2005) examined the Hill cognitive-experiential model with clients who were spiritually oriented. In this study, clients gained more spiritual insight and had greater increases in existential well-being when therapists provided spiritual interpretations of their dreams in the insight stage than when therapists offered waking life interpretations. These findings suggest that there may be some value in therapists addressing spiritual and existential concerns with clients who are spiritually oriented.

C. Client Dreams about Therapists

Although many therapists, particularly of a psychoanalytic orientation, have written about the clinical importance of client dreams about therapists (e.g., Eyre, 1988), only a few empirical studies have investigated this phenomenon. Harris (1962) and Rosenbaum (1965) reported that about 10% of client dreams reported in sessions were manifestly about the therapists, and Rohde *et al.* (1992) found that 33% of clients who were themselves therapists had dreams in which their own therapists appeared in undisguised form. Hence, these data indicate that some clients, particularly those in psychodynamic therapy, do have dreams about their therapists.

In terms of the content, Harris (1962) indicated that client dreams about therapists reflected transference, but Rosenbaum (1965) reported no such evidence. Harris also reported that the manifest content ranged from wish fulfillment to a reflection of anxiety, whereas Rohde *et al.* found themes of separation-rejection, seduction-antagonism, protectiveness-responsiveness, and praise in dreams. Thus, it appeared that client dreams about therapists covered a range of topics, although many appeared to be negative, with the therapist/analyst treating the client badly. Methodological problems plagued these studies, however: Harris used his own clients and did his own data analyses from case notes; Rosenbaum surveyed a small non-representative sample of analysts and relied on his own judgment to analyze the data; Rohde *et al.* used trained judges and a larger sample size, but their sample consisted of psychotherapists and thus their findings might not generalize to clients who are not therapists. A study that we are currently conducting in a clinic setting examining client dreams about their therapists might provide some further evidence about this topic.

D. Therapist Dreams about Clients

We found three empirical studies about therapists' dreams about clients. In a survey of members of the Canadian Psychoanalytic Society (Lester *et al.*, 1989), 78% of participants reported having had countertransference dreams (i.e., dreams where the client appeared in undisguised form in the manifest content of the dream). These dreams most often occurred at difficult points in the therapy (when there was a strong erotic transference, 46%; when therapists were not understanding their clients, 46%; when clients were angry, 32%), although they also occasionally occurred when progress was being made (26%), or when therapists were introducing something new into the therapy (14%). Most therapists reported having gained insight into their dreams about clients (76%), although a few indicated guilt (22%) or embarrassment (20%). Male therapists had more sadistic/erotic, competitive, and sadistic dreams and fewer identification/closeness dreams than did female therapists.

Kron and Avny (2003) studied dreams of 22 Israeli therapists about 31 clients. The majority of the dreams (65%) were characterized by negative emotions, in that therapists felt betrayed, abandoned, and forsaken by clients who were characterized as aggressive, neglectful, abandoning, or invading of their personal space. Kron and Avny speculated that the dreams reflected therapists' unresolved issues, a projection of clients' difficulties, or problems in the therapeutic relationship.

Spangler et al. (2009) qualitatively examined eight experienced therapists' dreams about their clients. Therapists' dreams reflected either particularly challenging clients or an extreme amount of stress in the therapists' life. The dreams typically involved negative interpersonal content (e.g., awkwardness, boundary violations, aggression), although there were a few positive interactions.

In sum, therapists' dreams about clients are most often negative, reflecting difficult or challenging interactions, although some involved positive interactions. A caveat across these studies, however, is that all were retrospective (collected using a survey format or interviews) from selective samples of therapists. In addition, recall bias may have played a role, in that more salient or more negative dreams may have been remembered more often. We are currently conducting a study where therapists keep dream journals, and thus may be able to obtain a clearer picture of the frequency and types of dreams therapists have about clients.

E. TRAINING THERAPISTS TO DO DREAM WORK

Three studies were found that examined training in dream work, all using a retrospective survey method (i.e., asking practicing therapists about their training). In all three studies (Crook and Hill, 2003; Fox, 2002; Keller et al., 1995), most therapists indicated that they had at least minimal graduate training in dream work. In addition, Fox (2002) found that the more training therapists had in dream work, the more likely they were to perceive themselves as competent in working with dreams and to consider dream work to be effective. Similarly,

Crook and Hill (2003) found that the more training therapists had, the more likely they were to feel competent in working with dreams, to have had clients who brought up dreams in therapy, to have spent time in therapy working on dreams, and to have used many activities for working with dreams. These findings suggest that therapists feel more competent and engage in more dream work when they have had training in dream work. These studies were correlational, however, so we cannot rule out the possibility that those therapists who felt more competent in working with dreams sought out more dream training. To address the issue of the effects of dream training, experimental work is needed.

Ullman (1994) presented an experiential group approach for teaching therapists how to make connections between dream images and waking life experiences. In this method, he stressed the importance of dialogue between the dreamer and therapist, with the therapist listening to and questioning the dreamer to elicit relevant client information. He also stressed the importance of safety to help the client feel free to engage in the discovery process. Unfortunately, there is yet no empirical evidence regarding Ullman's training method.

Crook (2004) developed a training model for the Hill cognitive-experiential approach in which therapists read about the model, participate in discussions of the model, and then practice the model in group and dyadic settings. In a recent empirical study with s small sample and only one trainer, Crook-Lyon et al. (2009) found evidence that therapists felt more self-efficacy for working with dreams, had more positive attitudes toward dreams, and had higher self-reported competence for working with dreams as a result of training. In addition, there was some preliminary evidence that feedback from supervisors about their performance in sessions and practice doing sessions with clients both led to higher levels of self-efficacy, attitudes toward dreams, and ability to conduct dream sessions, but these findings await replication with larger samples.

V. Future Directions

Given the potential effectiveness of dream work, it seems appropriate for therapists to incorporate such content into psychotherapy, especially after being adequately trained in how to work with dreams. Therapists would ideally be trained by experts to use approaches that have received empirical support, but, alternatively, therapists can learn methods for working with dreams by reading texts and practicing on their own.

In terms of research, we need more empirical investigations of the efficacy and effectiveness of different dream models, including direct comparisons of various dream models. For example, Hill's cognitive-experiential model and Krakow's IRT have quite different approaches to affect in dream work: Hill recommends re-experiencing and processing the affective material, whereas Krakow recommends avoidance of exposure to the dream images. Both of these approaches appear effective, so it would be important to compare the two directly, and also to determine if each is more effective with certain types of clients.

Furthermore, work is needed to determine the effectiveness of various components of the different models. More work is needed, as well, on the best methods for including dream work in therapy and for training therapists.

We hope that this review is helpful in encouraging therapists and researchers to pay more attention to dreams in psychotherapy. In a similar review of dream work, say 20 years from now, we hope that there will be many more approaches to dream work and that these approaches will have received substantial empirical attention so that we will know more about when, with whom, and how to use dreams in psychotherapy effectively.

References

Adler, A. (1936). On the interpretation of dreams. *Int. J. Individ. Psychol.* **2**, 3–16.

Adler, A. (1938). Social Interest: Challenge to Mankind. Faber & Faber, London.

Adler, A. (1958). What Life Should Mean to You. Capricorn, New York.

Baumann, E., and Hill, C. E. (2008). The attainment of insight in the insight stage of the Hill dream model: the influence of client reactance and therapist interventions. *Dreaming* **18**, 127–137.

Beck, A. T. (1971). Cognitive patterns in dreams and daydreams. In: Scientific proceedings of the American Academy of Psychoanalysis: Vol. 19. Dream dynamics: Sciene and psychoanalysis (J. H. Masserman, ed.), Grune & Stratton, New York, pp. 2–7.

Beebe, J. (1993). A Jungian approach to working with dreams. In: New Directions in Dream Interpretation (G. Delaney, ed.), SUNY Press, Albany, NY, pp. 77–102.

Bird, B. E. I. (2005). Understanding dreams and dreamers: an Adlerian perspective. *J. Individ. Psychol.* **61**, 200–216.

Bonime, W. (1987). Collaborative dream interpretation. In: Dreams in a New Perspective: The Royal Road Revisited (M. L. Glucksman and S. L. Warner, ed.), Human Sciences Press, New York.

Bosnak, R. (1988). A Little Course in Dreams: A Basic Handbook of Jungian Dreamwork. Shambhala, Boston.

Boss, M. (1958). The Analysis of Dreams. Philosophical Library, New York.

Boss, M. (1963). Psychoanalysis and Daseinsanalysis. Basic Books, New York.

Bradlow, P. A., and Bender, E. P. (1997). First dreams in psychoanalysis: a case study. *J. Clin. Psychoanal.* **6**, 107–122.

Caroppo, E., Dimaggio, G. G., Popolo, R., Salvatore, G., and Ruggeri, G. (1997). Recurrent oneiric themes: a clinical research on dream evaluation during psychotherapy. *New Trends Exp. Clin. Psychiatr.* **8**, 275–278.

Celucci, A., and Lawrence, P. (1978). The efficacy of systematic desensitization in reducing nightmares. *J. Behav. Ther. Exp. Psychiatr.* **9**, 109–114.

Cowen, D., and Levin, R. (1995). The use of the Hartmann boundary questionnaire with an adolescent population. *Dreaming* **5**, 105–114.

Craig, E., and Walsh, S. J. (1993). The clinical use of dreams. In: New Directions in Dream Interpretation (G. Delaney, ed.), SUNY Press, Albany, NY, pp. 103–154.

Crook, R. E. (2004). Training therapists to work with dreams in therapy. In: Dream Work in Therapy: Facilitating Exploration, Insight, and Action (C. E. Hill, ed.), American Psychological Association, Washington, DC.

Crook, R. E., and Hill, C. E. (2003). Working with dreams in psychotherapy: the therapists'perspective. Dreaming 13, 83–93.

Crook Lyon, R. E., and Hill, C. E. (2004). Client reactions to working with dreams in psychotherapy. Dreaming 14, 207–219.

Crook-Lyon, R. E., Hill, C. E., Wimmer, C. L., Hess, S. A., and Goates-Jones, M. K. (2009). Therpist training, feedback, and practice for dream work: a pilot study. Pycholl Rep. 105, 87–98.

Davis, T. L. (2004). Incorporating spirituality into dream work. In: Dream Work in Therapy: Facilitating Exploration, Insight, and Action (C. E. Hill, ed.), American Psychological Association, Washington, DC, pp. 149–168.

Davis, T. L., and Hill, C. E. (2005). Including spirituality in the Hill model of dream interpretation. J. Couns. Dev. 83, 492–503.

Delaney, G. (1991). Breakthrough Dreaming. Bantam Books, New York.

Delaney, G. (1993). The dream interview. In: New Directions in Dream Interpretation (G. Delaney, ed.), SUNY Press, Albany, NY, pp. 195–240.

Diemer, R., Lobell, L., Vivino, B., and Hill, C. E. (1996). A comparison of dream interpretation, event interpretation, and unstructured sessions in brief psychotherapy. J. Coun. Psychol. 43, 99–112.

Dimaggio, G. G., Popolo, R., Serio, A. V., and Ruggeri, G. (1997). Deam emotional experience changes and psychotherapeutic process: an experimental contribution. New Trends Exp. Clin. Psychiatr. 8, 271–273.

Eudell-Simmons, E. M., and Hilsenroth, M. J. (2005). A review of empirical research supporting four conceptual uses of dreams in psychotherapy. Clin. Psychol. Psychother. 12, 255–269.

Eyre, D. (1988). The use of the analyst as a dream symbol. Br. J. Psychother. 5, 5–18.

Falk, D. R., and Hill, C. E. (1995). The process and outcome of dream interpretation groups for divorcing women. Dreaming 5, 29–42.

Faraday, A. (1972). Dream Power. Coward, McCann & Geoghegan, New York.

Faraday, A. (1974). The Dream Game. Harper & Row, New York.

Fosshage, J. L. (1983). The psychoanalytic function of dreams: a revised psychoanalytic perspective. Psychoanal. Contemp. Thought 6, 641–669.

Fosshage, J. L. (1987). New vistas in dream interpretation. In: Dreams in a New Perspective: The Royal Road Revisited (M. L. Glucksman and S. L. Warner, eds.), Human Sciences Press, New York, pp. 23–44.

Fox, S. A. (2002). A survey of mental health clinicians' use of dream interpretation in psychotherapy. Diss. Abstr. Int. B Sci. Eng. 62(7-B), 3376.

Freeman, A., and White, B. (2004). Dreams and the dream image: using dreams in cognitive therapy. In: Cognitive Therapy and Dreams (R. I. Rosner, W. J. Lyddon, and A. Freeman, eds.), Springer Publishing Company, New York, pp. 69–87.

Freud, S. (1966). The Interpretation of Dreams. Avon, New York (Original work published 1900).

Garma, A. (1987). Freudian approach. In: Dream Interpretation: A Comparative Study (J. L. Fosshage and C. A. Loew, eds.), PMA, New York, pp. 16–51.

Gendlin, E. (1986). Let Your Body Interpret Your Dream. Chiron, Wilmette, IL.

Germain, A., and Nielson, T. (2003). Impact of imagery rehearsal treatment on distressing dreams, psychological distress, and sleep paramters in nightmare patients. Behav. Sleep Med. 1, 140–154.

Germain, A., Krakow, B., Faucher, B., Zadra, A., Nielson, T., Hollifield, M., Warner, T. D., and Koss, M. (2004). Increased masstery elements associated with imagery rehearsal treatment for nightmares in sexual assault survivors with PTSD. Dreaming 14, 195–206.

Gillman, R. D. (1993). Dreams in which the analyst appears as himself. In: The Dream in Clinical Practice (J. Natterson, ed.), Jason Aronson, Inc, Northvale, NJ, pp. 29–44.

Glucksman, M. L. (1988). The use of successive dreams to facilitate and document change during treatment. *J. Am. Acad. Psychoanal.* **16**, 47–69.

Glucksman, M. L., and Warner, S. L. eds. (1987). Dreams in a New Perspective: The Royal Road Revisited. Human Sciences Press, New York.

Hall, C., and Van de Castle, R. (1966). The Content Analysis of Dreams. Appleton-Century-Crofts, New York.

Halliday, G. (1992). Effect of encouragement on dream recall. *Dreaming* **2**, 39–44.

Harris, I. (1962). Dreams about the analyst. *Int. J. Psychoanal.* **43**, 151–158.

Heaton, K. J., Hill, C. E., Petersen, D., Rochlen, A. B., and Zack, J. (1998). A comparison of therapist-facilitated and self-guided dream interpretation sessions. *J. Coun. Psychol.* **45**, 115–121.

Hill, C. E. (1996). Working with Dreams in Psychotherapy. Guilford Press, New York.

Hill, C. E. ed. (2004). Dream Work in Therapy: Facilitating Exploration, Insight, and Action. American Psychological Association, Washington, DC.

Hill, C. E., Crook-Lyon, R. E., Hess, S., Goates-Jones, M. K., Roffman, M., Stahl, J., Sim, W., and Johnson, M. (2006). Prediction of session process and outcome in the Hill dream model: contributions of client dream-related characteristics and the process of the three stages. *Dreaming* **16**, 159–185.

Hill, C. E., Diemer, R., and Heaton, K. J. (1997). Dream interpretation sessions: who volunteers, who benefits, and what volunteer clients view as most and least helpful. *J. Coun. Psychol.* **44**, 53–62.

Hill, C. E., and Goates, M. K. (2004). Research on the Hill cognitive-experiential dream model. In: Dream Work in Therapy: Facilitating Exploration, Insight, and Action (C. E. Hill, ed.), American Psychological Association, Washington, DC, pp. 245–288.

Hill, C. E., Kelley, F. A., Davis, T. L., Crook, R. E., Maldonado, L. E., Turkson, M. A., Wonnell, T. L., Suthakaren, V., Zack, J. S., Rochlen, A. B., Kolchakian, M. R., and Codrington, J. N. (2001). Predictors of outcome of dream interpretation sessions: volunteer client characteristics, dream characteristics, and type of interpretation. *Dreaming* **11**, 53–72.

Hill, C. E., Knox, S., Hess, S., Crook-Lyon, R., Goates-Jones, M., and Sim, W. (2007a). The attainment of insight in the Hill dream model: a single case study. In: Insight in Psychotherapy (L. Castonguay and C. E. Hill, eds.), American Psychological Association, Washington, DC.

Hill, C. E., Liu, J., Spangler, P., Sim, W., and Schottenbauer, M. (2008). Working with dreams in psychotherapy: what do psychodynamic therapists report that they do? *Psychoanal. Psychol.* **25**, 565–573.

Hill, C. E., Nakayama, E., and Wonnell, T. (1998). A comparison of description, association, and combined description/association in exploring dream images. *Dreaming* **8**, 1–13.

Hill, C. E., Rochlen, A. B., Zack, J. S., McCready, T., and Dematatis, A. (2003). Working with dreams using the Hill cognitive-experiential model: a comparison of computer-assisted, therapist empathy, and therapist empathy + input conditions. *J. Coun. Psychol.* **50**, 211–220.

Hill, C. E., and Spangler, P. (2007). Dreams and psychotherapy. In: The New Science of Dreaming (D. Barrett and P. McNamara, eds.), Greenwood Publishers, Westport, CT, pp. 159–186.

Hill, C. E., Spangler, P., Sim, W., and Baumann, E. (2007b). Interpersonal content of dreams in relation to the process and outcome of single sessions using the Hill dream model. *Dreaming* **17**, 1–19.

Hill, C. E., Tien, H. S., Sheu, H., Sim, W., Ma, Y., Choi, K., and Tashiro, T. (2007c). Predictors of outcome of dream work for East Asian volunteer clients: dream factors, anxious attachment, Asian values, and therapist input. *Dreaming* **17**, 208–226.

Hill, C. E., Zack, J., Wonnell, T., Hoffman, M. A., Rochlen, A., Goldberg, J., Nakayama, E., Heaton, K. J., Kelley, F., Eiche, K., Tomlinson, M., and Hess, S. (2000). Structured brief therapy with a focus on dreams or loss for clients with troubling dreams and recent losses. *J. Coun. Psychol.* **47**, 90–101.

Hillman, D. J. (1990). The emergence of grassroots dreamwork movement. In: Dream Time and Dreamwork: Decoding the Language of the Night (S. Krippner, ed.), Jeremy P. Tarcher, Inc, Los Angeles, CA, pp. 13–20.

Huermann, R., Crook-Lyon, R. E., Heath, M. A., Fischer, L., and Potkar, K. (2009). Dream work with children: perceptions and practices of school mental health practitioners. *Dreaming* **19**, 85–96.

Johnson, R. (1986). Inner Work. Harper & Row, San Francisco.

Jung, C. G. ed. (1964). Man and His Symbols. Dell, New York.

Jung, C. G. (1974). Dreams (R. F. C. Hull, Trans.). Princeton University Press, Princeton, NJ.

Keller, J. W., Brown, G., Maier, K., Steinfurth, K., Hall, S., and Piotrowski, C. (1995). Use of dreams in therapy: a survey of clinicians in private practice. *Psycholl Rep.* **76**, 1288–1290.

Kellner, R., Neidhardt, J., Krakow, B., and Pathak, D. (1992). Changes in chronic nightmares after one session of desensitization or rehearsal of instructions. *Am. J. Psychiatr.* **149**, 659–663.

Knox, S., Hill, C. E., Hess, S., and Crook-Lyon, R. (2008). The attainment of insight in the Hill dream model: replication and extension. *Psychother. Res.* **18**, 200–215.

Kolchakian, M. R., and Hill, C. E. (2002). Working with unmarried couples with dreams. *Dreaming* **12**, 1–16.

Krakow, B. (2004). Imagery rehearsal therapy for chronic posttraumatic nightmares: a mind's eye view. In: Cognitive Therapy and Dreams (R. I. Rosner, W. J. Lyddon, and A. Freeman, eds.), Springer Publishing Company, New York, pp. 89–109.

Krakow, B., Hollifield, M., Johnston, L., Koss, M., Schrader, R., Warner, T. D., Tandberg, D., Lauriello, J., McBride, L., Cutchen, L., Cheng, D., Emmons, S., Germain, A., Melendrez, D., Sandoval, D., and Prince, H. (2001). Imagery rehearsal therapy for chronic nightmares in sexual assault survivors with posttraumatic stress disorder. *J. Am. Med. Assoc.* **286**, 537–545.

Krakow, B., Hollifield, M., Schrader, R., Koss, M., Tandberg, D., Lauriello, J., McBride, L., Warner, T. D., Cheng, D., Edmond, T., and Kellner, R. (2000). A controlled study of imagery rehearsal for chronic nightmares in sexual assault survivors with PTSD: a preliminary report. *J. Trauma. Stress.* **13**, 589–609.

Krakow, B., Johnston, L., Melendrez, D., Warner, T. D., Chavez-Kennedy, D., and Herlan, M. J. (2001). An open-label trial of evidence-based cognitive-behavior therapy for nightmares and incomnia in crime victims with PTSD. *Am. J. Psychiatr.* **158**, 2043–2047.

Krakow, B., Sandoval, D., Schrader, R., Keuhne, B., McBride, L., Yau, C. L., and Tandberg, D. (2001). Treatment of chronic nightmares in adjudicated adolescent girls in a residential facility. *J. Adolesc. Health* **29**, 94–100.

Krakow, B., and Zadra, A. (2006). Clinical management of chronic nightmares: imagery rehearsal therapy. *Behav. Sleep Med.* **4**, 45–70.

Kron, T., and Avny, N. (2003). Psychotherapists' dreams about their patients. *J. Anal. Psychol.* **48**, 317–339.

Lester, E. P., Jodoin, R. M., and Robertson, B. M. (1989). Countertransference dreams reconsidered: a survey. *Int. Rev. Psychoanal.* **16**, 305–314.

Lippman, P. (2000). Nocturnes: On Listening to Dreams. The Analytic Press, Hillsdale, NJ.

Lombardi, D. N., and Elcock, L. E. (1997). Freud versus Adler on dreams. *Am. Psychol.* **52**, 572–573.

Mahrer, A. R. (1990). Dream Work in Psychotherapy and Self-Change. Norton, New York.

Miller, W. R., and DiPlato, M. (1983). Treatment of nightmares via relaxation and desensitization: a controlled evaluation. *J. Consult. Clin. Psychol.* **51**, 870–877.

Natterson, J. M. (1980). The dream in group psychotherapy. In: The Dream in Clinical Practice (J. M. Natterson, ed.), Jason Aronson, New York, pp. 434–443.

Natterson, J. M. (1993). Dreams: the gateway to consciousness. In: New Directions in Dream Interpretation (G. Delaney, ed.), SUNY Press, Albany, NY, pp. 41–76.

Oremland, J. D. (1973). A specific dream during the termination phase of successful psychoanalyses. *J. Am. Psychoanal. Assoc.* **21**, 285–302.

Perls, F. (1969). Gestalt Therapy Verbatim. Bantam, New York.

Polster, E., and Polster, M. (1973). Gestalt Therapy Integrated. Random House, New York.

Popp, C., Diguer, L., Luborsky, L., Johnson, S., Morris, M., Schaffer, N., Schaffer, P., and Schmidt, K. (1998). The parallel of the CCRT from waking narratives with the CCRT from dreams: a further validation. In: Understanding Transference: The Core Conflictual Relationship Theme Method, 2nd ed. (L. Luborsky and P. Crits-Christoph, ed.), American Psychological Association, Washington, DC.

Popp, C., Luborsky, L., and Crits-Christoph, P. (1998). The parallel of the CCRT from waking narratives with the CCRT from dreams. In: Understanding Transference: The Core Conflictual Relationship Theme Method, 2nd ed. (L. Luborsky and P. Crits-Christoph, eds.), American Psychological Association, Washington, DC.

Rochlen, A. B. (2004). Using dreams to work with male clients. In: Dream Work in Therapy: Facilitating Exploration, Insight, and Action (C. E. Hill, ed.), American Psychological Association, Washington, DC, pp. 187–201.

Rochlen, A. B., Hill, C. E., et al. (2005). Gender role conflict and the process and outcome of dream work with men. *Dreaming* **15**, 227–239.

Rohde, A. B., Geller, J. D., and Farber, B. A. (1992). Dreams about the therapist: mood, interactions, and themes. *Psychother. Theory Res. Pract. Train.* **29**, 536–544.

Rosenbaum, M. (1965). Dreams in which the analyst appears undisguised—a clinical and statistical study. *Int. J. Psychoanal.* **46**, 429–437.

Schredl, M. (2000). Gender differences in dream recall. *J. Ment. Imagery.* **24**, 169–176.

Schredl, M., Bohusch, C., Kahl, J., Mader, A., and Somesan, A. (2000). The use of dreams in psychotherapy: a survey of psychotherapists in private practice. *J. Psychother. Pract. Res.* **9**, 81–87.

Schredl, M., and Piel, E. (2005). Gender differences in dreaming: are they stable over time? *Pers. Individ. Dif.* **39**, 309–316.

Schwartz, W. (1990). A psychoanalytic approach to dreamwork. In: Dreamtime and Dreamwork: Decoding the Language of the Night (S. Krippner, ed.), Tarcher, Los Angeles, pp. 49–58.

Shuttleworth-Jordan, A. B., and Saayman, G. S. (1989). Differential effects of alternative strategies on psychotherapeutic process in group dream work. *Psychotherapy* **26**, 514–519.

Sim, W., Hill, C. E., Chowdhury, S., Huang, T., Zaman, N., and Talavera, P. (2010). Problems and action ideas discussed by first- and second-generation female East Asian students during dream sessions. *Dreaming.* **20**, 42–59

Sirois, F. (1994). Dreaming about the session. *Psychoanal. Q.* **63**, 332–345.

Spangler, P., Hill, C. E., Mettus, C., Guo, A. H., and Heymsfield, L. (2009). Therapist perspectives on the dreams about clients: a qualitative investigation. *Psychother. Res.* **19**, 81–95.

Taylor, J. (1992). Where People Fly and Water Runs Uphill: Using Dreams to Tap the Wisdom of the Unconscious. Warner, New York.

Taylor, J. (1998). The Living Labyrinth: Exploring Universal Themes in Myths, Dreams, and the Symbolism of Waking Life. Paulist Press, Mahwah, NJ.

Tien, H. S., Lin, C., and Chen, S. (2006). Dream interpretations sessions for college students in Taiwan: who benefits and what volunteer clients view as most and least helpful. *Dreaming* **16**, 246–257.

Ullman, M. (1987). The experiential dream group. In: Handbook of Dreams (B. B. Wolman, ed.), Van Nostrand, New York, pp. 407–423.

Ullman, M. (1994). The experiential dream group: its applications in the training of therapists. *Dreaming* **4**, 223–229.

Van de Castle, R. L. (1994). Our Dreaming Mind. Ballantine Books, New York.

Weiss, L. (1986). Dream Analysis in Psychotherapy. Pergamon, New York.

Weiss, L. (1999). Practical Dreaming: Awakening the Power of Dreams in Your Life. New Harbinger Press, Oakland, CA.

Wolk, D. J. (1996). The psychodramatic reenactment of a dream. *J. Group Psychother. Psychodrama Sociom.* **49**, 3–9.

Wonnell, T. L. (2004). Working with dreams in groups. In: Dream Work in Therapy: Facilitating Exploration, Insight, and Action (C. E. Hill, ed.), American Psychological Association, Washington, DC.

Wonnell, T., and Hill, C. E. (2000). The effects of including the action stage in dream interpretation. *J. Coun. Psychol.* **47**, 372–379.

Wonnell, T. L., and Hill, C. E. (2005). Predictors of intention to act and implementation of action in dream sessions: therapist skills, level of difficulty of action plan, and client involvement. *Dreaming* **15**, 129–141.

Zack, J., and Hill, C. E. (1998). Predicting dream interpretation outcome by attitudes, stress, and emotion. *Dreaming* **8**, 169–185.

INDEX

CONTENTS OF RECENT VOLUMES